160   Approximation by Algebraic Numbers

Approximation by Algebraic Numbers

YANN BUGEAUD

Université Louis Pasteur, Strasbourg

CAMBRIDGE UNIVERSITY PRESS
Cambridge, New York, Melbourne, Madrid, Cape Town, Singapore, São Paulo

Cambridge University Press
The Edinburgh Building, Cambridge CB2 8RU, UK

Published in the United States of America by Cambridge University Press, New York

www.cambridge.org
Information on this title: www.cambridge.org/9780521823296

First published 2004
This digitally printed version 2007

A catalogue record for this publication is available from the British Library

Library of Congress Cataloguing in Publication data

Bugeaud Yann, 1971–
Approximation by algebraic numbers/Yann Bugeaud.
p.   cm. – (Cambridge tracts in mathematics; 160)
Includes bibliographical references and index.
ISBN 0 521 82329 3
1. Approximation theory.   2. Algebraic number theory.   I. Title.   II. Series.
QA221.B78 2004
512.7′4 – dc22   2003063884

ISBN 978-0-521-82329-6 hardback
ISBN 978-0-521-04567-4 paperback

Contents

Preface

Und alles ist mir dann immer wieder zerfallen, auf dem Konzentrationshöhepunkt
ist mir dann immer wieder alles zerfallen.

Thomas Bernhard

(...) il faut continuer, je vais continuer.

Samuel Beckett

The central question in Diophantine approximation is: how well can a given
real number ξ be approximated by rational numbers, that is, how small can the
difference $|\xi - p/q|$ be made for varying rational numbers p/q? The accuracy
of the approximation of ξ by p/q is being compared with the 'complexity' of
the rational number p/q, which is measured by the size of its denominator q. It
follows from the theory of continued fractions (or from Dirichlet's Theorem)
that for any irrational number ξ there exist infinitely many rational numbers
p/q with $|\xi - p/q| < q^{-2}$. This can be viewed as the first general result in
this area.

There are two natural generalizations of the central question. On the one
hand, one can treat rational numbers as algebraic numbers of degree one and
study, for a given positive integer n, how well ξ can be approximated by algebraic numbers of degree at most n. On the other hand, $\xi - p/q$ can be viewed
as $q\xi - p$, that is as $P(\xi)$, where $P(X)$ denotes the integer polynomial $qX - p$.
Thus, for a given positive integer n, one may ask how small $|P(\xi)|$ can be made
for varying integer polynomials $P(X)$ of degree at most n. To do this properly,
one needs to define a notion of size, or 'complexity', for algebraic numbers α
and for integer polynomials $P(X)$, and we have to compare the accuracy of

ix

the approximation of ξ by α (*resp.* the smallness of $|P(\xi)|$) with the size of α (*resp.* of $P(X)$). In both cases, we use for the size the naive height H: the height $H(P)$ of $P(X)$ is the maximum of the absolute values of its coefficients and the height $H(\alpha)$ of α is that of its minimal polynomial over $\mathbb{Z}$.

In 1932, Mahler proposed to classify the real numbers (actually, the complex numbers) into several classes according to the various types of answers to the latter question, while in 1939 Koksma introduced an analogous classification based on the former question. In both cases, the algebraic numbers form one of the classes. Let ξ be a real number and let n be a positive integer. According to Mahler, we denote by $w_n(\xi)$ the supremum of the real numbers w for which there exist infinitely many integer polynomials $P(X)$ of degree at most n satisfying

$$0 < |P(\xi)| \leq H(P)^{-w},$$

and we divide the set of real numbers into four classes according to the behaviour of the sequence $(w_n(\xi))_{n \geq 1}$. Following Koksma, we denote by $w_n^*(\xi)$ the supremum of the real numbers w for which there exist infinitely many real algebraic numbers α of degree at most n satisfying

$$0 < |\xi - \alpha| \leq H(\alpha)^{-w-1}.$$

It turns out that both classifications coincide, inasmuch as each of the four classes defined by Mahler corresponds to one of the four classes defined by Koksma. However, they are not strictly equivalent, since there exist real numbers ξ for which $w_n(\xi)$ and $w_n^*(\xi)$ differ for any integer n at least equal to 2. In addition, it is a very difficult (and, often, still open) question to determine to which class a given 'classical' number like π, e, $\zeta(3)$, $\log 2$, etc. belongs.

The present book is mainly concerned with the following problem: given two non-decreasing sequences of real numbers $(w_n)_{n \geq 1}$ and $(w_n^*)_{n \geq 1}$ satisfying some necessary restrictions (e.g. $w_n^* \leq w_n \leq w_n^* + n - 1$), does there exist a real number ξ with $w_n(\xi) = w_n$ and $w_n^*(\xi) = w_n^*$ for all positive integers n? This question is very far from being solved, although we know (see Chapter 4) that almost all (in the sense of the Lebesgue measure on the line) real numbers share the same approximation properties, namely they satisfy $w_n(\xi) = w_n^*(\xi) = n$ for all positive integers n.

There are essentially two different points of view for investigating such a problem. We may try to construct explicitly (or semi-explicitly) real numbers with the required properties (Chapter 7) or, if this happens to be too difficult, we may try to prove the existence of real numbers with a given property by showing that the set of these numbers has positive Hausdorff dimension

(Chapters 5 and 6). Most often, however, we are unable to exhibit a single explicit example of such a number.

The content of the present book is as follows. Chapter 1 is devoted to the approximation by rational numbers. We introduce the notion of continued fractions and establish their main properties needed to prove the celebrated metric theorem of Khintchine saying that, for a continuous function $\Psi : \mathbb{R}_{\geq 1} \to \mathbb{R}_{>0}$ such that $x \mapsto x\Psi(x)$ is non-increasing, the equation $|q\xi - p| < \Psi(q)$ has infinitely many integer solutions (p, q) with q positive for either almost no or almost all real numbers ξ, according to whether the sum $\sum_{q=1}^{+\infty} \Psi(q)$ converges or diverges.

In Chapter 2, we briefly survey the approximation to algebraic numbers by algebraic numbers, and recall many classical results (including Roth's Theorem and Schmidt's Theorem). We make a clear distinction between effective and ineffective statements.

Mahler's and Koksma's classifications of real numbers are defined in Chapter 3, where we show, following ideas of Wirsing, how closely they are related. Some links between simultaneous rational approximation and these classifications are also mentioned, and we introduce four other functions closely related to w_n and w_n^*.

In Chapter 4, we establish Mahler's Conjecture to the effect that almost all real numbers ξ satisfy $w_n(\xi) = w_n^*(\xi) = n$ for all positive integers n. This result, first proved by Sprindžuk in 1965, has been refined and extended since that time and we state the most recent developments, essentially due to a new approach found by Kleinbock and Margulis.

Exceptional sets are investigated from a metric point of view in Chapters 5 and 6. To this end, we introduce a classical powerful tool for discriminating between sets of Lebesgue measure zero, namely the notion of Hausdorff dimension. We recall the basic definitions and some well-known results useful in our context. This allows us to prove the theorem of Jarník and Besicovitch saying that for any real number $\tau \geq 1$ the Hausdorff dimension of the set of real numbers ξ with $w_1^*(\xi) \geq 2\tau - 1$ is equal to $1/\tau$. We also establish its generalization to any degree n (with $w_1^*(\xi) \geq 2\tau - 1$ replaced by $w_n^*(\xi) \geq (n + 1)\tau - 1$) obtained in 1970 by A. Baker and Schmidt. Chapter 6 is devoted to refined statements and contains general metric theorems on sets of real numbers which are very close to infinitely many elements of a fixed set of points which are, in some sense, evenly distributed.

In Chapter 7, we prove, following ideas of Schmidt, that the class formed by the real numbers ξ with $\limsup_{n\to+\infty} w_n(\xi)/n$ infinite and $w_n(\xi)$ finite for all positive integers n is not empty. At the same time, we show that there exist real numbers ξ for which the quantities $w_n(\xi)$ and $w_n^*(\xi)$ differ by preassigned values, a result due to R. C. Baker. The real numbers ξ with the required property are obtained as limits of sequences of algebraic numbers. This illustrates the importance of results on approximation of algebraic numbers by algebraic numbers. The remaining part of Chapter 7 is concerned with some other (simpler) explicit constructions.

Mahler's and Koksma's classifications emphasize the approximation by algebraic numbers of bounded degree. We may as well exchange the roles played by degree and height or let both vary simultaneously. We tackle this question in Chapter 8 by considering the classification introduced by Sprindžuk in 1962 and the so-called 'order functions' defined by Mahler in 1971. Further, some recent results of Laurent, Roy, and Waldschmidt expressed in terms of a more involved notion of height (namely, the absolute logarithmic height) are given.

In Chapter 9, we briefly discuss approximation in the complex field, in the Gaussian field, in p-adic fields, and in fields of formal Laurent series.

Chapter 10, which begins by a brief survey on the celebrated Littlewood Conjecture, offers a list of open questions. We hope that these will motivate further research.

Finally, there are two appendices. Appendix A is devoted to lemmas on zeros of polynomials: all proofs are given in detail and the statements are the best known at the present time. Appendix B lists classical auxiliary results from the geometry of numbers.

The Chapters are largely independent of each other.

We deliberately do not give proofs to all the theorems quoted in the main part of the text. We have clearly indicated when this is the case (see below). Furthermore, we try, in the end-of-chapter notes, to be as exhaustive as possible and to quote less-known papers, which, although interesting, did not yield up to now to further research. Of course, exhaustivity is an impossible task, and it is clear that the choice of the references concerning works at the border of the main topic of this book reflects the personal taste and the limits of the knowledge of the author.

The purpose of the exercises is primarily to give complementary results, thus they are often an adaptation of an original research work to which the reader is directed.

There exist already many textbooks dealing, in part, with the subject of the present one, e.g., by Schneider [517], Sprindžuk [539, 540], A. Baker [44], Schmidt [510, 512], Bernik and Melnichuk [90], Harman [273], and Bernik and Dodson [86]. However, the intersection rarely exceeds two or three chapters. Special mention should be made to the wonderful book of Koksma [332], which contains an impressive list of references which appeared before 1936.

Maurice Mignotte and Michel Waldschmidt encouraged me constantly. Many colleagues sent me comments, remarks, and suggestions. I am very grateful to all of them. Special thanks are due to Guy Barat and Damien Roy, who carefully read several parts of this book.

The present book will be regularly updated on my institutional Web page:
`http://www-irma.u-strasbg.fr/~bugeaud/Book`

The following statements are not proved in the present book:
Theorems 1.13 to 1.15, 1.17, 1.20, 2.1 to 2.8, 3.7, 3.8, 3.10, 3.11, 4.4 to 4.7, Proposition 5.1, Theorems 5.7, 5.9, 5.10, Proposition 6.1, Theorems 6.3 to 6.5, 8.1, 8.5, 8.8 (partially proved), 9.1 to 9.8, Lemma 10.1, Theorems 10.1, B.3, and B.4.

The following statements are left as exercises:
Theorems 1.16, 1.18, 1.19, 5.4, 5.6, 6.2, 6.9, 6.10, 7.2, 7.3, 7.6, 8.4, 8.6, 8.12, and Proposition 8.1.

Frequently used notation

deg   degree.

positive   strictly positive.

$\mathcal{N}$ infinite set of integers.

$[\cdot]$ integer part.

$\{\cdot\}$ fractional part.

$||\cdot||$ distance to the nearest integer.

An empty sum is equal to 0 and an empty product is equal to 1.

ϕ the Euler totient function.

$\log_i$ the i-fold iterated logarithm.

$\ll, \gg$ means that there is an implied constant.

Card   the cardinality (of a finite set).

c, c_i, κ constants.

$c(\mathrm{var}_1, \ldots, \mathrm{var}_m)$ constant depending (at most) on the variables $\mathrm{var}_1, \ldots, \mathrm{var}_m$.

$\mathbb{A}_n$ the set of real algebraic numbers of degree at most n.

H   naive height, Ch. 2.

Λ size, Ch. 8.

h   absolute height, Ch. 8.

M Mahler's measure, App. A.

λ the Lebesgue measure on the real line.

$\lambda(I) = |I|$ the Lebesgue measure of an interval I.

vol   the n-dimensional Lebesgue measure.

μ a measure (not the Lebesgue one).

ξ the number we approximate.

d the degree of ξ (when ξ is algebraic).

α the algebraic approximant of ξ.

n (an upper bound for) the degree of the approximant α.

H (an upper bound for) the height of the approximant α.

$\underline{x}$ the n-tuple $(x_1, \ldots, x_n)$.

$\mathcal{B}$ set of badly approximable real numbers, Ch. 1.

$\mathcal{B}_M$ set of real numbers whose partial quotients are at most equal to M, Ch. 5.

$\Psi, x \mapsto x^{\tau}$ approximation functions, Ch. 1, 5, 6.

$\mathcal{K}_n^*(\Psi), \mathcal{K}_n(\tau), \mathcal{K}_n^*(\tau), \mathcal{K}_n^+(\tau), \mathcal{K}_{\mathcal{S}}^*(\Psi)$ Ch. 1, 5, 6.

$w_n(\xi, H), w_n^*(\xi, H), w_n^{*c}(\xi, H), w_n(\xi), w_n^*(\xi), w_n^{*c}(\xi), w(\xi), w^*(\xi)$ Ch. 3

$w_n'(\xi), \hat{w}_n'(\xi), \hat{w}_n(\xi), \hat{w}_n^*(\xi)$ Ch. 3

A-, S-, T-, U-, A^*-, S^*-, T^*-, U^*-numbers   Ch. 3.

U_m-numbers   Ch. 7.

$t(\xi), t^*(\xi)$ the type and the $*$-type of ξ, Ch. 3.

Π_+ Ch. 4.

$w_n^+(\xi)$ Ch. 5.

$\mathcal{W}_n(\tau), \mathcal{W}_n^*(\tau), \mathcal{W}_n^+(\tau), \mathcal{W}_n^{\geq}(\tau)$ Ch. 5.

dim   Hausdorff dimension, Ch. 5.

$\mathcal{H}^f$ Hausdorff $\mathcal{H}^f$-measure, Ch. 5.

$\mathcal{H}^s$ Hausdorff s-measure, Ch. 5.

$\prec$ order between dimension functions, Ch. 5.

Φ, Ξ increasing functions, Ch. 5.

$\mathcal{S}$ infinite set of real numbers, Ch. 5, 6.

λ lower order at infinity, Ch. 6.

$\mathcal{M}_\infty^f$ net measure, Ch. 6.

$\tilde{w}(\xi, H), \tilde{w}(\xi), \tilde{\mu}(\xi, H), \tilde{\mu}(\xi), \tilde{w}^*(\xi, H), \tilde{w}^*(\xi)$ Ch. 8.

$\tilde{A}$-, $\tilde{S}$-, $\tilde{T}$-, $\tilde{U}$-numbers   Ch. 8.

$O(u \mid \xi), O^*(u \mid \xi)$ order functions, Ch. 8.

$\asymp$ order between order functions, Ch. 8.

$\tau(\xi)$ transcendence type of ξ, Ch. 8.

Res   the resultant of two polynomials, App. A.

Disc   the discriminant of an integer polynomial, App. A.

$\mathcal{C}$ a bounded convex body, App. B.

$\lambda_1, \ldots, \lambda_n$ the successive minima of a bounded convex body, App. B.

1

Approximation by rational numbers

Throughout the present Chapter, we are essentially concerned with the following problem: for which functions $\Psi : \mathbb{R}_{\geq 1} \to \mathbb{R}_{\geq 0}$ is it true that, for a given real number ξ, or for all real numbers ξ in a given class, the equation $|\xi - p/q| < \Psi(q)$ has infinitely many solutions in rational numbers p/q? We begin by stating the results on rational approximation obtained by Dirichlet and Liouville in the middle of the nineteenth century. In Section 1.2, we define the continued fraction algorithm and recall the main properties of continued fractions expansions. These are used in Section 1.3 to give a full proof of a metric theorem of Khintchine. The next two Sections are devoted to the Duffin–Schaeffer Conjecture and to some complementary results on continued fractions.

1.1 Dirichlet and Liouville

Every real number ξ can be expressed in infinitely many ways as the limit of a sequence of rational numbers. Furthermore, for any positive integer b, there exists an integer a with $|\xi - a/b| \leq 1/(2b)$, and one may hope that there are infinitely many integers b for which $|\xi - a/b|$ is in fact much smaller than $1/(2b)$. For instance, this is true when ξ is irrational, as follows from the theory of continued fractions. In order to measure the accuracy of the approximation of ξ by a rational number a/b (written in its lowest terms), we have to compare the difference $|\xi - a/b|$ with the *size*, or complexity, of a/b. A possible definition for this notion is, for example, the number of digits of a plus the number of digits of b. However, as usual, we define the size, or the height, of a/b as the maximum of the absolute values of its denominator and numerator: this definition is more relevant and can be easily extended (see Definition 2.1).

The first statement of Theorem 1.1 is often referred to as Dirichlet's Theorem, although it is not explicitly stated under this form in [196], a paper which appeared in 1842. However, it follows easily from the proof of the main result of [196], which actually provides an extension of the second assertion of Theorem 1.1 to linear forms and to systems of linear forms.

THEOREM 1.1. *Let ξ and Q be real numbers with $Q \geq 1$. There exists a rational number p/q, with $1 \leq q \leq Q$, such that*

$$\left| \xi - \frac{p}{q} \right| < \frac{1}{qQ}.$$

Furthermore, if ξ is irrational, then there exist infinitely many rational numbers p/q such that

$$\left| \xi - \frac{p}{q} \right| < \frac{1}{q^2}, \tag{1.1}$$

and if $\xi = a/b$ is rational, then for any rational $p/q \neq a/b$ with $q > 0$ we have

$$\left| \xi - \frac{p}{q} \right| \geq \frac{1}{|b|q}.$$

PROOF. Let t denote the integer part of Q. If ξ is the rational a/b, with a and b integers and $1 \leq b \leq t$, it is sufficient to set $p = a$ and $q = b$. Otherwise, the $t + 2$ points $0, \{\xi\}, \ldots, \{t\xi\}$, and 1 are pairwise distinct and they divide the interval $[0, 1]$ into $t + 1$ subintervals. Clearly, at least one of these has its length at most equal to $1/(t + 1)$. This means that there exist integers k, ℓ and m_k, m_ℓ with $0 \leq k < \ell \leq t$ and

$$|(\ell\xi - m_\ell) - (k\xi - m_k)| \leq \frac{1}{t + 1} < \frac{1}{Q}.$$

We conclude by setting $p := m_\ell - m_k$ and $q := \ell - k$, and by noticing that q satisfies $1 \leq q \leq t \leq Q$. Instead of reasoning with the lengths of the intervals, we could as well use an argument dating back to Dirichlet [196], now called Dirichlet's *Schubfachprinzip* (or pigeon-hole principle, or *principe des tiroirs*, or *principio dei cassetti*, or *principio de las cajillas*, or *principiul cutiei*, or *skatulya-elv*, or *lokeroperiaate*, or *zasada pudełkowa*). It asserts that at least two among the $t + 2$ points $0, \{\xi\}, \ldots, \{t\xi\}$, and 1 lie in one of the $t + 1$ intervals $[j/(t + 1), (j + 1)/(t + 1)]$, where $j = 0, \ldots, t$; hence, the existence of integers k, ℓ, m_k, and m_ℓ as above.

Suppose now that ξ is irrational and let Q_0 be a positive integer. By the first assertion of the theorem, there exists an integer q with $1 \leq q \leq Q_0$ such that

$$\left| \xi - \frac{p}{q} \right| < \frac{1}{qQ_0} \leq \frac{1}{q^2}$$

holds for some integer p. We may assume that q is the smallest integer between 1 and Q_0 with this property. By the first assertion of the theorem applied with $Q = 1/|\xi - p/q|$, there exists a rational number p'/q' with $1 \le q' \le 1/|\xi - p/q|$ such that

$$\left| \xi - \frac{p'}{q'} \right| < \frac{1}{q'} \left| \xi - \frac{p}{q} \right| < \frac{1}{q'Q_0} \quad \text{and} \quad \left| \xi - \frac{p'}{p'} \right| < \frac{1}{q^2}.$$

Our choice of q ensures that q' is strictly larger than q and we proceed inductively to get an infinite sequence of distinct rational numbers satisfying (1.1), thus the second assertion is proved. The third one is immediate.

Theorem 1.1 provides a useful criterion of irrationality: a real number having infinitely many *good* rational approximants must be irrational.

Recall that a complex number ξ is an algebraic number if it is root of a non-zero integer polynomial $P(X)$. Otherwise, ξ is a transcendental number. In 1844, two years after Dirichlet's paper, Liouville [368, 369] was the first to prove that transcendental numbers exist, and, moreover, he constructed explicit examples of such numbers. Thirty years later, Cantor [152] gave an alternative proof of the existence of real transcendental numbers: he showed that the set of real algebraic numbers is countable and that, given a countable set of real numbers, any real interval of positive length contains points *not* belonging to that set. Cantor's proof, however, does not yield any explicit example of a real trancendental number.

A detailed proof of Theorem 1.2 is given in [370] and includes the case $n = 1$ (that is, the last assertion of Theorem 1.1). The main idea, however, already appeared in Liouville's note [369].

THEOREM 1.2. *Let ξ be a real root of an irreducible integer polynomial $P(X)$ of degree $n \ge 2$. There exists a positive constant $c_1(\xi)$ such that*

$$\left| \xi - \frac{p}{q} \right| \ge \frac{c_1(\xi)}{q^n} \tag{1.2}$$

for all rational numbers p/q. A suitable choice for $c_1(\xi)$ is

$$c_1(\xi) := \frac{1}{1 + \max_{|t-\xi| \le 1} |P'(t)|}.$$

PROOF. With $c_1(\xi)$ defined as above, inequality (1.2) is true when $|\xi - p/q| \ge 1$. Let p/q be a rational number satisfying $|\xi - p/q| < 1$. Since $P(X)$ is irreducible and has integer coefficients, we have $P(p/q) \ne 0$ and $|q^n P(p/q)| \ge 1$. By Rolle's Theorem, there exists a real number t lying between ξ and p/q such that

$$|P(p/q)| = |P(\xi) - P(p/q)| = |\xi - p/q| \times |P'(t)|.$$

Hence, we have $|t - \xi| \leq 1$ and

$$\left| \xi - \frac{p}{q} \right| \geq \frac{1}{q^n \, |P'(t)|} \geq \frac{c_1(\xi)}{q^n},$$

as claimed.

COROLLARY 1.1. *The number $\xi := \sum_{n \geq 1} 10^{-n!}$ is transcendental.*

PROOF. Since its decimal expansion is not ultimately periodic, ξ is irrational. For any integer $n \geq 2$, set $q_n = 10^{(n-1)!}$ and $p_n = q_n(10^{-1!} + \ldots + 10^{-(n-1)!})$. Then we have

$$\left| \xi - \frac{p_n}{q_n} \right| = \sum_{m \geq n} \frac{1}{10^{m!}} \leq \frac{2}{10^{n!}} = \frac{2}{q_n^n}$$

and ξ is not algebraic of degree greater than or equal to 2, by Theorem 1.2. Consequently, ξ is a transcendental number.

Corollary 1.1 illustrates how Theorem 1.2 can be applied to prove the transcendence of a large class of real numbers, which are now called Liouville numbers.

DEFINITION 1.1. *Let ξ be a real number. If for any positive real number w there exists a rational number p/q such that*

$$0 < \left| \xi - \frac{p}{q} \right| < \frac{1}{q^w},$$

then ξ is called a Liouville number.

An easy modification of the proof of Corollary 1.1 shows that any real number $\sum_{n \geq 1} a_n 10^{-n!}$ with a_n in $\{1, 2\}$ is a Liouville number. Hence, there exist uncountably many Liouville numbers. Furthermore, Theorem 1.2 provides a useful transcendence criterion, see Exercise 1.1.

Combining the theorems of Liouville and Dirichlet, we see that the problem of rational approximation of real quadratic numbers is, in some sense, solved.

COROLLARY 1.2. *Let ξ be a real quadratic algebraic number. Then there exists a positive real number $c_2(\xi)$ such that*

$$\left| \xi - \frac{p}{q} \right| \geq \frac{c_2(\xi)}{q^2} \quad \text{for all rationals } p/q \tag{1.3}$$

whereas

$$\left| \xi - \frac{p}{q} \right| < \frac{1}{q^2} \quad \text{for infinitely many rationals } p/q.$$

One may ask whether there exist real numbers, other than quadratic irrationalities, for which the property (1.3) is satisfied. The answer is affirmative, and a way to prove it is to use the theory of continued fractions.

1.2 Continued fractions

Let $x_0, x_1, \ldots$ be real numbers with $x_1, x_2, \ldots$ positive. A *finite continued fraction* denotes any expression of the form

$$[x_0; x_1, x_2, \ldots, x_n] = x_0 + \cfrac{1}{x_1 + \cfrac{1}{x_2 + \cfrac{1}{\ldots + \cfrac{1}{x_n}}}}.$$

More generally, we call any expression of the above form or of the form

$$[x_0; x_1, x_2, \ldots] = x_0 + \cfrac{1}{x_1 + \cfrac{1}{x_2 + \cfrac{1}{\ldots}}} = \lim_{n \to +\infty} [x_0; x_1, x_2 \ldots, x_n]$$

a *continued fraction*, provided that the limit exists.

The aim of this Section is to show how a real number ξ can be expressed as $\xi = [x_0; x_1, x_2, \ldots]$, where x_0 is an integer and x_n a positive integer for any $n \geq 1$. We first deal with the case of a rational number ξ, then we describe an algorithm which associates to any irrational ξ an infinite sequence of integers $(a_n)_{n \geq 0}$, with $a_n \geq 1$ for $n \geq 1$, and we show that the sequence of rational numbers $[a_0; a_1, a_2, \ldots, a_n]$ converges to ξ.

For more results on continued fractions or/and different points of view on this theory, the reader may consult, for example, a text of Van der Poorten [462] and the books of Cassels [155], Dajani and Kraaikamp [174], Hardy and Wright [271], Iosifescu and Kraaikamp [286], Perron [454], Rockett and Szüsz [474], Schmidt [512], and Schweiger [519].

LEMMA 1.1. *Any rational number r has exactly two different continued fraction expansions. These are $[r]$ and $[r - 1; 1]$ if r is an integer and, otherwise, one of them has the form $[a_0; a_1, \ldots, a_{n-1}, a_n]$ with $a_n \geq 2$, and the other one is $[a_0; a_1, \ldots, a_{n-1}, a_n - 1, 1]$.*

PROOF. Let r be a rational number and write $r = u/v$ with v positive and u and v coprime. We argue by induction on v. If $v = 1$, then $r = u = [u]$, and if $r = [a_0; a_1, \ldots, a_n]$ with $n \geq 1$, we have $r = a_0 + 1/[a_1; a_2, \ldots, a_n]$.

Since $a_1 \geq 1$ and r is an integer, we deduce that $n = 1$ and $a_1 = 1$, thus $r = a_0 + 1 = [r - 1; 1]$.

We assume $v \geq 2$ and that the lemma holds true for any rational of denominator positive and at most equal to $v - 1$. Performing the Euclidean division of u by v, there exist integers q and c with $u = qv + c$ and $1 \leq c \leq v - 1$. Thus, $u/v = q + c/v$ and, by our inductive hypothesis, the rational v/c has exactly two expansions in continued fractions, which we denote by $[a_1; a_2, \ldots, a_{n-1}, a_n]$ with $a_n \geq 2$, and $[a_1; a_2, \ldots, a_{n-1}, a_n - 1, 1]$. Setting a_0 equal to q, the desired result follows for u/v.

Unless otherwise explicitly stated, by 'the' continued fraction expansion of a rational number p/q, we mean [1] if $p/q = 1$, and if not, the unique expansion which does not end with 1.

The following algorithm allows us to associate to any irrational real number ξ an infinite sequence of integers. Let us define the integer a_0 and the real number $\xi_1 > 1$ by

$$a_0 = [\xi] \quad \text{and} \quad \xi_1 = 1/\{\xi\}.$$

We then have $\xi = a_0 + 1/\xi_1$. For any positive integer n, we define inductively the integer a_n and the real number $\xi_{n+1} > 1$ by

$$a_n = [\xi_n] \quad \text{and} \quad \xi_{n+1} = 1/\{\xi_n\},$$

and we observe that $\xi_n = a_n + 1/\xi_{n+1}$. We point out that the algorithm does not stop since ξ is assumed to be irrational. Thus, we have associated to any irrational real number ξ an infinite sequence of integers $a_0, a_1, a_2, \ldots$ with a_n positive for all $n \geq 1$.

If ξ is rational, the same algorithm terminates and associates to ξ a *finite* sequence of integers. Indeed, the ξ_js are then rational numbers and, if we set $\xi_j = u_j/v_j$, with u_j, v_j positive and $\gcd(u_j, v_j) = 1$, an easy induction shows that we get $u_j > u_{j+1}$, for any positive integer j with $v_j \neq 1$. Consequently, there must be some index n for which $v_{n-1} \neq 1$ and ξ_n is an integer. Thus, $a_{n+1}, a_{n+2}, \ldots$ are not defined. We have $\xi = [a_0; a_1, \ldots, a_n]$, and this corresponds to the Euclidean algorithm.

DEFINITION 1.2. *Let ξ be an irrational number (resp. a rational number). Let $a_0, a_1, \ldots$ (resp. $a_0, a_1, \ldots, a_N$) be the sequence of integers associated to ξ by the algorithm defined above. For any integer $n \geq 1$ (resp. $n = 1, \ldots, N$), the rational number*

$$\frac{p_n}{q_n} := [a_0; a_1, \ldots, a_n]$$

*is called the n-th convergent of ξ and a_n is termed the n-th partial quotient of
ξ. Further, for any integer $n \geq 1$ (resp. $n = 1, \ldots, N - 1$), there exists a real
number η_n in $]0, 1[$ such that*

$$\xi = [a_0; a_1, \ldots, a_{n-1}, a_n + \eta_n].$$

We observe that the real numbers η_n occurring in Definition 1.2 are exactly the
real numbers $1/\xi_{n+1}$ given by the algorithm.

In all of what follows until the end of Theorem 1.5, unless otherwise explicitly stated, we assume that ξ is a real irrational number and we associate to
ξ the sequences $(a_n)_{n\geq 0}$ and $(p_n/q_n)_{n\geq 1}$ as given by Definition 1.2. However,
the statements below remain true for rational numbers ξ provided that the a_ns
and the p_n/q_ns are well-defined.

The integers p_n and q_n can be easily expressed in terms of a_n, p_{n-1}, p_{n-2},
q_{n-1}, and q_{n-2}.

THEOREM 1.3. *Setting*

$$p_{-1} = 1, \quad q_{-1} = 0, \quad p_0 = a_0, \quad and \quad q_0 = 1,$$

we have, for any positive integer n,

$$p_n = a_n p_{n-1} + p_{n-2} \quad and \quad q_n = a_n q_{n-1} + q_{n-2}.$$

PROOF. We proceed by induction. Since $p_1/q_1 = a_0 + 1/a_1 = (a_0 a_1 + 1)/a_1$, the definitions of p_{-1}, q_{-1}, p_0, and q_0 show that the theorem is true
for $n = 1$. Assume that it holds true for a positive integer n and denote by
$p'_0/q'_0, \ldots, p'_n/q'_n$ the convergents of the rational number $[a_1; a_2, \ldots, a_{n+1}]$.
For any integer j with $0 \leq j \leq n + 1$ we have

$$\frac{p_j}{q_j} = [a_0; a_1, \ldots, a_j] = a_0 + \frac{1}{[a_1; a_2, \ldots, a_j]} = a_0 + \frac{q'_{j-1}}{p'_{j-1}},$$

thus

$$p_j = a_0 p'_{j-1} + q'_{j-1} \quad and \quad q_j = p'_{j-1}. \tag{1.4}$$

It follows from (1.4) with $j = n + 1$ and the inductive hypothesis applied to
the rational $[a_1; a_2, \ldots, a_{n+1}]$ that

$$p_{n+1} = a_0(a_{n+1} p'_{n-1} + p'_{n-2}) + a_{n+1} q'_{n-1} + q'_{n-2}$$
$$= a_{n+1}(a_0 p'_{n-1} + q'_{n-1}) + a_0(p'_{n-2} + q'_{n-2})$$

and

$$q_{n+1} = a_{n+1} p'_{n-1} + p'_{n-2},$$

whence, by (1.4) with $j = n$ and $j = n - 1$, we get $q_{n+1} = a_{n+1}q_n + q_{n-1}$ and $p_{n+1} = a_{n+1}p_n + p_{n-1}$, as claimed.

THEOREM 1.4. *For any non-negative integer n, we have*

$$q_n p_{n-1} - p_n q_{n-1} = (-1)^n \qquad (1.5)$$

and, for all $n \geq 1$,

$$q_n p_{n-2} - p_n q_{n-2} = (-1)^{n-1} a_n. \qquad (1.6)$$

PROOF. These equalities are clearly true for $n = 0$ and $n = 1$. It then suffices to argue by induction, using Theorem 1.3.

LEMMA 1.2. *For any irrational number ξ and any non-negative integer n, the difference $\xi - p_n/q_n$ is positive if, and only if, n is even.*

PROOF. We easily check that this is true for $n = 0, 1$, and 2 and we proceed by induction. Let $n \geq 4$ be an even integer. Then $\xi = [a_0; a_1, [a_2; a_3, \ldots, a_n + \eta_n]]$ and the inductive hypothesis implies that $[a_2; a_3, \ldots, a_n + \eta_n] > [a_2; a_3, \ldots, a_n]$. Since $[a_0; a_1, u] > [a_0; a_1, v]$ holds for all positive real numbers $u > v$, we get that $\xi > [a_0; a_1, [a_2; a_3, \ldots, a_n]] = p_n/q_n$. We deal with the case n odd in exactly the same way.

As a corollary of Lemma 1.2, we get a result of Vahlen [575].

COROLLARY 1.3. *Let p_n/q_n and p_{n+1}/q_{n+1} be two consecutive convergents of the continued fraction expansion of an irrational number ξ. Then at least one of them satisfies*

$$\left| \xi - \frac{p}{q} \right| < \frac{1}{2q^2}.$$

PROOF. We infer from Lemma 1.2 that ξ is an inner point of the interval bounded by p_n/q_n and p_{n+1}/q_{n+1}. Thus, using (1.5) and the inequality $a^2 + b^2 > 2ab$, valid for any distinct real numbers a and b, we get

$$\frac{1}{2q_n^2} + \frac{1}{2q_{n+1}^2} > \frac{1}{q_n q_{n+1}} = \left| \frac{p_n}{q_n} - \frac{p_{n+1}}{q_{n+1}} \right| = \left| \xi - \frac{p_n}{q_n} \right| + \left| \xi - \frac{p_{n+1}}{q_{n+1}} \right|,$$

and the claimed result follows.

The next two theorems show that the real irrational numbers are in a one-to-one correspondence with the set of integer sequences $(a_i)_{i \geq 0}$ with a_i positive for $i \geq 1$.

THEOREM 1.5. *The convergents of even order of any real irrational ξ form a strictly increasing sequence and those of odd order a strictly decreasing sequence. The sequence of convergents $(p_n/q_n)_{n \geq 0}$ converges to ξ, and we set*

$$\xi = [a_0; a_1, a_2, \ldots].$$

Any irrational number has a unique expansion in continued fractions.

PROOF. It follows from (1.6) that for any integer n with $n \geq 2$ we have

$$\frac{p_{n-2}}{q_{n-2}} - \frac{p_n}{q_n} = \frac{(-1)^{n-1} a_n}{q_n q_{n-2}},$$

and, since the a_ns are positive, we deduce that the convergents of even order of the real irrational ξ form a strictly increasing sequence and those of odd order a strictly decreasing sequence. To conclude, we observe that, by Lemma 1.2, we have $p_{2n}/q_{2n} < \xi < p_{2n+1}/q_{2n+1}$ for all $n \geq 0$, and, by (1.5), the difference $p_{2n}/q_{2n} - p_{2n+1}/q_{2n+1}$ tends to 0 when n tends to infinity. Uniqueness is clear. Indeed, if $(b_i)_{i \geq 0}$ is a sequence of integers with b_i positive for $i \geq 1$ and such that $\lim_{n \to +\infty} [b_0; b_1, b_2, \ldots]$ exist, then this limit cannot be equal to $\lim_{n \to +\infty} [a_0; a_1, a_2, \ldots]$ as soon as there exists a non-negative integer i with $a_i \neq b_i$.

THEOREM 1.6. *Let $a_0, a_1, \ldots$ be integers with $a_1, a_2, \ldots$ positive. Then the sequence of rational numbers $[a_0; a_1, \ldots, a_i]$, $i \geq 1$, converges to the irrational number whose partial quotients are precisely $a_0, a_1, \ldots$*

PROOF. For any positive integer n, denote by p_n/q_n the rational number $[a_0; a_1, \ldots, a_n]$. The recurrence relations obtained in Theorems 1.3 and 1.4 hold true in the present context. As in the proof of Theorem 1.5, we deduce from (1.5) and (1.6) that the sequences $(p_{2n}/q_{2n})_{n \geq 1}$ and $(p_{2n+1}/q_{2n+1})_{n \geq 1}$ are adjacent. Hence, they converge to the same limit, namely to the irrational number $[a_0; a_1, a_2, \ldots]$, whose partial quotients are precisely $a_0, a_1, \ldots$, by Theorem 1.5.

We observe that for any irrational number ξ the sequences $(a_n)_{n \geq 0}$ and $(\xi_n)_{n \geq 1}$ given by the algorithm defined below Lemma 1.1 satisfy $\xi_n = [a_n; a_{n+1}, a_{n+2}, \ldots]$ for all positive integers n.

THEOREM 1.7. *Let n be a positive integer and $\xi = [a_0; a_1, a_2 \ldots]$ be an irrational number. We then have*

$$\xi = [a_0; a_1, \ldots, a_n, \xi_{n+1}] = \frac{p_n \xi_{n+1} + p_{n-1}}{q_n \xi_{n+1} + q_{n-1}}$$

and

$$q_n\xi - p_n = \frac{(-1)^n}{q_n\xi_{n+1} + q_{n-1}} = \frac{(-1)^n}{q_n} \cdot \frac{1}{\xi_{n+1} + [0; a_n, a_{n-1}, \ldots, a_1]}.$$

Furthermore, the set of real numbers having a continued fraction expansion whose $n + 1$ first partial quotients are $a_0, a_1, \ldots, a_n$ is precisely the closed interval bounded by $(p_{n-1} + p_n)/(q_{n-1} + q_n)$ and p_n/q_n, which are equal to $[a_0; a_1, \ldots, a_n, 1]$ and $[a_0; a_1, \ldots, a_n]$, respectively.

PROOF. We proceed by induction, using Theorem 1.3 and noticing that we have $\xi_n = a_n + 1/\xi_{n+1}$ and $q_n/q_{n-1} = [a_n; a_{n-1}, \ldots, a_1]$ for all positive integers n. The last assertion of the theorem follows immediately, since the admissible values of ξ_{n+1} run exactly through the interval $]1, +\infty[$.

COROLLARY 1.4. *For any irrational number ξ and any non-negative integer n, we have*

$$\frac{1}{q_n(q_n + q_{n+1})} < \left| \xi - \frac{p_n}{q_n} \right| < \frac{1}{q_n q_{n+1}}.$$

PROOF. Writing $\xi = [a_0; a_1, a_2, \ldots]$ and $\xi_{n+1} = [a_{n+1}; a_{n+2}, \ldots]$, we observe that $a_{n+1} < \xi_{n+1} < a_{n+1} + 1$, and we get from Theorem 1.7 that

$$\frac{1}{q_n\big((a_{n+1} + 1)q_n + q_{n-1}\big)} < \left| \xi - \frac{p_n}{q_n} \right| < \frac{1}{q_n(a_{n+1}q_n + q_{n-1})}.$$

The corollary follows then from Theorem 1.3.

The following result of Legendre [359] provides a partial converse to Corollary 1.3.

THEOREM 1.8. *Let ξ be a real number. Any non-zero rational number a/b with*

$$\left| \xi - \frac{a}{b} \right| < \frac{1}{2b^2}$$

is a convergent of ξ.

PROOF. We assume that $\xi \neq a/b$ and we write $\xi - a/b = \varepsilon\theta/b^2$, with $\varepsilon = \pm 1$ and $0 < \theta < 1/2$. By Lemma 1.1, setting $a_{n-1} = 1$ if necessary, we may write $a/b = [a_0; a_1, \ldots, a_{n-1}]$, with n given by $(-1)^{n-1} = \varepsilon$, and we denote by $p_1/q_1, \ldots, p_{n-1}/q_{n-1}$ the convergents of a/b. Let ω be such that

$$\xi = \frac{p_{n-1}\omega + p_{n-2}}{q_{n-1}\omega + q_{n-2}} = [a_0; a_1, \ldots, a_{n-1}, \omega].$$

We check that

$$
\frac{\varepsilon\theta}{b^2} = \xi - \frac{a}{b} = \xi - \frac{p_{n-1}}{q_{n-1}} = \frac{1}{q_{n-1}}(\xi q_{n-1} - p_{n-1})
$$

$$
= \frac{1}{q_{n-1}} \cdot \frac{(-1)^{n-1}}{\omega q_{n-1} + q_{n-2}},
$$

whence

$$
\omega = \frac{1}{\theta} - \frac{q_{n-2}}{q_{n-1}}
$$

and $\omega > 1$. Denote by $[a_n; a_{n+1}, a_{n+2}, \dots]$ the continued fraction expansion of ω. Since $a_j \geq 1$ for all $j \geq 1$, we have

$$
\xi = [a_0; a_1, \dots, a_{n-1}, \omega] = [a_0; a_1, a_2, \dots],
$$

and Theorem 1.6 yields that $a/b = p_{n-1}/q_{n-1}$ is a convergent of ξ.

A less-known result of Fatou [239] (see Grace [259] for a complete proof) provides a satisfactory converse to Theorem 1.1. It asserts that if the real number ξ and the rational number a/b satisfy $|\xi - a/b| < 1/b^2$, then there exists an integer n such that

$$
\frac{a}{b} \quad \text{belongs to} \quad \left\{ \frac{p_n}{q_n}, \frac{p_{n+1} + p_n}{q_{n+1} + q_n}, \frac{p_{n+2} - p_{n+1}}{q_{n+2} - q_{n+1}} \right\}.
$$

DEFINITION 1.3. *The real number ξ is said to be badly approximable if there exists a positive constant $c_3(\xi)$ such that*

$$
\left| \xi - \frac{p}{q} \right| \geq \frac{c_3(\xi)}{q^2} \quad \text{for any rational } p/q \text{ distinct from } \xi.
$$

The set of badly approximable real numbers is denoted by $\mathcal{B}$.

Theorem 1.1 and Corollary 1.2 show that rational and quadratic real numbers are badly approximable. However, many other real numbers share this property, as follows from Theorem 1.9.

THEOREM 1.9. *An irrational real number ξ is badly approximable if, and only if, the sequence of its partial quotients is bounded. Consequently, the set $\mathcal{B}$ is uncountable.*

PROOF. Assume that ξ is badly approximable, and let c_4 be a positive real number such that $|\xi - p/q| > c_4/q^2$ for any rational p/q. Let n be a positive integer. It follows from Corollary 1.4 that $q_n \leq q_{n-1}/c_4$. Since Theorem 1.3 yields that $q_n \geq a_n q_{n-1}$, we get $a_n \leq 1/c_4$, and the sequence of partial quotients of ξ is bounded.

Conversely, if the partial quotients of ξ are not greater than a constant M, we then have $q_{n+1} \leq (M+1)q_n$ for any non-negative integer n. Furthermore, if a/b satisfies $|\xi - a/b| < 1/(2b^2)$, then, by Theorem 1.8 and Corollary 1.4, there exists a non-negative integer n such that $a/b = p_n/q_n$ and

$$\left| \xi - \frac{a}{b} \right| > \frac{1}{b(b+q_{n+1})} \geq \frac{1}{(M+2)b^2}.$$

This shows that ξ is badly approximable. The last assertion of the theorem follows from Theorem 1.6.

1.3 The theorem of Khintchine

In this Section, we aim to prove Theorem 1.10, due to Khintchine [317], by using the theory of continued fractions, as in [317] and in his book [323]. This is one of the first metric results in Diophantine approximation. We denote by λ the Lebesgue measure on the real line and, if I is a bounded real interval, we often simply write $|I| = \lambda(I)$ for its length. A set of Lebesgue measure zero is called a *null* set; the complement of a null set is termed a set of *full* measure, usually simply called *full*. As usual, we say that *almost no* points belong to a set if this set is null, while a full set contains *almost all* points.

THEOREM 1.10. *Let* $\Psi : \mathbb{R}_{\geq 1} \rightarrow \mathbb{R}_{>0}$ *be a continuous function such that* $x \mapsto x^2\Psi(x)$ *is non-increasing. Then, the set*

$$\mathcal{K}_1^*(\Psi) := \left\{ \xi \in \mathbb{R} : \left| \xi - \frac{p}{q} \right| < \Psi(q) \text{ for infinitely many rational} \right.$$

$$\left. \text{numbers } \frac{p}{q} \right\}$$

has Lebesgue measure zero if the sum $\sum_{q=1}^{+\infty} q\Psi(q)$ *converges and has full Lebesgue measure otherwise.*

The function Ψ occurring in Theorem 1.10 is called an *approximation function*. Throughout this book, we assume for commodity that Ψ is continuous on $\mathbb{R}_{\geq 1}$, although it only requires to be defined for every sufficiently large integer. The elements of $\mathcal{K}_1^*(\Psi)$ are termed Ψ-approximable, or approximable at order Ψ.

Since $x \mapsto \Psi(x)$ is non-increasing in Theorem 1.10, for any ξ in $\mathcal{K}_1^*(\Psi)$ there are infinitely many *reduced* rational numbers p/q with $|\xi - p/q| < \Psi(q)$. Thus, we may add in the definition of $\mathcal{K}_1^*(\Psi)$ the extra assumption 'with p and q coprime' without any change in the conclusion. Furthermore, we point out that the assumption '$x \mapsto x^2\Psi(x)$ is non-increasing' can be removed in the convergence half of Theorem 1.10.

An alternative proof of Theorem 1.10, valid under slightly less restrictive conditions on the function Ψ, is given in Chapter 6. It is based on an improvement of the divergence half of the Borel–Cantelli Lemma.

An immediate consequence of Theorem 1.10 is that almost all real numbers ξ have infinitely many convergents with $|\xi - p/q| < q^{-2}(\log q)^{-1}$, and almost all real numbers ξ have only finitely many convergents with $|\xi - p/q| < q^{-2}(\log q)^{-2}$.

Throughout the proof of Theorem 1.10, we often need the following auxiliary result, referred to as the (convergence half of the) Borel–Cantelli Lemma.

LEMMA 1.3. *Let* $(E_n)_{n \geq 0}$ *be a sequence of Borelian real subsets such that the sum* $\sum_{n \geq 0} \lambda(E_n)$ *converges. We then have*

$$\lambda \left(\bigcap_{N \geq 1} \bigcup_{n \geq N} E_n \right) = 0,$$

that is, almost all real numbers belong to only a finite number of sets E_n.

PROOF. Let ε be a positive real number. Since $\sum_{n \geq 0} \lambda(E_n)$ converges, there exists N_0 such that, for any integer $N \geq N_0$, we have $\lambda(\bigcup_{n \geq N} E_n) < \varepsilon$; hence, in particular, $\lambda(\bigcap_{N \geq 1} \bigcup_{n \geq N} E_n) < \varepsilon$.

Before turning to the proof of Theorem 1.10, we state an application of Lemma 1.2.

DEFINITION 1.4. *The real number* ξ *is said to be* very well approximable *if there exists a positive real number* τ *such that*

$$\left| \xi - \frac{p}{q} \right| < \frac{1}{q^{2+\tau}} \quad \text{for infinitely many rational numbers } \frac{p}{q}.$$

As pointed out in [86], it would seem more appropriate to use the terminology 'well approximable' instead of 'very well approximable'. To avoid confusion, we follow current usage.

COROLLARY 1.5. *The Lebesgue measure of the set of very well approximable numbers is equal to zero.*

PROOF OF COROLLARY 1.5 AND OF THE CONVERGENCE HALF OF THEOREM 1.10. Since $\mathcal{K}_1^*(\Psi)$ is invariant by translation by 1, it is plainly enough to prove the conclusion of Theorem 1.10 for the set $\mathcal{K}_1^*(\Psi) \cap [0, 1]$. For any positive integer n, denote by E_n the union of the intervals $[m/n - \Psi(n), m/n + \Psi(n)] \cap [0, 1]$ for $m = 0, \ldots, n$. Observe that $\mathcal{K}_1^*(\Psi) \cap [0, 1]$ is contained in $\bigcap_{N \geq 1} \bigcup_{n \geq N} E_n$. If the sum $\sum_{q=1}^{+\infty} q\Psi(q)$ converges, then $\sum_{n \geq 1} \lambda(E_n)$ also converges, and Lemma 1.2 yields that $\mathcal{K}_1^*(\Psi) \cap [0, 1]$ has

Lebesgue measure zero. In particular, taking for Ψ the functions $x \mapsto x^{-2-1/k}$, where k is a positive integer, we get Corollary 1.5.

The divergence half of Theorem 1.10 is much more difficult to prove. We begin with some preliminary results on continued fractions. As a standard matter, if ξ is a real number in $[0, 1[$, we denote its continued fraction expansion by $[0; a_1, a_2, \ldots]$ and the sequence of its convergents by $(p_n/q_n)_{n \geq 0}$. Let $n \geq 1$ be an integer and $\underline{k} = (k_1, \ldots, k_n)$ be a n-tuple of positive integers. We denote by $F_{\underline{k}}$ the set of real numbers in $[0, 1]$ whose partial quotients satisfy $a_i = k_i$ for $i = 1, \ldots, n$. By Theorem 1.7, $F_{\underline{k}}$ is a closed interval bounded by p_n/q_n and $(p_{n-1} + p_n)/(q_{n-1} + q_n)$. We can write $F_{\underline{k}}$ as the union

$$F_{\underline{k}} = \bigcup_{s \geq 1} F_{(\underline{k}, s)},$$

where $F_{(\underline{k}, s)}$ denotes the sub-interval of $F_{\underline{k}}$ composed by the real numbers with $a_{n+1} = s$. This union is not disjoint but the intersection of two sets $F_{(\underline{k}, s)}$ and $F_{(\underline{k}, s')}$ with $s' > s \geq 1$ consists of at most one point, which is a rational number. Since

$$\lambda(F_{(\underline{k}, s)}) = \left| \frac{sp_n + p_{n-1}}{sq_n + q_{n-1}} - \frac{(s+1)p_n + p_{n-1}}{(s+1)q_n + q_{n-1}} \right|$$

$$= \frac{1}{s^2 \left(q_n + \frac{q_{n-1}}{s} \right) \left(\frac{q_{n-1}}{s} + \left(1 + \frac{1}{s} \right) q_n \right)},$$

and

$$\lambda(F_{\underline{k}}) = \frac{1}{q_n(q_n + q_{n-1})},$$

we get

$$\frac{\lambda(F_{(\underline{k}, s)})}{\lambda(F_{\underline{k}})} = \frac{1}{s^2} \cdot \frac{1 + \frac{q_{n-1}}{q_n}}{1 + \frac{q_{n-1}}{sq_n}} \cdot \frac{1}{1 + \frac{1}{s} + \frac{q_{n-1}}{sq_n}},$$

and thus

$$\frac{1}{3s^2} < \frac{\lambda(F_{(\underline{k}, s)})}{\lambda(F_{\underline{k}})} < \frac{2}{s^2}. \tag{1.7}$$

Two intermediate results are needed towards the proof of Theorem 1.10. The first one is the Borel–Bernstein theorem [110, 98, 111, 99].

THEOREM 1.11. *Let $(u_n)_{n \geq 1}$ be a sequence of positive real numbers. If the sum $\sum_{n \geq 1} u_n^{-1}$ diverges, then, for almost all $\xi = [0; a_1, a_2, \ldots]$ in $[0, 1[$, there exist infinitely many integers n such that $a_n \geq u_n$. Further, if this sum* *converges, then, for almost all $\xi = [0; a_1, a_2, \ldots]$ in $[0, 1[$, there exist only a finite number of integers n such that $a_n \geq u_n$.*

PROOF. Let m, n, and j be positive integers with $n \geq 2$ and $1 \leq j \leq n$. Let $\underline{k}$ be a $(m + j)$-tuple of positive integers, and set $\underline{k} := (k_1, \ldots, k_m, k_{m+1}, \ldots, k_{m+j})$. Let $X > 1$ be a real number. We infer from (1.7) that

$$\sum_{s<X} \lambda(F_{(\underline{k},s)}) = \sum_{s\geq 1} \lambda(F_{(\underline{k},s)}) - \sum_{s\geq X} \lambda(F_{(\underline{k},s)}) \leq \left(1 - \frac{1}{3}\sum_{s\geq X}\frac{1}{s^2}\right)\lambda(F_{\underline{k}})$$

$$\leq \left(1 - \frac{1}{3(1+X)}\right)\lambda(F_{\underline{k}}).$$

(1.8)

Denote by $F_{m,j}$ the union of the intervals $F_{\underline{k}}$ over all the $(m + j)$-tuples $\underline{k}$ with $k_i \geq 1$ for $i = 1, \ldots, m$ and $k_{m+i} < u_{m+i}$ for $i = 1, \ldots, j$. By choosing $X = u_{m+n}$ in (1.8), we get the upper bound

$$\lambda(F_{m,n}) \leq \left(1 - \frac{1}{3(1+u_{m+n})}\right)\lambda(F_{m,n-1}),$$

whence, by induction,

$$\lambda(F_{m,n}) \leq \prod_{i=2}^{n}\left(1 - \frac{1}{3(1+u_{m+i})}\right)\lambda(F_{m,1}).$$

The divergence of the sum $\sum_{n\geq 1} u_n^{-1}$ implies that, for any positive integer m, the product

$$\prod_{i=2}^{n}\left(1 - \frac{1}{3(1+u_{m+i})}\right)$$

and thus $\lambda(F_{m,n})$ tend to 0 when n tends to infinity. Since the set B_m of real numbers satisfying $a_{m+i} < u_{m+i}$ for all $i \geq 1$ is contained in every set $F_{m,n}$, its Lebesgue measure is zero. Consequently, if ξ belongs to none of the sets B_m, and that is indeed the case for almost all ξ, then there exist infinitely many integers n such that $a_n \geq u_n$. This proves the first part of the theorem.

The other part is easier. Assume that the sum $\sum_{n\geq 1} u_n^{-1}$ converges. Let $n \geq 1$ be an integer and denote by E_n the set of real numbers satisfying $a_n \geq u_n$. The right inequality of (1.7) yields that

$$\sum_{s\geq u_{n+1}} \lambda(F_{(\underline{k},s)}) < \sum_{s\geq u_{n+1}} \frac{2}{s^2}\lambda(F_{\underline{k}}) \leq \frac{4\lambda(F_{\underline{k}})}{u_{n+1}}.$$

(1.9)

By (1.9), we get $\lambda(E_{n+1}) < 4/u_{n+1}$, thus the sum $\sum_{n \geq 1} \lambda(E_n)$ converges and we infer from Lemma 1.2 that the Lebesgue measure of the set of real numbers belonging to infinitely many sets E_n is zero.

The next statement, due to Borel [110], is a direct consequence of Theorem 1.11.

COROLLARY 1.6. *The Lebesgue measure of the set $\mathcal{B}$ of badly approximable real numbers is equal to zero.*

PROOF. Applying Theorem 1.11 to the divergent sequence $(1/n)_{n \geq 1}$, we get that, for almost all real numbers $\xi = [a_0; a_1, a_2, \dots]$, there exist infinitely many positive integers n such that $a_n \geq n$. Hence, almost all real numbers have unbounded partial quotients.

The set $\mathcal{B}$ is not too small in the sense that its Hausdorff dimension is equal to 1, see Exercise 5.1.

THEOREM 1.12. *There exists a positive real number B and, for almost all real numbers ξ, an integer $n_0(\xi)$ such that $q_n < e^{Bn}$ for all integers $n \geq n_0(\xi)$.*

PROOF. For a real number $g \geq 1$ and an integer $n \geq 2$, denote by $E_n(g)$ the set of real numbers ξ in $[0, 1[$ such that $a_1 \dots a_n \geq g$. By (1.5), Theorem 1.3, and Theorem 1.7, this set is a union of intervals, each of which being of length

$$\left| \frac{p_n}{q_n} - \frac{p_n + p_{n-1}}{q_n + q_{n-1}} \right| < \frac{1}{q_n^2} < \frac{1}{(a_1 \dots a_n)^2}.$$

Consequently, we get

$$\lambda(E_n(g)) < \sum_{a_1 \dots a_n \geq g} \frac{1}{(a_1 \dots a_n)^2}.$$

In order to estimate the above summation, we compare it with an integral and we notice that, for any positive integer a, we have

$$\frac{1}{a^2} \leq 2 \int_a^{a+1} \frac{dx}{x^2}.$$

Thus, if

$$J_n(g) := \int \dots \int_{\substack{x_i \geq 1 \\ x_1 \dots x_n \geq g}} \frac{dx_1}{x_1^2} \dots \frac{dx_n}{x_n^2},$$

we get $\lambda(E_n(g)) < 2^n J_n(g)$. For $g \leq 1$, the integral is on the whole domain $x_1 \geq 1, \dots, x_n \geq 1$, thus $J_n(g) = 1$. For $g > 1$, an easy induction on n yields that

$$J_n(g) = \frac{1}{g} \sum_{i=0}^{n-1} \frac{(\log g)^i}{i!}. \tag{1.10}$$

Consequently, if we choose $g = e^{An}$, where $A > 1$ is a real number which will be fixed afterwards, we deduce from (1.10) that

$$\lambda\left(E_n(e^{An})\right) < 2^n e^{-An} \sum_{i=0}^{n-1} \frac{(An)^i}{i!} \le 2^n e^{-An} n \frac{(An)^n}{n!} < 2^n e^{-An} n (Ae)^n.$$

If A is large enough, then the sum $\sum_{n\ge 1} \lambda(E_n(e^{An}))$ converges and we infer from Lemma 1.2 that almost all real numbers ξ in $[0, 1]$ belong to only a finite number of sets $E_n(e^{An})$. In other words, for almost all ξ in $[0, 1]$ and for any sufficiently large integer n (in terms of ξ), we have $a_1 \dots a_n < e^{An}$ and then, by Theorem 1.3,

$$q_n < 2 a_n q_{n-1} < \dots < 2^n e^{An}.$$

Since for any real number ξ the denominators of the convergents of ξ and $\xi + 1$ are the same, the theorem is proved.

Khintchine [317] proved that any real number B strictly greater than $\log 2 + \exp(\sqrt{2 \log 2})$ satisfies the conclusion of Theorem 1.12. Actually, a much stronger statement holds true: there exists a real number ℓ such that for almost all real numbers ξ the sequence $(\sqrt[n]{q_n})_{n\ge 1}$ converges to ℓ. This was established in 1936 by Khintchine [322] and, the same year, Lévy [366] proved that $\ell = \exp(\pi^2/(12 \log 2))$.

We are now ready to complete the proof of Theorem 1.10, the main result of this Section.

COMPLETION OF THE PROOF OF THEOREM 1.10. We assume that the sum $\sum_{q\ge 1} q\Psi(q)$ diverges and we aim to prove that the set $\mathcal{K}_1^*(\Psi) \cap [0, 1]$ has full measure. For any $x > 0$, set $\Phi(x) := e^{2Bx} \Psi(e^{Bx})$, where B is the real number given by Theorem 1.12. By assumption, the function Φ is non-increasing. Further, for any positive real numbers a and A with $a < A$, we have

$$\int_a^A \Phi(x)\,\mathrm{d}x = \frac{1}{B} \int_{e^{Ba}}^{e^{BA}} u\Psi(u)\,\mathrm{d}u,$$

which diverges when A tends to infinity. Consequently, the sum $\sum_{q\ge 1} \Phi(q)$ diverges. Theorem 1.11 asserts then that for almost all real numbers ξ in $[0, 1]$ we have $a_{n+1} \ge 1/\Phi(n)$ for infinitely many integers n. Thus, for all these integers n we have

$$\left| \xi - \frac{p_n}{q_n} \right| \le \frac{1}{q_n q_{n+1}} < \frac{1}{a_{n+1} q_n^2} \le \frac{\Phi(n)}{q_n^2}.$$

However, by Theorem 1.12, there exists a real number $B > 0$ such that, for almost all ξ, we have $q_n < e^{Bn}$, that is, $n > (\log q_n)/B$, provided that n is large enough in terms of ξ. Since Φ is non-increasing, we get that, for almost all ξ in $[0, 1]$, there exist infinitely many integers n such that

$$\left| \xi - \frac{p_n}{q_n} \right| < \frac{1}{q_n^2} \, \Phi\left(\frac{\log q_n}{B} \right) = \Psi(q_n),$$

which concludes the proof of the theorem.

1.4 The Duffin–Schaeffer Conjecture

In the statement of Theorem 1.10, the assumption on the approximation function Ψ is quite restrictive, and it would be desirable to weaken it. For instance, the alternative proof of Theorem 1.10 given in Chapter 6 requires only that Ψ is non-increasing. The result generally conjectured involves the function φ, the Euler totient function, defined for all positive integers q by

$$\varphi(q) = \mathrm{Card}\{1 \le x \le q : \gcd(x, q) = 1\},$$

and has been proposed by Duffin and Schaeffer at the end of [213].

CONJECTURE (DUFFIN–SCHAEFFER). *Let* $\Psi : \mathbb{R}_{\ge 1} \to \mathbb{R}_{\ge 0}$ *be some continuous function. Then the set*

$$\left\{ \xi \in \mathbb{R} : \left| \xi - \frac{p}{q} \right| < \Psi(q) \text{ for infinitely many rationals } \frac{p}{q} \right.$$
$$\left. \text{with } gcd(p, q) = 1 \right\}$$

has full Lebesgue measure if the sum $\sum_{q=1}^{+\infty} \varphi(q)\Psi(q)$ *diverges.*

When the above sum converges, the above set has Lebesgue measure zero, as easily follows from Lemma 1.2.

As observed after the statement of Theorem 1.10, Khintchine's result holds also when we demand that the integers p and q are coprime. This is a consequence of the assumption made on Ψ and this remark does not apply for a general function Ψ. Thus, we must distinguish between approximation by rationals not necessarily reduced and approximation by reduced rationals. According to [213], '(...) the more natural formulation of this problem is in terms of reduced fractions'.

The notorious difficulties in the Duffin–Schaeffer Conjecture are due to the fact that it is very difficult to control the pairwise intersections of the intervals $]p/q - \Psi(q), p/q + \Psi(q)[$, where p and $q \ge 1$ are integers. However, it has been proved that the Conjecture holds true under some additional hypotheses. We choose to quote only two results, a first one due to Duffin and Schaeffer

[213], and a second one to Erdös [228]; the interested reader can find an exhaustive survey of that problem (including detailed proofs of Theorems 1.13 to 1.15 below) in Harman [273] (see also his survey [274]).

THEOREM 1.13. *The Duffin–Schaeffer Conjecture holds true if we assume that there exists a real number c such that the function* $x \mapsto x^c \Psi(x)$ *is decreasing.*

THEOREM 1.14. *The Duffin–Schaeffer Conjecture holds true if there exists a positive real number* κ *such that, for every positive integer q, we have* $\Psi(q) = 0$ *or* κ/q^2. *In particular, if* $(n_i)_{i \geq 1}$ *is a strictly increasing sequence of positive integers, then, for almost all real numbers* ξ, *it contains infinitely many (resp. a finite number of) denominators of convergents to* ξ *if the sum* $\sum_{i \geq 1} \varphi(n_i)/n_i^2$ *diverges (resp. converges).*

In the opposite direction, the next result, due to Duffin and Schaeffer [213], asserts that the divergence of the sum $\sum_{q \geq 1} q \Psi(q)$ is certainly not sufficient to ensure the existence of infinitely many solutions to $|\xi - p/q| < \Psi(q)$ for almost all ξ.

THEOREM 1.15. *There exists a positive function* Ψ *such that the sum* $\sum_{q \geq 1} q \Psi(q)$ *diverges and*

$$\left| \xi - \frac{p}{q} \right| < \Psi(q)$$

has only a finite number of solutions for almost all ξ.

1.5 Complementary results on continued fractions

In this Section, we list a few complementary results on continued fractions. The reader is referred to the books quoted at the beginning of Section 1.2 for deeper statements and proofs (notice that proofs of Theorems 1.16, 1.18, and 1.19 are also given in Exercises 1.2 and 1.3).

Corollary 1.2 asserts that quadratic numbers are badly approximable numbers, thus, by Theorem 1.9, their partial quotients are bounded. However, much more is known: the sequence of their partial quotients is ultimately periodic.

THEOREM 1.16. *The real irrational number* $\xi = [a_0; a_1, a_2, \dots]$ *has a periodic continued fraction expansion (that is, there exist integers* $k \geq 0$ *and* $n \geq 1$ *such that* $a_{m+n} = a_m$ *for all* $m \geq k$) *if, and only if,* ξ *is a quadratic irrationality.*

The 'only if' part is due to Euler [230], and the 'if' part was established by Lagrange [347] in 1770.

The next result dates back to (at least) 1877 and can be found in Serret [520].

THEOREM 1.17. *Let* $\xi = [a_0; a_1, a_2, \ldots]$ *and* $\eta = [b_0; b_1, b_2, \ldots]$ *be two irrational numbers. There exist integers* u *and* v *such that* $a_{u+n} = b_{v+n}$ *for every positive integer* n *if, and only if, there exist integers* a, b, c, *and* d *such that* $|ad - bc| = 1$ *and* $\eta = (a\xi + b)/(c\xi + d)$.

Two real numbers satisfying the equivalent conditions of Theorem 1.17 are called *equivalent*.

Hurwitz [284] improved the second assertion of Theorem 1.1.

THEOREM 1.18. *For any real irrational number* ξ, *there exist infinitely many rationals* p/q *satisfying*

$$\left| \xi - \frac{p}{q} \right| < \frac{1}{\sqrt{5}q^2}.$$

Further, the constant $\sqrt{5}$ *cannot be replaced by a larger real number.*

We observe that the Golden Section $(1 + \sqrt{5})/2 = [1; 1, 1, \ldots, 1, \ldots]$ is, up to equivalence, the irrational number which is the most badly approximable by rational numbers. As shown by Hurwitz [284], Theorem 1.18 can be improved if, besides the rationals, we also exclude the numbers which are equivalent to the Golden Section.

THEOREM 1.19. *For any irrational real number* ξ *which is not equivalent to* $(1 + \sqrt{5})/2$, *there exist infinitely many rationals* p/q *satisfying*

$$\left| \xi - \frac{p}{q} \right| < \frac{1}{\sqrt{8}q^2}.$$

The constant $\sqrt{8}$ *cannot be replaced by a larger real number.*

In other words, the assumptions of Theorem 1.19 are satisfied by any irrational ξ having infinitely many partial quotients greater than or equal to 2. The limiting case is obtained with real numbers equivalent to $(1 + \sqrt{2})/2$.

More generally, for a real number ξ we define

$$\nu(\xi) = \liminf_{q \to +\infty} q \|q\xi\|,$$

where $\| \cdot \|$ denotes the function distance to the nearest integer. Clearly, $\nu(\xi) = 0$ holds for any rational ξ. The set of values taken by the function ν is called the Lagrange spectrum. By Theorem 1.18, it is contained in $[0, 1/\sqrt{5}]$.

THEOREM 1.20. *There exists a sequence of numbers*

$$\nu_1 = \frac{1}{\sqrt{5}} > \nu_2 = \frac{1}{\sqrt{8}} > \nu_3 > \nu_4 > \cdots$$

tending to $1/3$ *such that, for all* v_i, *there is, up to equivalence, a finite number of real numbers* ξ *satisfying* $v(\xi) = v_i$.

More results on the Lagrange spectrum can be found in Cusick and Flahive [173].

1.6 Exercises

EXERCISE 1.1. Use Liouville's Theorem 1.2 to prove that if the denominators q_n of the convergents of ξ satisfy

$$\limsup_{n \to +\infty} \frac{\log \log q_n}{n} = +\infty,$$

then ξ is transcendental.

EXERCISE 1.2. (Proof of Theorem 1.16). Show that any irrational real number having a periodic continued fraction expansion is a quadratic irrationality. Prove the converse, following Steinig [543]. Let ξ be a quadratic real number and define the sequences $(a_n)_{n \geq 0}$ and $(\xi_n)_{n \geq 1}$ by the algorithm given in Section 1.2. Set $\xi_0 := \xi$.

1) Show, by induction, that for each non-negative integer n there is an integer polynomial $f_n(x) := A_n x^2 + B_n x + C_n$, with $B_n^2 - 4A_n C_n$ positive and not a square, such that $f_n(\xi_n) = 0$. Prove that A_{n+1}, B_{n+1} and C_{n+1} are given by

$$A_{n+1} = a_n^2 A_n + a_n B_n + C_n, \quad B_{n+1} = 2a_n A_n + B_n, \quad \text{and} \quad C_{n+1} = A_n.$$

$$(1.11)$$

2) Observe that

$$D := B_0^2 - 4A_0 C_0 = B_n^2 - 4A_n C_n, \quad \text{for any } n \geq 0, \tag{1.12}$$

and deduce from (1.11) that there exists an infinite set $\mathcal{N}$ of distinct positive integers such that $A_n A_{n-1}$ is negative for any n in $\mathcal{N}$.

3) Deduce from (1.12) that

$$|B_n| < \sqrt{D}, \quad |A_n| \leq D/4, \quad \text{and} \quad |C_n| \leq D/4,$$

for any n in $\mathcal{N}$. Conclude.

EXERCISE 1.3. (Proof of Hurwitz Theorems 1.18 and 1.19, following Forder [248].)

Let ξ be an irrational number and assume that its convergents p_{n-1}/q_{n-1}, p_n/q_n, and p_{n+1}/q_{n+1} satisfy $|\xi - p_j/q_j| \geq 1/(\sqrt{5}q_j^2)$ for $j = n-1, n, n+1$.

1) Prove that we have

$$\frac{1}{q_{n-1}q_n} \geq \frac{1}{\sqrt{5}}\left(\frac{1}{q_{n-1}^2} + \frac{1}{q_n^2}\right),$$

and deduce that q_n/q_{n-1} and q_{n-1}/q_n belong to the interval $](\sqrt{5}-1)/2$, $(\sqrt{5}+1)/2[$. Prove that the same conclusion also holds for q_n/q_{n+1} and q_{n+1}/q_n. Conclude.

2) With a suitable adaptation of the above proof, show that if the convergents p_n/q_n of an irrational number ξ satisfy $|\xi - p_n/q_n| \geq 1/(\sqrt{8}q_n^2)$ from some integer n_0 onwards then $q_{n+1} < 2q_n + q_{n-1}$ holds for any sufficiently large integer n. Conclude.

EXERCISE 1.4. Improvement of Dirichlet's Theorem 1.1.

Theorem 1.1 asserts that if ξ is a given real number, then, for any positive integer Q, there exist integers p and q satisfying

$$1 \leq q \leq Q \quad \text{and} \quad |q\xi - p| < c/Q, \tag{1.13}$$

with $c = 1$. Show that this result holds true for a real number ξ with a constant c strictly less than 1 if, and only if, ξ is badly approximable.

A stronger statement is due to Davenport and Schmidt [183]:

Let $\xi = [a_0; a_1, a_2, \ldots]$ be a real number, and set

$$c_5(\xi) = \liminf_{n\to+\infty} [0; a_n, a_{n-1}, \ldots, a_1] \times [0; a_{n+1}, a_{n+2}, \ldots].$$

If $c > 1/(1+c_5(\xi))$, then (1.13) has a solution for all Q sufficiently large, and this is not true if $c < 1/(1+c_5(\xi))$.

Use Theorem 1.7 to prove this result and show that the largest possible value for $c_5(\xi)$ is $(3-\sqrt{5})/2$.

EXERCISE 1.5. A result of Jarník [292], Satz 6, on approximation order.

Let $\Psi : \mathbb{R}_{\geq 1} \to \mathbb{R}_{>0}$ be a non-increasing function such that $\Psi(x) = o(x^{-2})$ as x tends to infinity. Using the theory of continued fractions, prove that there exists an uncountable set of real numbers ξ which are approximable at order Ψ, but not at any order $c\Psi$, with $0 < c < 1$.

EXERCISE 1.6. In 1962, Erdős [227] showed that every real number (*resp.* every non-zero real number) is the sum (*resp.* product) of two Liouville numbers. He gave two proofs: a first one is direct and constructive, while a second one rests on Baire's Theorem. His result has been extended by Rieger [471], Schwarz [518], and Alniaç k and Saias [19], who established the following statement (recall that a G_δ-set is a countable intersection of open sets):

Let I be a real interval of positive length. Let G be a real G_δ-set dense in I and $(f_n)_{n\geq 0}$ be a sequence of continuous, nowhere locally constant, real functions on I. Then, the set $\cap_{n\geq 0} f_n^{-1}(G)$ is a G_δ-set dense in I.

Prove Alniaç k and Saias' assertion. Show that the set of Liouville numbers is a dense G_δ-set and deduce Erdös' result.

1.7 Notes

• In 1844 Liouville [368] used a result of Lagrange on continued fractions to prove that, for any n, the n-th partial quotient of an algebraic number of degree $d \geq 2$ is less than some number (independent of n) times the $(d-2)$-th power of the denominator of the $(n-1)$-th convergent. Then, he observed that if ξ is a real number such that, for any n, its n-th partial quotient a_n is defined in terms of the denominator q_{n-1} of its $(n-1)$-th convergent by taking $a_n = q_{n-1}^{q_{n-1}}$ or $a_n = q_{n-1}^n$, then the number ξ must be transcendental. These numbers are, historically, the first examples of transcendental numbers. At the end of [368], Liouville mentioned that analogous results exist for ordinary sums, including $\sum_{m\geq 1} a^{-m!}$ for any integer $a \geq 2$. In 1851, he gave [370] a complete proof of this last assertion. In a second note of 1844, Liouville [369] simplified the proof given in his first note [368] by removing the use of Lagrange's result. The reader wishing more information is directed to Chapter XII of Lützen [374].

• Dirichlet's Theorem 1.1 has two natural multidimensional extensions (as also follows from Theorem B.2). Let $n \geq 2$ be an integer and $\xi_1, \ldots, \xi_n$ be real numbers. On the one hand (approximation of linear forms), for any positive integer Q, there exist integers $p_1, \ldots, p_n, q$ not all zero satisfying

$$|p_1\xi_1 + \ldots + p_n\xi_n + q| < Q^{-n}, \quad \max\{|p_1|, \ldots, |p_n|\} \leq Q.$$

On the other hand (simultaneous approximation), for any positive integer Q, there exist integers $p_1, \ldots, p_n, q$ not all zero satisfying

$$\max\{|q\xi_1 - p_1|, \ldots |q\xi_n - p_n|\} < Q^{-1}, \quad |q| \leq Q^n.$$

Davenport and Schmidt [184] (see also [183] and Schmidt [509] for a further generalization) proved that for almost every n-tuple $(\xi_1, \ldots, \xi_n)$ none of these two forms of Dirichlet's Theorem can be improved (in the same sense as in Exercise 1.4). Raisbeck [466] established a theorem in the opposite direction (best possible for $n = 2$), later improved upon by Kaindl [306] for $n \geq 3$ (see also Tichy [559] and Langmayr [350] for systems of linear forms).

Further extensions of Dirichlet's Theorem, where the approximants must satisfy some restriction, have been considered by Obreškov [441], Rogers [475], Schmidt [511], and Thurnheer [554, 555, 556, 557, 558].

Khintchine [320] proved that, given any function $\varepsilon(Q)$ which tends to 0 as Q tends to infinity, no matter how rapidly, there exists a pair (ξ_1, ξ_2) of real numbers, with $1, \xi_1, \xi_2$ linearly independent over the rationals, such that the inequalities

$$|p_1\xi_1 + p_2\xi_2 + q| < \varepsilon(Q), \quad 1 \le \max\{|p_1|, |p_2|\} \le Q$$

are soluble in integers for every Q sufficiently large (see Theorem XIV of Chapter V of [155]).

• Let ξ be a real irrational number and k be a positive real number. Extending Fatou's result [239] quoted after the proof of Theorem 1.8, Worley [601] has expressed the rational solutions a/b of $|\xi - a/b| < k/b^2$ in terms of the convergents of ξ.

• An n-tuple $(\xi_1, \ldots, \xi_n)$ of real numbers is said to be badly approximable if there exists a positive constant c_6 such that $\max_{1 \le i \le n} \|q\xi_i\| > c_6\, q^{-1-1/n}$ for any positive integer q. This extends Definition 1.3. The first proof of the existence of continuum-many badly approximable pairs of distinct real numbers is due to Davenport [175]. For further references and an exposition of the theory of α-β games, see Chapter III of [512]. Schmidt [514] proved the existence of infinite sets of real numbers whose finite subsets do not have good simultaneous rational approximations.

• A very general form of Khintchine's Theorem 1.10 has been obtained by Groshev [260], extending earlier work of Khintchine [319] to systems of linear forms. Let m and n be positive integers and $\Psi : \mathbb{R}_{\ge 1} \to \mathbb{R}_{>0}$ be continuous and such that $x \mapsto x^{m-1}\,\Psi(x)^n$ is non-increasing. Groshev proved that for either almost all or almost no matrices M in $\mathbb{R}^{mn}$ there are infinitely many integer vectors $\underline{q}$ in $\mathbb{Z}^m$ such that $|\langle \underline{q}M \rangle| < \Psi(|\underline{q}|)$, depending on whether the series $\sum_{x=1}^{+\infty} x^{m-1}\,\Psi(x)^n$ diverges or converges. Here, for a vector $\underline{\xi} = (\xi_1, \ldots, \xi_n)$, we set $|\underline{\xi}| = \max\{|\xi_1|, \ldots, |\xi_n|\}$ and $\langle \underline{\xi} \rangle = (\langle \xi_1 \rangle, \ldots, \langle \xi_n \rangle)$, where $\langle \underline{\xi} \rangle$ lies in $]-1/2, 1/2]$ and differs from ξ_i to within an integer.

• Theorem 1.10 asserts in the divergence case the existence of infinitely many rational numbers p/q with $|\xi - p/q| < \Psi(q)/q$, but the method of proof does not give any information regarding the asymptotic behaviour of their number. An asymptotic formula has been obtained by Erdös [226] and LeVeque [364], and, with a good error term, by Schmidt [500, 502], see also Chapter III of [512], Chapter 1 of [540], Chapter 4 of [273], and the survey of Harman [274].

• Harman [272] established an analogue of Theorem 1.10 where numerators and denominators are restricted to the set of prime numbers. See also Chapter 6 of his book [273]. An asymptotic formula for the number of solutions in the divergence case, and under some extra hypothesis on the function Ψ, has been established by Jones in the second chapter of his Ph.D. thesis [302]. Furthermore, Jones [303] extended the results of [272] to simultaneous approximation.

• For inhomogeneous approximation, we refer the reader to Cassels [155], Hardy and Wright [271], Khintchine [323], and Gruber and Lekkerkerker [262]. An inhomogeneous analogue of Theorem 1.10, due to Schmidt [502] (who also obtained an asymptotic formula for the number of solutions in the divergence case), follows from Exercise 6.2 and Theorem 6.1.

• Sullivan [547] established a variant of Khintchine's Theorem 1.10 under a weaker assumption. His proof is of a geometric nature.

• Theorems 1.18 and 1.19 are closely related to the works of Korkine and Zolotareff [337] and Markoff [406], where indefinite, binary quadratic forms are considered. Hurwitz was first to give these results in the form of a statement about Diophantine approximation.

• Let $b \geq 2$ be an integer. A real number is said to be *simply normal* in base b if each digit $0, 1, \ldots, b - 1$ occurs in its expansion in base b with frequency $1/b$. It is said to be *normal* if it is simply normal in any base $b \geq 2$. R. C. Baker (see Montgomery [427], page 203, and Queffélec and Ramaré [465]) observed that there exist badly approximable numbers which are normal. Using a similar method, that is, combining the results of Bluhm [102, 103] (inspired by a work of Kaufman [315]) with a Theorem of Davenport, Erdös and LeVeque [177], Bugeaud [126] proved that there exist Liouville numbers which are normal. Conversely, an explicit example of a Liouville number which is simply normal in no base has been given by Martin [408] (see also [126]).

• Schmidt [507] established that, for any irrational number ξ which is not a Liouville number, there exists an irrational number α such that ξ/α is a Liouville number. Burger [144] extended Erdös' result [227] quoted in Exercise 1.6, and he proved that there exists a Liouville number ξ such that $\log \xi$ is also a Liouville number. Alniaçk [16] established the existence of Liouville numbers with special properties.

• Hall [270] proved that any real number in [0, 1] can be expressed as a sum of two real numbers having partial quotients less than or equal to 4. See also related works of Astels [34, 35, 36].

• From the topological point of view, Gruber [261] proved that the set of badly approximable real numbers is 'small' (precisely, it is a meager set), while the set of Liouville numbers is 'large' (its complement is a meager set).

• Kargaev and Zhigljavsky [309] established metric results related to the approximation of real numbers by rational numbers with bounded denominators.

• A different metric aspect of rational approximation has been investigated by Kühnlein [345].

• Viola [580] studied rational Diophantine approximation in short intervals.

• R. C. Baker [50] proved that real numbers having k distinct rational approximations with given denominators $q_1, \ldots, q_k$ are quite sparse.

• The constant $c_5(\xi)$ occurring in Exercise 1.4 allows us to define the so-called Dirichlet spectrum, studied, for example, by Ivanov [287] and Kopetzky [335]. Diviš [197] and Burger [145] considered closely related problems.

2

Approximation to algebraic numbers

In this book, we study various questions related to classifications of real numbers and we mainly focus our attention on the approximation of real *transcendental* numbers by algebraic numbers. In the present Chapter however, we briefly review the most important results which have followed Liouville's Theorem 1.2 and deal with algebraic approximation to algebraic numbers. Since a broad literature is available on this topic, we omit most of the proofs and refer the reader to, for example, the monographs of Mahler [388], Schmidt [510, 512], A. Baker [44], and Feldman and Nesterenko [244] for further information.

Completeness is not the only reason for making this survey. Indeed, some results of the present Chapter will be used in subsequent parts of the book. For instance, Theorem 2.7 (or Theorem 2.6) is crucial for proving the main result of Chapter 7, namely the existence of T-numbers. These real transcendental numbers with very specific properties of approximation by algebraic numbers are defined in Chapter 3.

We divide our exposition into four main Sections, dealing respectively with rational approximation, effective rational approximation, algebraic approximation to algebraic numbers, and effective algebraic approximation to algebraic numbers. A broad variety of methods are needed for the proofs of the results below. In a fifth Section, we briefly mention various applications to irrationality and transcendence statements.

2.1 Rational approximation

Let ξ be a real algebraic number. We say that μ is an *irrationality measure* for ξ if there exists a positive constant $c_1(\xi, \mu)$ such that, for any rational number

p/q distinct from ξ and with $q > 0$, we have

$$\left| \xi - \frac{p}{q} \right| > \frac{c_1(\xi, \mu)}{q^\mu}. \tag{2.1}$$

When the constant $c_1(\xi, \mu)$ can be explicitly computed, we say that μ is an *effective irrationality measure* for ξ.

Theorem 1.1 asserts that the rational numbers are precisely the numbers for which 1 is an irrationality measure and that any irrationality measure of any irrational number is at least equal to 2. This fact can be rephrased as an irrationality criterion, as already mentioned in Chapter 1. In the opposite direction, Liouville's Theorem states that the degree of any real algebraic number ξ is an irrationality measure for ξ. In particular, when ξ is a real algebraic number of degree at most 2, its smallest irrationality measure is equal to its degree. On the other hand, when ξ is of degree at least 3, our knowledge was, a century ago, not very satisfactory. Fortunately, since the early twentieth century, many mathematicians succeeded in strengthening Liouville's result.

The first significant improvement is due to Thue [553], who proved that, for any real algebraic number ξ of degree d at least 3, every real number greater than $d/2 + 1$ is an irrationality measure for ξ. As an immediate application, it follows that the Diophantine equation

$$F(x, y) = m, \quad \text{in integers } x, y, \tag{2.2}$$

where $F(X, Y)$ is an irreducible, homogeneous binary form of degree at least 3 with integer coefficients and m is a non-zero integer, has only finitely many solutions. Equation (2.2) is now called the *Thue Equation*.

In 1921, Siegel [524] sharpened considerably Thue's result by showing that every real number μ with

$$\mu > \min_{1 \le j \le d} \left(\frac{d}{j+1} + j \right)$$

is an irrationality measure for ξ. Later, Dyson [223] and independently Gelfond [256], slightly refined Siegel's estimate by proving that every μ greater than $\sqrt{2d}$ is indeed an irrationality measure for ξ. All these results have been superseded by Roth's [477], which can be stated as follows.

THEOREM 2.1. *Let ξ be a real algebraic number of degree $d \ge 2$. Then, for any positive real number ε, there exists a positive constant $c_2(\xi, \varepsilon)$ such that*

$$\left| \xi - \frac{p}{q} \right| > \frac{c_2(\xi, \varepsilon)}{q^{2+\varepsilon}} \tag{2.3}$$

for any rational number p/q with $q > 0$.

Khintchine's Theorem 1.10 may suggest that Theorem 2.1 could be further refined: a conjecture of Lang claims that the map $q \mapsto q^{2+\varepsilon}$ occurring in (2.3) could be replaced by $q \mapsto q^2(\log q)^{1+\varepsilon}$.

The common feature of all the results mentioned in the present Section, except Liouville's Theorem, is that they all are ineffective. This means that the methods used by Thue, Siegel, Dyson, Gelfond, and Roth did not allow them to compute explicitly the constant $c_1(\xi, \mu)$ occurring in (2.1).

However, as established by Davenport and Roth [179], Theorem 2.1 provides a criterion for proving that a real number with 'too large' partial quotients cannot be algebraic, which improves the criterion deduced from Liouville's Theorem 1.2 (see Exercise 1.1).

THEOREM 2.2. *Let ξ be a real number and, for any integer $n \geq 1$, denote by q_n the denominator of the n-th convergent in its continued fraction expansion. If*

$$\limsup_{n \to +\infty} \frac{(\log \log q_n)(\log n)^{1/2}}{n} = +\infty,$$

then ξ is transcendental.

In 1957, Ridout [470] obtained an important extension of Theorem 2.1, which can be formulated as follows.

THEOREM 2.3. *Let ξ be a non-zero algebraic number and let $p_1, \ldots, p_r, q_1, \ldots, q_s$ be distinct rational prime numbers. Let μ, ν, and c be real numbers with $0 \leq \mu \leq 1$, $0 \leq \nu \leq 1$ and $c > 0$. Let p and q be restricted to integers of the form $p = p^* p_1^{a_1} \ldots p_r^{a_r}$, $q = q^* q_1^{b_1} \ldots q_s^{b_s}$, where $a_1, \ldots, a_r, b_1, \ldots, b_s$ are non-negative integers and p^*, q^* are non-zero integers satisfying $|p^*| \leq cp^\mu$ and $|q^*| \leq cq^\nu$. Then, if $\kappa > \mu + \nu$, there are at most a finite number of solutions of the inequality*

$$0 < \left| \xi - \frac{p}{q} \right| < \frac{1}{q^\kappa}.$$

We observe that, taking $\mu = \nu = c = 1$ in Theorem 2.3, we recover Roth's Theorem. For more references on this subject, the reader is directed to [388], pages 73 to 76. We postpone some applications of Theorems 2.2 and 2.3 to Section 2.5.

2.2 Effective rational approximation

Let ξ be a real algebraic number of degree d greater than or equal to 3. Roth's Theorem 2.1 asserts that every real number strictly larger than 2 is an irrationality measure for ξ, while d is an *effective* irrationality measure for ξ, by

Liouville's Theorem 1.2. There is a big gap between these results and we may ask whether ξ admits smaller effective irrationality measures. As an immediate application, this would imply an effectively computable upper bound for the size of the Thue equation (2.2) which is polynomial in $|m|$.

A general result has been obtained by Feldman [241] (see also Chapter 9 of his book [242]) by means of A. Baker's theory of linear forms in the logarithms of algebraic numbers [43].

THEOREM 2.4. *Let ξ be a real algebraic number of degree $d \geq 3$. There exist effectively computable positive numbers $c_3(\xi)$ and $\tau(\xi)$ such that*

$$\left| \xi - \frac{p}{q} \right| \geq \frac{c_3(\xi)}{q^{d-\tau(\xi)}}$$

for any rational number p/q.

However, $\tau(\xi)$ is very small and, denoting by R_ξ the regulator of the number field $\mathbb{Q}(\xi)$ (see, for example, [435], page 106, for definition), we may take

$$\tau(\xi) = \left(3^{d+26} d^{15d+20} R_\xi \log \max\{e, R_\xi\} \right)^{-1},$$

as proved by Bugeaud and Győry [136], who used essentially the same method as Feldman and A. Baker. Notice that R_ξ is smaller than $(2d^2 H \log(dH))^{d-1}$, where $H \geq 3$ is an upper bound for the absolute values of the coefficients of the minimal polynomial of ξ over $\mathbb{Z}$ (see [136]).

An alternative approach, which yields Theorem 2.4 but does not rest on the theory of linear forms in logarithms, has been successfully worked out by Bombieri [104, 105] and developed by Bombieri and several co-authors [106, 108]. It gives a slightly better value for $\tau(\xi)$, as far as the dependence on the regulator is concerned. As noticed in [120], a combination of both methods yields the value

$$\tau(\xi) = \left(10^{26d} d^{14d} R_\xi \right)^{-1},$$

which is, at present, the best known general result for $d \geq 4$. Furthermore, for cubic irrationalities, the sharpest estimate is due to A. Baker and Stewart [46].

A short and self-contained proof of Theorem 2.4 has been worked out by Bilu and Bugeaud [101] (see also [591], Corollary 10.18).

Better effective irrationality measures are known for classes of algebraic numbers, including k-th roots of rational numbers, see [244] (Sections 3.5 and 3.6 of Chapter 1 and Section 4.6 of Chapter 2) for references.

Results like Theorem 2.4 have many applications to the resolution of Diophantine equations.

2.3 Approximation by algebraic numbers

Up to now, we were only interested in the approximation of a real number ξ by rational numbers a/b. To this end, we have compared the difference $|\xi - a/b|$ with a natural way to measure the size of a/b, namely with $\max\{|a|, |b|\}$. We have now to define a notion of size to evaluate the complexity of an algebraic number α, which, if possible, coincides with $\max\{|a|, |b|\}$ when α is the rational a/b. Many definitions have been proposed (see Appendix A), and we choose the most natural one, usually termed the *naive height*, which we simply call *height*.

DEFINITION 2.1. *The height of a complex polynomial $P(X)$, denoted by $H(P)$, is the maximum of the moduli of its coefficients. The height of an algebraic number α, denoted by $H(\alpha)$, is the height of its minimal polynomial over $\mathbb{Z}$.*

We point out that, in all questions investigated up to Chapter 7, the degree of the approximant is fixed, and the height is allowed to vary. In this case, and for these kind of questions, the choice of the size has no particular significance. This does not remain true in Chapter 8, where degree and height are allowed to vary simultaneously, and some of the results presented there are given in terms of the absolute logarithmic height or in terms of the Mahler measure.

The next theorem, asserted by Roth [477] and proved by LeVeque [363], extends Theorem 2.1 to number fields.

THEOREM 2.5. *Let $\mathbb{K}$ be a number field and ξ be a real algebraic number not in $\mathbb{K}$. Then, for any $\varepsilon > 0$, there exists an (ineffective) positive constant $c_4(\xi, \mathbb{K}, \varepsilon)$ such that*

$$|\xi - \alpha| > c_4(\xi, \mathbb{K}, \varepsilon) \, H(\alpha)^{-2-\varepsilon}, \tag{2.4}$$

for every α in $\mathbb{K}$.

An extension of Dirichlet's Theorem 1.1 shows that the exponent $2 + \varepsilon$ in (2.4) cannot be replaced by a real number strictly less than 2.

Instead of studying approximation of an algebraic number by elements of a fixed number field, we now consider approximation by algebraic numbers of degree less than or equal to n. This would generalize Theorem 2.1 in another direction than Theorem 2.5. Wirsing [599] has given a satisfactory answer to this problem, by obtaining an exponent depending only on n.

THEOREM 2.6. *Let ξ be a real algebraic number and $n \geq 1$ be an integer. For any $\varepsilon > 0$, there exists a positive (ineffective) constant $c_5(\xi, n, \varepsilon)$ such that*

$$|\xi - \alpha| > c_5(\xi, n, \varepsilon) \, H(\alpha)^{-2n-\varepsilon}, \tag{2.5}$$

for any algebraic number $\alpha \neq \xi$ of degree at most n.

Theorem 2.6 reduces to Roth's Theorem for $n = 1$; the proof of Wirsing uses the main steps of that of Roth. However, a metric argument suggests that it could be possible to replace the exponent $-2n - \varepsilon$ in (2.5) by $-n - 1 - \varepsilon$. This conjecture has been proved by Schmidt (see [504] for $n = 2$ and [506] for $n \geq 3$), nearly at the same time as Wirsing published Theorem 2.6, but with very different arguments.

THEOREM 2.7. *Let ξ be a real algebraic number and $n \geq 1$ be an integer. Let ε be a positive real number. Then there exists a positive (ineffective) constant $c_6(\xi, n, \varepsilon)$ such that*

$$|P(\xi)| > c_6(\xi, n, \varepsilon) \, \mathrm{H}(P)^{-n-\varepsilon}$$

for any integer polynomial $P(X)$ of degree at most n and such that $P(\xi) \neq 0$. Furthermore, there exists a positive (ineffective) constant $c_7(\xi, n, \varepsilon)$ such that

$$|\xi - \alpha| > c_7(\xi, n, \varepsilon) \, \mathrm{H}(\alpha)^{-n-1-\varepsilon}$$

for any algebraic number $\alpha \neq \xi$ of degree at most n.

The first assertion of Theorem 2.7 is a corollary to the celebrated Subspace Theorem of Schmidt, a remarkable result having many important applications in Diophantine approximation. The proof, very involved, is given with many details in [512], its main lines being also in [510]. The second assertion of Theorem 2.7 follows from the first one, as proved in Theorem 3.5. Notice that while it is used in Chapter 7 in order to prove that T-numbers (see Definition 3.1) do exist, Theorem 2.6 however is sufficient to get the same conclusion. Clearly, Theorem 2.7 (and, thus, Theorem 2.8 below) is interesting only for n at most equal to the degree of ξ minus two, since otherwise it is superseded by Liouville's inequality (Corollary A.2).

THEOREM 2.8. *Let $n \geq 1$ be an integer and ξ be a real algebraic number of degree d. Then, for any positive real number ε, there exist an ineffective constant $c_8(\xi, n, \varepsilon)$ and infinitely many algebraic numbers α of degree at most n such that*

$$|\xi - \alpha| < c_8(\xi, n, \varepsilon) \, \mathrm{H}(\alpha)^{-\min\{n+1, d\} + \varepsilon}.$$

Theorem 2.8 follows from the first assertion of Theorem 2.7, as proved in Theorem 3.5. In the next Section, we show that Theorem 2.8 can be improved upon when $n \geq d - 1$.

2.4 Effective approximation by algebraic numbers

Corollary 1.2 asserts that real quadratic algebraic numbers are badly approximable by rational numbers. This is a particular case of a more general phenomenon: for any integers $d \geq 2$ and $n \geq d - 1$, real algebraic numbers of degree d are badly approximable by algebraic numbers of degree n. Theorem 2.9 covers the case $n = d - 1$, while Theorem 2.11 deals with the case $n \geq d$.

THEOREM 2.9. *Let ξ be a real algebraic number of degree $d \geq 2$. There exist effectively computable positive constants $c_9(\xi)$ and $c_{10}(\xi)$ such that*

$$|\xi - \alpha| \geq c_9(\xi)\,\mathrm{H}(\alpha)^{-d} \quad \textit{for any algebraic number } \alpha \textit{ of degree} \\ \textit{at most } d - 1,$$

and

$$|\xi - \alpha| \leq c_{10}(\xi)\,\mathrm{H}(\alpha)^{-d} \quad \textit{for infinitely many real algebraic numbers} \\ \alpha \textit{ of degree } d - 1.$$

The first part of Theorem 2.9 is a restatement of Corollary A.2, while the second part originates in the work of Wirsing [598]. Bombieri and Mueller [107] have given an alternative proof, which we reproduce here. They proved Theorem 2.10 below, which implies the second part of Theorem 2.9.

THEOREM 2.10. *Let ξ with $|\xi| \leq 1/2$ be a real algebraic number of degree $d \geq 2$. For every $H \geq 2$, there exists an algebraic number α of degree at most $d - 1$ such that*

$$|\xi - \alpha| \leq \frac{d!(d - 1)}{H^d}$$

and

$$\mathrm{H}(\alpha) \leq 2^{d-1}\big(d\sqrt{d+1}\,\mathrm{H}(\xi)\big)^{(d-1)^2/d}\,H. \tag{2.6}$$

Moreover, α is real as soon as $H > 2^d\,(d+1)^{(3d+5)/2}\,\mathrm{H}(\xi)^{d-1}$.

PROOF. For any real number $H \geq 2$, let $\mathcal{C}(H)$ denote the convex symmetric body in $\mathbb{R}^d$ defined by

$$|x_0 + x_1\xi + \ldots + x_{d-1}\xi^{d-1}| \leq H^{-d+1},$$

$$|x_1|, \ldots, |x_{d-1}| \leq H.$$

Let $\lambda_i = \lambda_i(H)$, $i = 1, \ldots, d$, be the successive minima of $\mathcal{C}(H)$ (see Appendix B for the definition). We denote by $(x_0^{(i)}, \ldots, x_{d-1}^{(i)})$, $i = 1, \ldots, d$, linearly independent points at which the successive minima λ_i are attained and we set

$$P_i(X) = x_0^{(i)} + x_1^{(i)}X + \ldots + x_{d-1}^{(i)}X^{d-1}.$$

We first show that there exists some integer i for which

$$|P_i'(\xi)| \geq \frac{\lambda_i}{d!} H \qquad (2.7)$$

holds. To this end, we set

$$M = \max_{1 \leq i \leq d} \frac{|P_i'(\xi)|}{\lambda_i},$$

and we observe that $\lambda_1, \ldots, \lambda_d$ are as well the successive minima of the convex body

$$\mathcal{C}(H, M) := \mathcal{C}(H) \cap \{|x_1 + 2\xi x_2 + \ldots + (d-1)\xi^{d-2}x_{d-1}| \leq M\}.$$

By Theorem B.3, we have the inequalities

$$\frac{2^d}{d!} \leq \lambda_1 \ldots \lambda_d \, \mathrm{vol}(\mathcal{C}(H, M)) \qquad (2.8)$$

and

$$\lambda_1 \ldots \lambda_d \leq 1, \qquad (2.9)$$

since $\mathrm{vol}(\mathcal{C}(H)) \leq 2^d$. Further, the volume of $\mathcal{C}(H, M)$ is at most equal to the volume of the convex body defined by the inequalities

$$|x_0 + x_1\xi + \ldots + x_{d-1}\xi^{d-1}| \leq H^{-d+1},$$
$$|x_1 + 2x_2\xi + \ldots + (d-1)x_{d-1}\xi^{d-2}| \leq M,$$
$$|x_2|, \ldots, |x_{d-1}| \leq H.$$

Hence, we get

$$\mathrm{vol}(\mathcal{C}(H, M)) \leq 2^d M/H.$$

Combined with (2.8) and (2.9), this implies that $M \geq H/d!$, thus (2.7) holds for some index i.

For such an i, we have by construction $|P_i(\xi)| \leq \lambda_i H^{-d+1}$ and $H(P_i) \leq \lambda_i H$, since $H \geq 2$ and $|\xi| \leq 1/2$. Hence, Lemma A.5 and (2.7) yield that the polynomial $P_i(X)$ has a root α with

$$|\xi - \alpha| \leq (d-1)\frac{|P_i(\xi)|}{|P_i'(\xi)|} \leq \frac{d!(d-1)}{H^d}, \qquad (2.10)$$

and

$$H(\alpha) \le 2^{d-1} H(P_i) \le 2^{d-1} \lambda_i H, \qquad (2.11)$$

by Lemma A.3.

It remains for us to bound λ_i. We infer from (2.9) that

$$\lambda_i \le \lambda_d \le \lambda_1^{-d+1}, \qquad (2.12)$$

and we are led to bound λ_1 from below. By Theorem A.1, we get

$$|P_1(\xi)| \ge d^{-d+1} (d+1)^{-(d-1)/2} H(P_1)^{-d+1} H(\xi)^{-d+1}. \qquad (2.13)$$

Furthermore, by definition, the polynomial $P_1(X)$ satisfies $|P_1(\xi)| \le \lambda_1 H^{-d+1}$ and $H(P_1) \le \lambda_1 H$, hence

$$|P_1(\xi)| \le \lambda_1^d H(P_1)^{-d+1}. \qquad (2.14)$$

Combining (2.13) and (2.14), we get

$$\lambda_1 \ge \left(d\sqrt{d+1} \, H(\xi) \right)^{-(d-1)/d}, \qquad (2.15)$$

and (2.6) follows from (2.10), (2.11), (2.12), and (2.15).

If the algebraic number α constructed above is non-real, then, denoting by $P_\alpha(X)$ its minimal polynomial over $\mathbb{Z}$ and by $\overline{\alpha}$ its complex conjugate, Lemma A.6 and (2.10) imply

$$\begin{aligned} |P_\alpha(\xi)| &\le 2^{d-1} \sqrt{d} \, H(\alpha) \, |\xi - \alpha| \cdot |\xi - \overline{\alpha}| \\ &\le 2^{d-1} \sqrt{d} \, (d-1)^2 \, (d!)^2 \, H(\alpha) \, H^{-2d}. \end{aligned} \qquad (2.16)$$

On the other hand, Theorem A.1 yields

$$|P_\alpha(\xi)| \ge d^{-d+1} (d+1)^{-(d-1)/2} H(\alpha)^{-d+1} H(\xi)^{-d+1},$$

which, combined with (2.16) and (2.6), gives the upper bound $H \le 2^d (d+1)^{(3d+5)/2} H(\xi)^{d-1}$. This completes the proof of the theorem.

We are now concerned with the approximation of algebraic numbers by algebraic numbers of same or larger degree. Apparently, Theorem 2.11 did not appear previously. Its first part is a restatement of Corollary A.2, while its second part originates in a work of Davenport and Schmidt [182]. Theorem 2.11 is used in the proof of Theorem 7.5. We recall that an algebraic integer is an algebraic number whose minimal polynomial over $\mathbb{Z}$ is monic.

THEOREM 2.11. *Let ξ be a real algebraic number of degree $d \geq 2$. Let $n \geq d$ be an integer. There exist positive, effectively computable, constants $c_{11}(\xi, n)$ and $c_{12}(\xi, n)$ such that*

$$|\xi - \alpha| \geq c_{11}(\xi, n)\, H(\alpha)^{-d} \quad \text{for any algebraic number } \alpha \neq \xi \text{ of degree at most } n,$$

and

$$|\xi - \alpha| \leq c_{12}(\xi, n)\, H(\alpha)^{-d} \quad \text{for infinitely many real algebraic integers } \alpha \text{ of degree } n.$$

PROOF. Without loss of generality, we may assume that $|\xi| \leq 1/2$. Theorem A.1 implies that for any non-zero integer polynomial $P(X)$ of degree at most $d - 1$ we have

$$|P(\xi)| > (d + 1)^{-3(d-1)/2}\, H(\xi)^{-d+1}\, H(P)^{-d+1}. \tag{2.17}$$

Set $\kappa = (d + 1)^{3(d-1)^2/(2d)}\, H(\xi)^{(d-1)^2/d}$ and let H be a real number with

$$H^d > 2^{n+5}\, (d + 1)^6\, (n + 1)^{5/2}\, \kappa^2. \tag{2.18}$$

It follows from (2.17) that the first minimum λ_1 of the compact convex $\mathcal{C}(H)$ defined by the inequalities

$$|x_{d-1}\xi^{d-1} + \ldots + x_1\xi + x_0| \leq H^{-d+1},$$

$$|x_1|, \ldots, |x_{d-1}| \leq H,$$

satisfies $\lambda_1 > \kappa^{-1/(d-1)}$. Consequently, by Theorem B.3, the d-th minimum λ_d of $\mathcal{C}(H)$ is smaller than κ. Thus, there exist d linearly independent integer polynomials $P_j(X) := x_{d-1}^{(j)} X^{d-1} + \ldots + x_1^{(j)} X + x_0^{(j)}$, for $j = 1, \ldots, d$, of degree at most $d - 1$ satisfying

$$|P_j(\xi)| \leq \kappa H^{-d+1} \quad \text{and} \quad H(P_j) \leq \kappa H, \tag{2.19}$$

for $j = 1, \ldots, d$, and $\delta := |\det(x_i^{(j)})| \leq d!$. Let p be a prime number which does not divide δ. Since the product of all prime numbers up to d^2 exceeds $d!$, we may assume that p is not larger than d^2. We argue exactly as Davenport and Schmidt [182] in order to construct a monic polynomial of degree n, small when evaluated at ξ.

Since p does not divide δ, there exists a superscript j such that p does not divide $x_0^{(j)}$. Without loss of generality, we assume that $j = 1$.

We consider the following linear system of d equations in the d real unknowns $\theta_1, \dots, \theta_d$:

$$\xi^n + p(\theta_1 P_1(\xi) + \dots + \theta_d P_d(\xi)) = p(d+1)\kappa H^{-d+1}$$
$$n\xi^{n-1} + p(\theta_1 P_1'(\xi) + \dots + \theta_d P_d'(\xi)) = pH + p \sum_{1 \le i \le d} |P_i'(\xi)| \quad (2.20)$$
$$\theta_1 x_m^{(1)} + \dots + \theta_d x_m^{(d)} = 0 \ (m = 2, \dots, d-1).$$

Since the polynomials $P_i(X)$, for $i = 1, \dots, d$, are linearly independent, this system has one and only one solution $(\theta_1, \dots, \theta_d)$. We take a d-tuple $(t_1, \dots, t_d)$ of integers such that $|\theta_i - t_i| \le 1$ for $i = 1, \dots, d$, and we set

$$x_m = t_1 x_m^{(1)} + \dots + t_d x_m^{(d)}, \ \text{for } m = 0, \dots, d-1.$$

We consider the polynomial

$$P(X) = X^n + p(x_{d-1}X^{d-1} + \dots + x_1 X + x_0)$$
$$= X^n + p(t_1 P_1(X) + \dots + t_d P_d(X)),$$

which, by a suitable choice of $(t_1, \dots, t_d)$, is irreducible. Indeed, by using Eisenstein's Criterion, it is sufficient to check that its constant coefficient, namely $p(t_1 x_0^{(1)} + \dots + t_d x_0^{(d)})$, is not divisible by p^2, since its leading coefficient is congruent to 1 modulo p. We fix a $(d-1)$-tuple $(t_2, \dots, t_d)$ and there remain two possible choices for t_1, which we denote by $t_{1,0}$ and $t_{1,1} = t_{1,0}+1$. Since p does not divide $x_0^{(1)}$, at least one of the integers $t_{1,0}x_0^{(1)} + \dots + t_d x_0^{(d)}$ or $t_{1,1}x_0^{(1)} + \dots + t_d x_0^{(d)}$ is not divisible by p. This enables us to choose t_1 such that the polynomial $P(X)$ is Eisensteinian with respect to the prime number p, hence, irreducible.

Furthermore, the polynomial $P(X)$ satisfies

$$P(\xi) = \xi^n + p(t_1 P_1(\xi) + \dots + t_d P_d(\xi)),$$

hence, by (2.18), (2.19), and the first equation of the system (2.20), we have

$$0 < p\kappa H^{-d+1} \le P(\xi) \le p(2d+1)\kappa H^{-d+1} < 1. \quad (2.21)$$

On the other hand, we have

$$P'(\xi) = n\xi^{n-1} + p(t_1 P_1'(\xi) + \dots + t_d P_d'(\xi));$$

thus, by (2.19) and the second equation of (2.20), we get

$$P'(\xi) \ge pH \quad (2.22)$$

and

$$P'(\xi) \le pH + 2p \sum_{i=1}^{d} |P_i'(\xi)| \le pH(1 + 4d(d-1)\kappa), \qquad (2.23)$$

since $|\xi| \le 1/2$ and $H(P_i') \le (d-1)\kappa H$ for $i = 1, \ldots, d$. Finally, by (2.19) and the last equations of (2.20), we get

$$|x_m| \le d\kappa H, \quad \text{for} \quad m = 2, \ldots, d-1, \qquad (2.24)$$

and we infer from (2.22), (2.23), (2.24), and $|\xi| \le 1/2$ that

$$|x_1| \le 1 + |P'(\xi)| + \sum_{j=2}^{d-1} j|x_j| \le (d^3 + 4d^2 p)\kappa H. \qquad (2.25)$$

It follows from (2.21), (2.24), and (2.25) that

$$|x_0| \le 1 + P(\xi) + |x_1| + \sum_{j=2}^{d-1} |x_j| \le (d^3 + d^2 + 4d^2 p)\kappa H,$$

and we finally derive from the upper bound $p \le d^2$ that

$$H(P) \le 6d^4 \kappa H. \qquad (2.26)$$

By (2.21), (2.22), and the first assertion of Lemma A.5, the polynomial $P(X)$ has a root α satisfying

$$|\xi - \alpha| \le n \frac{P(\xi)}{P'(\xi)} \le n(2d+1)\kappa H^{-d}, \qquad (2.27)$$

thus, by (2.26),

$$|\xi - \alpha| \le 6^{d+1} n (d+1)^{(3d^2+5d+5)/2} H(\xi)^{(d-1)^2} H(\alpha)^{-d}.$$

If α were non-real, then its conjugate would also satisfy (2.27) and, by (2.26), Lemma A.6, and $|\xi| \le 1/2$, we would then have

$$P(\xi) \le 2^{n+5} (n+1)^{5/2} (d+1)^6 \kappa^3 H^{1-2d}.$$

Together with (2.18), this would contradict the lower bound $P(\xi) \ge \kappa H^{-d+1}$ obtained in (2.21). Consequently, α is a real algebraic integer of degree n.

For an alternative proof of Theorem 2.9, it suffices to take $n = d - 1$ in the proof of Theorem 2.11: the Eisenstein Criterion ensures that the polynomial $P(X)$ is irreducible.

We could as well state analogues of Theorems 2.9 and 2.11, under the same hypothesis on ξ, with upper and lower bounds for $|P(\xi)|$, where $P(X)$ is an integer (or a monic integer) polynomial of degree d.

2.5 Remarks on irrationality and transcendence statements

It follows from Theorem 1.1 that a real number ξ is irrational if it has infinitely many *good* rational approximants. In a similar way, Roth's Theorem and its generalizations can be applied to establish that ξ is transcendental if it admits infinitely many *very good* rational approximants. Classical results, including the transcendence of the Champernowne number $0.1234567891011121314\ldots$, can be obtained by applying Theorems 2.1, 2.2, and 2.3 (although Mahler's original proof of this statement [380] used a weaker transcendence criterion), see Mahler [397] and Zhu [616]. For instance, Ferenczi and Mauduit [246] gave a combinatorial translation of Theorem 2.3 which shows in particular that real numbers whose expansion in a given base is Sturmian are transcendental. In most of the cases, an interesting question often remains open: after having shown that a given real number ξ has an irrationality measure strictly larger than 2 to conclude that ξ is transcendental, what can be said about its *smallest* irrationality measure?

Another typical problem is the algebraicity of real numbers defined in terms of their continued fraction expansion. By Corollary 1.6, real transcendental numbers with bounded partial quotients do exist and it is of interest to construct explicit examples. The first result of this type is due to Maillet [403] (see also Section 34 of Perron [454]), who found continued fractions with bounded partial quotients giving trancendental numbers by using a general form of Liouville's inequality (see Corollary A.2) which limits the precision of the approximation of an algebraic number by quadratic irrationalities. Such an argument is quite natural when we keep in mind that the quadratic irrationalities coincide with the real numbers having ultimately periodic continued fraction expansions (Theorem 1.16). Other examples of transcendental numbers with bounded partial quotients are due to A. Baker [37, 38] who applied Theorem 2.2, asserting that the partial quotients of the continued fraction expansion of an algebraic number cannot increase too rapidly.

An interesting application of the case $n = 2$ of Theorem 2.7 has been worked out by Queffélec [464] (the idea of using Schmidt's Theorem in a 'similar' context goes back to Davison [185]), who proved that the real number

$$\xi = [1; 2, 2, 1, 2, 1, 1, 2, 2, 1, 1, 2, 1, 2, 2, 1, \ldots],$$

with partial quotients being the Thue–Morse word on $\{1, 2\}$, that is, the fixed point of the substitution $1 \to 12$, $2 \to 21$, is transcendental. Again, although we know that ξ is too well approximable by quadratic numbers to be algebraic, we do not know precisely *how well* it is approximable. As we shall see in the course of the next Chapters, this question appears, in general, to be very difficult.

2.6 Notes

• Let ε be a positive real number. Theorem 2.1 implies that the n-th partial quotient a_n of a real algebraic number ξ of degree d satisfies $\log a_n < c_{13}(1 + \varepsilon)^n$ for any positive integer n and a suitable ineffective constant c_{13}, depending only on ε and ξ. Using Theorem 2.4, one can get an effective upper bound, but $1 + \varepsilon$ is then replaced by $d - 1 - \tau$, for a (very) small positive number τ. In the case of cubic algebraic numbers, this has been considerably improved by Wolfskill [600], who obtained the upper estimate $\log a_n < c_{14}(\sqrt{3} + \varepsilon)^n$ for any positive integer n and an effective constant c_{14}, depending only on ε and ξ.

• Using a method of Dyson [223], Mahler [383] proved that if $(p_n/q_n)_{n \geq 1}$ denotes the sequence of convergents of a quadratic or cubic real number, then the greatest prime factor of q_n tends to infinity with n. Actually, it follows from Ridout's Theorem 2.3 that this conclusion holds for any real irrational number. Shorey [523] established related quantitative results.

• Mahler [386] applied Ridout's Theorem 2.3 to get lower bounds for the distance to the nearest integer of powers of rational numbers.

• An interesting complement to Ridout's Theorem 2.3 has been given by Lagarias [346]. Stewart [544] used a similar construction to investigate a question on divisors of a product of consecutive integers.

• Mahler [390] (see also his book [388]) extended Theorem 2.5 and investigated approximation of algebraic numbers by algebraic integers lying in a given number field. He also proved an inhomogeneous result, later generalized by Schmidt, see Section 7.4 of [510].

• Allouche, Davison, Queffélec, and Zamboni [6], Davison [186], and Baxa [55] established extensions of Queffélec's result mentioned in Section 2.5. See also Liardet and Stambul [367] and Chapter 13 of Allouche and Shallit [7].

3

The classifications of Mahler and Koksma

The set of real numbers splits into algebraic and transcendental numbers, but these two subsets do not have the same *size*, the former being countable, while the latter has the power of continuum. Such a crude classification of real numbers seems to be rather unsatisfactory, and one aims to find some way to classify the transcendental numbers. First, we have to ask which requirements should satisfy a 'good' classification. Ideally, for a given real number ξ, we would like to have a simple criterion to determine the class to which ξ belongs. Furthermore, two algebraically dependent real numbers should belong to the same class. The first classification of transcendental real numbers has been proposed by Maillet [403, 404], and others were subsequently described by Perna [453] and by Morduchai-Boltovskoj [430], but none of these has proved to be relevant. For instance, Maillet's classification depends on the size of the partial quotients of the real numbers and, clearly, does not satisfy the second requirement.

An attempt towards a 'reasonable' classification was made in 1932 by Mahler [376], who proposed to subdivide the set of real numbers into four classes (including the class of algebraic numbers) according to, roughly speaking, their properties of approximation by algebraic numbers. Mahler's classification satisfies our second requirement: two algebraically dependent real numbers belong to the same class. However, if a real number is given (that is, for example, a Cauchy sequence of rational numbers), there is, in general, very little hope of determining to which class it belongs. Although non-trivial, Mahler's classification does not seem to be entirely satisfactory since almost all real numbers (in the sense of the Lebesgue measure) belong to the same class. Nevertheless, it has been widely studied, as has the closely related Koksma's classification, which was proposed a few years later [333].

In the present Chapter, we begin by defining Mahler's classification and we prove some of its properties. We then define Koksma's classification and

41

compare them. Further, we state the 'Main Problem', which motivates the next five Chapters. Finally, we discuss some links between algebraic approximation and simultaneous rational approximation, and we introduce four other exponents of Diophantine approximation.

3.1 Mahler's classification

Mahler's idea [376] consists in classifying the real numbers ξ according to the accuracy with which non-zero integer polynomials, evaluated at ξ, approach 0. For given positive integer n and real number $H \geq 1$, we define the quantity

$$w_n(\xi, H)$$
$$:= \min\{|P(\xi)| : P(X) \in \mathbb{Z}[X], \mathrm{H}(P) \leq H, \deg(P) \leq n, P(\xi) \neq 0\},$$

where $\mathrm{H}(P)$ denotes the height of the polynomial $P(X)$ (see Definition 2.1). Then, we set

$$w_n(\xi) = \limsup_{H \to +\infty} \frac{-\log w_n(\xi, H)}{\log H}$$

and

$$w(\xi) = \limsup_{n \to +\infty} \frac{w_n(\xi)}{n}.$$

In other words, $w_n(\xi)$ is the supremum of the real numbers w for which there exist infinitely many integer polynomials $P(X)$ of degree at most n satisfying

$$0 < |P(\xi)| \leq \mathrm{H}(P)^{-w}.$$

Further, it is an easy exercise (see Exercise 3.1) to check that for any positive integer n, for any real number ξ and any non-zero rational number a/b, we have $w_n(\xi) = w_n(\xi + a/b) = w_n(a\xi/b)$. This allows us in most of the proofs below to assume that ξ belongs to an interval of small length.

For $n \geq 1$, we have $0 \leq w_n(\xi) \leq +\infty$, since $w_n(\xi, H) \leq w_1(\xi, H) \leq 1$ for any H with $H \geq |\xi| + 1$. Thus, we get $0 \leq w(\xi) \leq +\infty$. Moreover, the sequence $(w_n(\xi))_{n\geq 1}$ is clearly non-decreasing. With these notations, Mahler [376] (actually, he used the Greek letter ω instead of w; hopefully, the present notation will not perturb the reader!) divided the set of real numbers into four disjoint classes.

DEFINITION 3.1. *Let ξ be a real number. We say that ξ is an*
A-number, if $w(\xi) = 0$;
S-number, if $0 < w(\xi) < +\infty$;

T-number, *if* $w(\xi) = +\infty$ *and* $w_n(\xi) < +\infty$ *for any* $n \geq 1$;
U-number, if $w(\xi) = +\infty$ *and* $w_n(\xi) = +\infty$ *from some n onwards.*

Theorem 3.1 below shows that the *A*-numbers are exactly the algebraic numbers. According to Stolarsky [546], the terminology '*S*-number' refers to Siegel, the letters *T* and *U* having then been chosen by alphabetical order.

In another paper of 1932, Mahler [377] proved that almost all (in the sense of Lebesgue measure) real numbers ξ satisfy $w_n(\xi) \leq 4n$ for any positive integer *n* (actually, his result is slightly stronger) and conjectured that it should be possible to replace 4 by 1. This has been confirmed by Sprindžuk [538] in 1965; Chapter 4 is devoted to this problem and related questions.

Mahler [375] had already introduced the terminology '*S*-number' in 1930, but this, however, does not match with Definition 3.1: the set of *S*-numbers in [375] is equal to the union of the sets of *A*- and *S*-numbers in [376].

The classes *S*, *T*, and *U* may be further subdivided into infinitely many subclasses, by introducing the notion of *type*.

DEFINITION 3.2. *We define the type* $t(\xi)$ *of an S-number* ξ *as the value of* $w(\xi)$, *that is,*

$$t(\xi) = \limsup_{n \to +\infty} \frac{w_n(\xi)}{n}.$$

Koksma [333] called 'Index der *S*-Zahl ξ' the quantity

$$\sup_{n \geq 1} \frac{w_n(\xi)}{n} \tag{3.1}$$

and he observed ([333], Satz 4) that any *S*-number is of index at least 1. Several authors, including Schneider [517], Kasch and Volkmann [311], A. Baker [38, 44], and Güting [269], used (3.1) to define the type of an *S*-number, instead of Definition 3.2. As observed by A. Baker and Schmidt [45] in 1970 in a footnote, it was at that time '*customary to use* $\sup w_n(\xi)/n$, *but the above would seem more natural*', where they refer to

$$\limsup_{n \to +\infty} \frac{w_n(\xi) + 1}{n + 1} \quad \text{and} \quad \sup_{n \geq 1} \frac{w_n(\xi) + 1}{n + 1}. \tag{3.2}$$

Indeed, the metric results stated in Chapter 5 show that one of the definitions in (3.2) should be preferred to (3.1). The first alternative in (3.2) coincides with Definition 3.2: in the author's opinion, the lim sup is much more relevant than the supremum. Notice however that the existence of S-numbers ξ with

$$\limsup_{n \to +\infty} \frac{w_n(\xi) + 1}{n + 1} < \sup_{n \geq 1} \frac{w_n(\xi) + 1}{n + 1}$$

remains an open problem. Actually, no matter how one defines the type, Theorem 5.8 asserts that there exist real S-numbers of arbitrary type greater than or equal to 1.

DEFINITION 3.3. *Let ξ be a T-number and, for any $n \geq 1$, define the real number τ_n by $w_n(\xi) = n^{\tau_n}$. The type of ξ, denoted by $t(\xi)$, is the quantity*

$$t(\xi) = \limsup_{n \to +\infty} \tau_n.$$

The type of a T-number takes its values in $[1, +\infty]$. Schmidt proved [507, 508] (see Chapter 7) that there exist T-numbers of arbitrary given type in $[3, +\infty]$, but the problem of the existence of T-numbers of type strictly less than 3 remains open.

DEFINITION 3.4. *Let ξ be a U-number. The type of ξ, denoted by $t(\xi)$, is the smallest positive integer n such that $w_n(\xi) = +\infty$.*

The U-numbers of type 1 are precisely the Liouville numbers (Definition 1.1). Furthermore, LeVeque [361] proved that there exist U-numbers of arbitrary given type (see Section 7.6). There are alternative definitions for the type of a U-number, see for instance Schmidt [508] and Section 7.1.

We conclude this Section by pointing out that the Lebesgue Density Theorem (see, for example, Riesz and Sz.-Nagy [472], page 13, or Mattila [415], Corollary 2.14) implies that the function w_n, which is invariant by rational translation, takes the same value at almost all real numbers, that is, for any real number w, the set of real numbers ξ with $w_n(\xi) = w$ has either measure zero or has full measure. Consequently, almost all real numbers belong to the same class and have the same type. This fact was first observed by Sprindžuk [530] (see also [539], page 23).

Furthermore, as soon as the set of real numbers is divided into disjoint measurable classes with the property that any two real numbers differing by a rational number belong to the same class (this is obviously true if any two algebraically dependent real numbers belong to the same class), then there exists a class to which almost all real numbers belong (see Exercise 6.5).

3.2 Some properties of Mahler's classification

Our first result gives a lower bound for $w(\xi)$ when ξ is a real transcendental number.

PROPOSITION 3.1. *Let n be a positive integer and ξ be a real number not algebraic of degree at most n. We then have $w_n(\xi) \geq n$. In particular, if ξ is transcendental, then we have $w(\xi) \geq 1$.*

PROOF. Without loss of generality, we may assume that $0 < \xi < 1/2$. Let $H \geq 2$ be real, and consider the system of inequalities

$$|a_n \xi^n + \ldots + a_1 \xi + a_0| \leq H^{-n}$$

$$|a_1|, \ldots, |a_n| \leq H.$$

By Theorem B.2, it has a non-zero integer solution $(a_0, a_1, \ldots, a_n)$. Further, $|a_n \xi^n + \ldots + a_1 \xi + a_0| < H^{-n}$ and $0 < \xi < 1/2$ imply that $|a_0| \leq H$. Since ξ is not algebraic of degree at most n, this means that there exists a non-zero integer polynomial $P_H(X)$, of height at most H, such that

$$0 < |P_H(\xi)| \leq \mathrm{H}(P_H)^{-n}.$$

By taking arbitrarily large values of H, we get a sequence of distinct integer polynomials $(P_k)_{k \geq 1}$, whose heights increase to infinity, and which satisfy $0 < |P_k(\xi)| \leq \mathrm{H}(P_k)^{-n}$ for $k \geq 1$. It immediately follows that $w_n(\xi) \geq n$. Consequently, $w(\xi) \geq 1$ if ξ is transcendental. We observe that we may alternatively use Dirichlet's *Schubfachprinzip* instead of Theorem B.2, see the proof of Lemma 8.1.

THEOREM 3.1. *The A-numbers are exactly the real algebraic numbers. Let ξ be an algebraic real number of degree d and let n be a positive integer. We then have $w_n(\xi) = \min\{n, d - 1\}$.*

PROOF. Let ξ be an algebraic number of degree d and let $P(X)$ be an integer polynomial of degree n and of height H such that $P(\xi) \neq 0$. We infer from Theorem A.1 that $w_n(\xi) \leq d - 1$, hence, we get that $w(\xi) = 0$. In view of the last assertion of Proposition 3.1, this shows that the sets of A-numbers and of real algebraic numbers coincide. By the first assertion of Schmidt's Theorem 2.7, the last inequality can be refined to $w_n(\xi) \leq \min\{n, d - 1\}$. Since $w_n(\xi) \geq w_{d-1}(\xi)$ holds if $n \geq d - 1$, we then infer from Proposition 3.1 that $w_n(\xi) = \min\{n, d - 1\}$.

It immediately follows from Theorem 3.1 that real numbers ξ with $0 < w(\xi) < 1$ do not exist.

COROLLARY 3.1. *Every real S-number is of type at least 1.*

Theorem 3.2, due to Mahler [375, 376], ensures that his classification satisfies the second requirement stated at the beginning of the Chapter.

THEOREM 3.2. *Two algebraically dependent real numbers ξ and η belong to the same class.*

PROOF. By Theorem 3.1, we may assume that ξ and η are transcendental. Let n be a positive integer. The constants c_1, c_2, and c_3 appearing below are positive, effectively computable, and depend only on ξ, η, and n. Let $H \geq 1$ be a real number and $A(X)$ be an integer polynomial of height at most H and degree at most n with $|A(\xi)| = w_n(\xi, H)$. Let

$$F(X, Y) = \sum_{h=0}^{M} \sum_{k=0}^{N} b_{hk} X^h Y^k = \sum_{h=0}^{M} B_h(Y) X^h$$

be a primitive, irreducible, integer polynomial vanishing at the point (ξ, η) and such that $B_M(Y)$ is not identically zero. For some values of y, the polynomials $A(X)$ and $F(X, y)$ have no common zero, hence the resultant (see Definition A.1)

$$R(Y) = \operatorname{Res}_X(A(X), F(X, Y))$$

is not the zero polynomial. Moreover,

$$\deg R(Y) \leq n \max_{0 \leq j \leq M} \deg B_j(Y) \leq nN,$$

and there exists a constant c_1 such that the height of the polynomial $R(Y)$ is not greater than $c_1 H^M$.

Classical properties of the resultant (see, for example, [349], Chapter IV, or [586], Chapter 5, or [156], Appendix A) imply that there exist two polynomials $g(X, Y)$ and $h(X, Y)$ satisfying

$$R(Y) = A(X)g(X, Y) + F(X, Y)h(X, Y),$$

and such that the height of the polynomial $g(X, Y)$ is not greater than $c_2 H^{M-1}$. Hence, we have $R(\eta) = A(\xi)g(\xi, \eta)$ and $|g(\xi, \eta)| \leq c_3 H^{M-1}$. Since $|R(\eta)| \geq w_{nN}(\eta, c_1 H^M)$, our choice of the polynomial $A(X)$ implies that

$$w_n(\xi, H) = |A(\xi)| \geq c_3^{-1} H^{-M+1} w_{nN}(\eta, c_1 H^M). \qquad (3.3)$$

Letting H tend to infinity, we obtain

$$w_n(\xi) \leq M - 1 + M w_{nN}(\eta). \qquad (3.4)$$

It follows that

$$w(\xi) \le \limsup_{n \to +\infty} \frac{(M-1)N + MNw_{nN}(\eta)}{nN} \le MNw(\eta) \qquad (3.5)$$

and, inverting the roles played by η and ξ, we get

$$w_n(\eta) \le N - 1 + Nw_{nM}(\eta) \quad \text{and} \quad w(\eta) \le MNw(\xi).$$

Consequently, $w(\xi)$ and $w(\eta)$ are simultaneously finite or infinite. Furthermore, $w_n(\xi)$ is finite for every positive integer n if, and only if, $w_n(\eta)$ is finite for every positive integer n. This completes the proof of the theorem.

We may ask whether inequalities (3.4) and (3.5) are optimal: how can the type of S-numbers ξ and η which are roots of a non-zero integer polynomial $P(X, Y)$ of bidegree (M, N) differ? It follows from (3.4) that for any real transcendental number ξ and any positive integers k and n, we have

$$w_n(\xi) + 1 \le k(w_n(\xi^k) + 1) \qquad (3.6)$$

and

$$w_n(\xi^k) \le w_{kn}(\xi). \qquad (3.7)$$

Inequality (3.6) is sharp for $n = 1$, since the difference $k(w_1(\xi^k) + 1) - 1 - w_1(\xi)$ can take any non-negative real value, see Exercise 3.6. Furthermore, it follows from Theorem 7.7 that for any integer $d \ge kn(kn + 2)$ and for ξ the positive real number with $\xi^{kn} = \sum_{j \ge 1} 2^{-(d+1)^j}$ we have $w_n(\xi^k) = w_{kn}(\xi) = d$. This shows that (3.7) is also sharp.

Moreover, it follows from (3.4) and (3.5) that two algebraically dependent T-numbers have same type. This is not the case for two algebraically dependent U-numbers, see Theorem 7.4.

3.3 Koksma's classification

Koksma's point of view [333] is close to that of Mahler, but instead of looking at the approximation of 0 by integer polynomials evaluated at the real number ξ, Koksma considered the approximation of ξ by algebraic numbers. For given positive integer n and real number $H \ge 1$, we define the quantity

$$w_n^*(\xi, H)$$
$$:= \min\{|\xi - \alpha| : \alpha \text{ real algebraic, } \deg(\alpha) \le n, \mathrm{H}(\alpha) \le H, \alpha \ne \xi\},$$

where $\mathrm{H}(\alpha)$ denotes the height of the algebraic number α (see Definition 2.1).

We then set

$$w_n^*(\xi) = \limsup_{H \to +\infty} \frac{-\log(H w_n^*(\xi, H))}{\log H} \qquad (3.8)$$

and

$$w^*(\xi) = \limsup_{n \to +\infty} \frac{w_n^*(\xi)}{n}.$$

In other words, $w_n^*(\xi)$ is the supremum of the real numbers w for which there exist infinitely many real algebraic numbers α of degree at most n satisfying

$$0 < |\xi - \alpha| \le H(\alpha)^{-w-1}. \qquad (3.9)$$

We have $w_n^*(\xi) = w_n^*(\xi + a/b) = w_n^*(a\xi/b)$ for any real number ξ, any positive integer n, and any non-zero rational number a/b (see Exercise 3.1). This implies that the function w_n^* takes the same value at almost all real numbers.

The factor H occurring in (3.8) (corresponding to the exponent -1 of $H(\alpha)$ in (3.9)) does not appear in the definition of $w_n(\xi)$, but it is natural to introduce it here. Indeed, if an integer polynomial $P(X)$ satisfies $|P(\xi)| < H(P)^{-w}$, then, by Lemma A.5, it has a root α with $|\xi - \alpha| < n H(P)^{-w} |P'(\xi)|^{-1}$. In general $|P'(\xi)|$ is likely to be comparable to $H(P)$ (or, equivalently, $P(X)$ is likely to have no other root close to ξ). Then, $|\xi - \alpha|$ is not much bigger than $H(\alpha)^{-w-1}$. Furthermore, we shall see in Chapter 4 that $w_n(\xi) = w_n^*(\xi) = n$ for almost all real numbers ξ.

For any real number ξ, any positive integer n, and any $H \ge |\xi| + 1$, we have $w_n^*(\xi, H) \le w_1^*(\xi, H) \le (|\xi| + 1)/H$, as can be seen by considering the rational number with denominator $[H]$ closest to ξ. Thus, we get $0 \le w_n^*(\xi) \le +\infty$. Moreover, the sequence $(w_n^*(\xi))_{n \ge 1}$ is clearly non-decreasing, and the functions w_1 and w_1^* are equal.

DEFINITION 3.5. *Let ξ be a real number. We say that ξ is an*

A^*-number, if $w^*(\xi) = 0$;
S^*-number, if $0 < w^*(\xi) < +\infty$;
T^*-number, if $w^*(\xi) = +\infty$ and $w_n^*(\xi) < +\infty$ for any $n \ge 1$;
U^*-number, if $w^*(\xi) = +\infty$ and $w_n^*(\xi) = +\infty$ from some n onwards.

Actually, Koksma [333] defined $w_n^*(\xi, H)$ as the minimum

$$\min\{|\xi - \alpha| : \alpha \text{ complex algebraic, } \deg(\alpha) \le n, H(\alpha) \le H\},$$

in such a way that ξ is transcendental if, and only if, $w_n^*(\xi, H)$ is positive for any $n \ge 1$ and $H \ge 1$. When Koksma introduced his classification (and also in 1957, when Schneider's book appeared [517]), it was not known that $w^*(\xi) = 0$ is a necessary and sufficient condition for ξ to be algebraic (a result due to Wirsing [598]). For this reason, Koksma only classified the *transcendental* numbers in three families (according to [333], an S^*-number ξ is a transcendental number with $w^*(\xi) < +\infty$), and, for consistency, he called A^*-numbers the set of algebraic numbers. Thus, our definition of A^*-numbers differs from Koksma's, but this is only apparent. We choose this way to proceed in order to match with Mahler's and Sprindžuk's classifications (see Chapter 8).

In [333] Koksma defined $w_n^*(\xi)$ for a complex transcendental number ξ by taking into account its complex approximants. Since we restrict our attention to the approximation of *real* numbers ξ, it is plainly natural to consider only *real* algebraic approximants. This definition is also taken by Schmidt in [512], page 279.

However, it is important to notice that it makes no difference if we also take into account complex non-real approximants. Indeed, for $n \geq 1$, $H \geq 1$, and a real number ξ, set

$$w_n^{*c}(\xi, H)$$
$$:= \min\{|\xi - \alpha| : \alpha \text{ algebraic}, \deg(\alpha) \leq n, H(\alpha) \leq H, \alpha \neq \xi\}$$

and

$$w_n^{*c}(\xi) = \limsup_{H \to +\infty} \frac{-\log(H w_n^{*c}(\xi, H))}{\log H}.$$

Clearly, we have the inequality

$$w_n^{*c}(\xi) \geq w_n^*(\xi),$$

which turns out to be an equality, as stated in the next lemma, extracted from [128].

LEMMA 3.1. *For any positive integer n and any real number ξ not algebraic of degree at most n, we have $w_n^{*c}(\xi) = w_n^*(\xi)$.*

PROOF. The case $n = 1$ is trivial. Let $n \geq 2$ be an integer, $H > 1$ be a real number, and ξ be a real number not algebraic of degree at most n. Let α_1 be an algebraic number of height at most H and of degree n_1 less than or equal to n such that $w_n^{*c}(\xi, H) = |\xi - \alpha_1|$. We may assume that α_1 is non-real, otherwise the lemma is clearly true. Then, the minimal polynomial of α_1 over $\mathbb{Z}$, denoted by $P_1(X)$, has two distinct roots α_1 and $\overline{\alpha}_1$ lying very close to ξ. Grace's complex version of Rolle's Theorem (see, for example, [114], page 25) asserts that its derivative $P_1'(X)$ has a root α_2 in the closed disc centered at

$(\alpha_1 + \overline{\alpha}_1)/2$ and of radius $|\alpha_1 - \overline{\alpha}_1| \cot(\pi/n_1)/2$. Observe that this closed disc reduces to the point $(\alpha_1 + \overline{\alpha}_1)/2$ if $n_1 = 2$. Consequently, we have

$$|\xi - \alpha_2| \leq |\xi - \mathfrak{Re}\,\alpha_1| + \frac{|\alpha_1 - \overline{\alpha}_1|}{2} \cdot \frac{n}{\pi} \leq n\,|\xi - \alpha_1|,$$

$$\deg(\alpha_2) \leq n_1 - 1, \quad \text{and} \quad \mathrm{H}(\alpha_2) \leq 2^n\,\mathrm{H}(P_1') \leq 2^n n H,$$

by Lemma A.3. If α_2 is non-real, we proceed further in the same way in order to construct an algebraic approximant α_3 of ξ whose degree is strictly less than the degree of α_2. We iterate this process until we end up with a real approximant. This always happens since the degrees of the algebraic numbers we construct form a strictly decreasing sequence. Consequently, there exists a real algebraic number α with

$$\mathrm{H}(\alpha) \leq (2^n n)^n\,H \quad \text{and} \quad |\xi - \alpha| \leq n^n\,|\xi - \alpha_1|,$$

and $\alpha \neq \xi$, since ξ is not algebraic of degree at most n. Thus, we have for any real number $H \geq 1$

$$w_n^*\big(\xi,\, 2^{n^2}\, n^n\, H\big) \leq n^n\, w_n^{*c}(\xi, H)$$

and

$$w_n^{*c}(\xi) \leq \limsup_{H \to +\infty} \frac{-\log\big(H n^{-n}\, w_n^*\big(\xi,\, 2^{n^2}\, n^n\, H\big)\big)}{\log H} = w_n^*(\xi)$$

holds, as asserted.

As for Mahler's classification, the classes S^*, T^*, and U^* may be further subdivided into infinitely many subclasses.

DEFINITION 3.6. *We define the $*$-type $t^*(\xi)$ of an S^*-number ξ as the value of $w^*(\xi)$, that is,*

$$t^*(\xi) = \limsup_{n \to +\infty} \frac{w_n^*(\xi)}{n}.$$

As for the S-numbers, this definition may not match with the quantity

$$\sup_{n \geq 1} \frac{w_n^*(\xi)}{n},$$

which Koksma [333] called 'Index der S^*-Zahl ξ'. The discussion following Definition 3.2 can be rephrased with w_n^* and S^* instead of w_n and S. Again, no matter how the $*$-type is defined, Theorem 5.6 asserts that there are real S^*-numbers of arbitrary $*$-type greater than or equal to 1.

DEFINITION 3.7. *Let ξ be a T^*-number and, for any positive integer n, define the real number τ_n^* by $w_n^*(\xi) = n^{\tau_n^*}$. The $*$-type of ξ, denoted by $t^*(\xi)$, is the quantity*

$$t^*(\xi) = \limsup_{n \to +\infty} \tau_n^*.$$

DEFINITION 3.8. *Let ξ be a U^*-number. The $*$-type of ξ, denoted by $t^*(\xi)$, is the smallest positive integer n such that $w_n^*(\xi) = +\infty$.*

The U^*-numbers of $*$-type 1 are precisely the Liouville numbers.

The analogue of Theorem 3.2 holds, namely two algebraically dependent real numbers belong to the same class. Moreover, they have same $*$-type if both are T^*-numbers. The proof is left as Exercise 3.6, where it is also shown that for any real transcendental number ξ and any positive integers k and n, we have

$$w_n^*(\xi) + 1 \le k\big(w_n^*(\xi^k) + 1\big)$$

and

$$w_n^*(\xi^k) \le w_{kn}^*(\xi). \tag{3.10}$$

As for inequality (3.7), Theorem 7.7 can be used to prove that (3.10) is sharp.

As in Mahler's classification, since the functions w_n^* are invariant by rational translations, almost all real numbers belong to the same class and have the same $*$-type. We end this Section with a metric result due to Koksma [333], Satz 13.

THEOREM 3.3. *Almost all real numbers are S^*-numbers of $*$-type less than or equal to 1.*

PROOF. For any positive integers H, n and k, let $E(H, n, k)$ denote the union of the intervals $[\alpha - H^{-n-1-1/k}, \alpha + H^{-n-1-1/k}]$ over all real algebraic numbers α of degree at most n and of height H. Since there are no more than $n(n+1)(2H+1)^n$ such algebraic numbers α, we get

$$\lambda\big(E(H, n, k)\big) \le 2n(n+1)(2H+1)^n H^{-n-1-1/k},$$

and the sum $\sum_{H \ge 1} \lambda(E(H, n, k))$ converges. It then follows from the Borel–Cantelli Lemma 1.2 that $\lambda(E(n, k)) = 0$, where $E(n, k)$ denotes the set of real numbers ξ such that the inequality

$$|\xi - \alpha| < H^{-n-1-1/k}$$

has infinitely many solutions in real algebraic numbers α of height at most H and of degree at most n. Since any S^*-number whose $*$-type is strictly greater than 1, any T^*-number and any U^*-number belong to some set $E(n, k)$, we conclude that almost all real numbers are S^*-numbers of $*$-type less than or equal to 1.

3.4 Comparison between both classifications

In this Section, we use a result of Wirsing [598] on the approximation of real numbers by algebraic numbers of bounded degree to show that the classifications of Mahler and Koksma are essentially equivalent (first assertion of Theorem 3.6). Actually, this fact had been established before Wirsing's result, but with a slightly more complicated proof, by Koksma [333] in the paper where he introduced its classification.

PROPOSITION 3.2. *For any positive integer n and any real number ξ, we have* $w_n(\xi) \geq w_n^*(\xi)$.

PROOF. We may assume that $w_n^*(\xi)$ is positive. Let w^* be real with $0 < w^* < w_n^*(\xi)$. By the definition of $w_n^*(\xi)$, there exist infinitely many non-zero algebraic numbers α of degree at most n such that $|\xi - \alpha| \leq H(\alpha)^{-1-w^*}$. Denoting by $P_\alpha(X)$ the minimal polynomial over $\mathbb{Z}$ of such an α, Rolle's Theorem implies that

$$|P_\alpha(\xi)| \leq |\xi - \alpha| \max_{t \in [\xi-1, \xi+1]} |P_\alpha'(t)| \leq c_4 \, H(P_\alpha) \, H(\alpha)^{-1-w^*},$$

for some positive constant c_4 depending only on ξ and n. Since $H(P_\alpha) = H(\alpha)$, this shows that $w_n(\xi) \geq w^*$. Letting w^* tend to $w_n^*(\xi)$, we get $w_n(\xi) \geq w_n^*(\xi)$, as claimed.

One may ask whether there exist real numbers ξ such that $w_n(\xi)$ and $w_n^*(\xi)$ are different. The answer is positive, and this will be proved in Chapter 7.

While it is easy to bound $w_n(\xi)$ from below in terms of $w_n^*(\xi)$, it is highly non-trivial to estimate $w_n^*(\xi)$ from below in terms of $w_n(\xi)$. Such a result was first obtained by Wirsing [598] in 1961 (and, later and independently, by Sprindžuk). He established various lower bounds for $w_n^*(\xi)$, when ξ is a transcendental real number or a real algebraic number of degree at least $n + 1$.

THEOREM 3.4. *Let n be a positive integer and ξ be a real number which is not algebraic of degree at most n. Then we have*

$$w_n^*(\xi) \geq w_n(\xi) - n + 1, \tag{3.11}$$

$$w_n^*(\xi) \geq \frac{w_n(\xi) + 1}{2}, \tag{3.12}$$

$$w_n^*(\xi) \geq \frac{w_n(\xi)}{w_n(\xi) - n + 1}, \tag{3.13}$$

and

$$w_n^*(\xi) \geq \frac{n}{4} + \frac{\sqrt{n^2 + 16n - 8}}{4}. \tag{3.14}$$

Inequality (3.11), under a hidden form, is due to Schneider [517], Hilfssatz 19. The lower bound (3.13) and a result slightly weaker than (3.14) (namely, the estimate $w_n^*(\xi) \geq (n+2+\sqrt{n^2 + 4n - 4})/4$, which follows by combining (3.12), (3.13), and Proposition 3.1) were obtained by Wirsing [598], while (3.14) is a result of Bernik and Tishchenko [95]. The proof of (3.13) given below is not the original one of Wirsing and can be found in [122].

We display an immediate consequence of (3.13) and Proposition 3.2.

COROLLARY 3.2. *Let n be a positive integer and ξ be a real number which is not algebraic of degree at most n. If $w_n(\xi) = n$, we then have $w_n^*(\xi) = n$.*

We are now able to show how the second assertion of Theorem 2.7 and Theorem 2.8 follow from the first assertion of Theorem 2.7.

THEOREM 3.5. *The A^*-numbers are exactly the algebraic numbers. Let ξ be a real algebraic number of degree d and let n be a positive integer. We then have $w_n^*(\xi) = \min\{n, d - 1\}$.*

PROOF. Inequality (3.12) of Theorem 3.4 combined with Proposition 3.1 shows that $w^*(\xi) \geq 1/2$ holds for any real transcendental number ξ. Conversely, let ξ be a real algebraic number of degree d and n be a positive integer. We infer from Corollary A.2 that $w_n^*(\xi) \leq d - 1$, hence we get that $w^*(\xi) = 0$, and the first assertion of the theorem is proved. Furthermore, we observe that for n at most equal to $d - 1$, Theorem 3.1 and Corollary 3.2 imply that $w_n^*(\xi) = n = \min\{n, d - 1\}$. In particular, for any n at least equal to $d - 1$, we have $w_n^*(\xi) \geq w_{d-1}^*(\xi) = d - 1$, hence $w_n^*(\xi) = d - 1$, by Corollary A.2. This completes the proof of the theorem.

It easily follows from Proposition 3.2 and Theorem 3.4 that the classifications of Mahler and of Koksma are equivalent.

THEOREM 3.6. *The classifications of Mahler and of Koksma coincide, in the sense that any S-number (resp. T-number, U-number) is an S^*-number (resp. T^*-number, U^*-number). Further, if ξ is an S-number, we then have $t^*(\xi) \leq t(\xi) \leq t^*(\xi) + 1$. If ξ is either a T-number or a U-number, then its type is equal to its $*$-type.*

PROOF. For a real transcendental number ξ, we infer from (3.11), (3.12), and Proposition 3.2 that the inequalities

$$w_n^*(\xi) \leq w_n(\xi) \leq \min\{n + w_n^*(\xi), 2w_n^*(\xi)\} \tag{3.15}$$

hold for any positive integer n. Consequently, $w_n(\xi)$ and $w_n^*(\xi)$ are simultaneously finite or infinite and $w^*(\xi) \leq w(\xi) \leq \min\{w^*(\xi) + 1, 2w^*(\xi)\}$. Thus, we have $t^*(\xi) \leq t(\xi) \leq t^*(\xi) + 1$ if ξ is an S-number. If ξ is a T-number, then $w_n(\xi) \geq n$ for infinitely many n and we infer from (3.15) that

$$\limsup_{n \to +\infty} \frac{\log w_n^*(\xi)}{\log n} \leq \limsup_{n \to +\infty} \frac{\log w_n(\xi)}{\log n} \leq \limsup_{n \to +\infty} \frac{\log 2w_n^*(\xi)}{\log n},$$

which implies that $t(\xi) = t^*(\xi)$.

Koksma [333], Satz 11, proved the first assertion of Theorem 3.6 without using Theorem 3.4 (see also Schneider [517]). Furthermore, he showed (Satz 10) that any S-number ξ is an S^*-number with $t^*(\xi) \leq t(\xi) \leq t^*(\xi) + 2$.

We now turn to the proof of Theorem 3.4. Clearly, the functions w_1 and w_1^* coincide. Let $n \geq 2$ be an integer and let ξ be a real number which is not algebraic of degree at most n. We first prove (3.11), and then inequalities (3.12), (3.14), and (3.13).

- Proof of the lower bound (3.11).
Let w be a real number with $w < w_n(\xi)$. We infer from the definition of $w_n(\xi)$ and Lemma A.3 that there exist infinitely many irreducible integer polynomials $P(X)$ of degree at most n satisfying

$$0 < |P(\xi)| \leq H(P)^{-w}.$$

By Lemma A.8, such a polynomial $P(X)$ has a root α with

$$|\xi - \alpha| \leq (2n)^n |P(\xi)| \cdot H(P)^{n-2}.$$

This implies the lower bound $w_n^{*c}(\xi) \geq w - n + 1$. We conclude by noticing that w can be taken arbitrarily close to $w_n(\xi)$ and by applying Lemma 3.1.

Without any loss of generality, we may assume in the sequel of the proof of Theorem 3.4 that $0 < \xi < 1/10$ holds and, by (3.11), that $w_n(\xi)$ is finite. In all what follows (apart from Lemma 3.2 and its proof), the notation $a \gg b$ means that there exists a real number λ, depending only on n, such that $a \geq \lambda b$.

- Proof of the lower bound (3.12).
Let ε be a real number with $0 < \varepsilon < 1/2$. We infer from the definition of $w_n(\xi)$ and Lemma A.3 that there exist infinitely many irreducible, primitive, integer polynomials $P(X)$ of degree at most n satisfying

$$0 < |P(\xi)| \leq H(P)^{-w_n(\xi)+\varepsilon}.$$

Let $P(X)$ be such a polynomial. If $R(X)$ is an integer polynomial of degree at most n which is a multiple of $P(X)$, then, again by Lemma A.3, there exists a positive constant c_5, depending only on n, with $c_5 < 1$ and $\mathrm{H}(R) \geq 2c_5 \mathrm{H}(P)$. By Theorem B.2, the system of inequalities

$$|b_n \xi^n + \ldots + b_0| \leq c_5^{-n} \mathrm{H}(P)^{-n}$$
$$|b_1|, \ldots, |b_n| \leq c_5 \mathrm{H}(P)$$

has a non-zero integer solution $(b_0, \ldots, b_n)$. Set $Q(X) = b_n X^n + \ldots + b_1 X + b_0$. If $\mathrm{H}(P) \geq 2c_5^{-1}$, it follows from the assumption $0 < \xi < 1/10$ that $\mathrm{H}(Q)$ is at most equal to $c_5 \mathrm{H}(P)$. Consequently, by our choice of c_5, the polynomials $P(X)$ and $Q(X)$ have no common factor.

Hence, one can build two sequences $(P_k)_{k \geq 1}$ and $(Q_k)_{k \geq 1}$ of non-zero integer polynomials of degree at most n, such that the height of $P_k(X)$ tends to infinity with k,

$$|P_k(\xi)| \leq \mathrm{H}(P_k)^{-w_n(\xi)+\varepsilon}, \quad \mathrm{H}(Q_k) \ll \mathrm{H}(P_k),$$
$$|Q_k(\xi)| \ll \mathrm{H}(P_k)^{-n} \quad (k \geq 1), \tag{3.16}$$

and

$$P_k(X) \text{ and } Q_k(X) \text{ are coprime } (k \geq 1).$$

To conclude, we need an auxiliary lemma, due to Wirsing [598]. Notice that in Lemma 3.2 below and in its proof, the constant implied in $\ll$ depends only on t.

LEMMA 3.2. *Let $t \geq 2$ be an integer and let $P(X)$ and $Q(X)$ be coprime polynomials with integer coefficients of degrees less than or equal to t. Let ξ be a real number with $|\xi| \leq 1$ and which is not algebraic of degree less than or equal to t. Assume that there exist a root of $P(X)$ and a root of $Q(X)$ in the open disc centered at ξ of radius 1. Then, we either have*

$$1 \ll \max\{P^2(\xi) \cdot \mathrm{H}(P)^{t-2} \cdot \mathrm{H}(Q)^t, Q^2(\xi) \cdot \mathrm{H}(Q)^{t-2} \cdot \mathrm{H}(P)^t\}, \tag{3.17}$$

or there exists a real root α of the polynomial $PQ(X)$ such that one of the following four cases holds:

$$|\xi - \alpha| \ll |P(\xi)| \cdot \mathrm{H}(P)^{-1}, \tag{3.18}$$
$$|\xi - \alpha| \ll |Q(\xi)| \cdot \mathrm{H}(Q)^{-1}, \tag{3.19}$$
$$|\xi - \alpha|^2 \ll P^2(\xi) \cdot |Q(\xi)| \cdot \mathrm{H}(P)^{t-2} \cdot \mathrm{H}(Q)^{t-1}, \tag{3.20}$$
$$|\xi - \alpha|^2 \ll |P(\xi)| \cdot Q^2(\xi) \cdot \mathrm{H}(P)^{t-1} \cdot \mathrm{H}(Q)^{t-2}. \tag{3.21}$$

PROOF. We denote by $\alpha_1, \ldots, \alpha_m$ the roots of $P(X)$ and by $\beta_1, \ldots, \beta_n$ those of $Q(X)$, numbered in such a way that, if $p_i := |\alpha_i - \xi|$ and $q_j := |\beta_j - \xi|$ for $i = 1, \ldots, m$ and $j = 1, \ldots, n$, we have $p_1 \leq \ldots \leq p_m$ and $q_1 \leq \ldots \leq q_n$. Without loosing any generality, we may assume $p_1 \leq q_1 < 1$. Denote by a_m the leading coefficient of $P(X)$ and by b_n that of $Q(X)$. Corollary A.1 applied with $\rho = 1$ gives

$$|P(\xi)| \ll H(P) \prod_{1 \leq i \leq m} \min\{1, p_i\} \ll |P(\xi)| \qquad (3.22)$$

and

$$|Q(\xi)| \ll H(Q) \prod_{1 \leq j \leq n} \min\{1, q_j\} \ll |Q(\xi)|. \qquad (3.23)$$

If $m = 1$ or if $m > 1$ and $p_2 > 1$ (*resp. if* $n = 1$ *or if* $n > 1$ *and* $q_2 > 1$), then α_1 (*resp.* β_1) is real and (3.22) (*resp.* (3.23)) implies that we are in the case (3.18) (*resp.* (3.19)). Thus, we can assume that $m \geq 2$, $n \geq 2$, and $p_2 \leq 1$, $q_2 \leq 1$. Denoting by R the resultant (see Definition A.1) of the polynomials $P(X)$ and $Q(X)$, we have

$$1 \leq |R| = |a_m|^n |b_n|^m \prod_{\substack{1 \leq i \leq m \\ 1 \leq j \leq n}} |\alpha_i - \beta_j|$$

$$\ll |a_m b_n|^t \prod_{\substack{1 \leq i \leq m \\ 1 \leq j \leq n}} \max\{p_i, q_j\} =: AB, \qquad (3.24)$$

where

$$A = \prod_{i:p_i \leq 1} \prod_{j:q_j \leq 1} \max\{p_i, q_j\}$$

and

$$B \leq |a_m b_n|^t \prod_{\substack{1 \leq i \leq m \\ 1 \leq j \leq n}} \left(\max\{1, p_i\} \max\{1, q_j\}\right) \ll H(P)^t H(Q)^t.$$

If $p_2 \leq q_1$, we have $p_1 \leq p_2 \leq q_1 \leq 1$, whence

$$A \leq \prod_{j:q_j \leq 1} q_j^2 \ll Q^2(\xi) H(Q)^{-2},$$

by (3.23). This gives (3.17).

Thus, we can assume that $q_1 < p_2$. If moreover q_2 is such that $p_1 \leq q_1 < p_2 \leq q_2$, then we get

$$A \ll q_1 \left(\prod_{i \geq 2: p_i \leq 1} p_i \right) \left(\prod_{j \geq 2: q_j \leq 1} q_j^2 \right).$$

By multiplying (3.24) by $p_1 q_1$, we infer from (3.22) that

$$p_1^2 \leq p_1 q_1 \ll \left(\prod_{i: p_i \leq 1} p_i \right) \left(\prod_{j: q_j \leq 1} q_j^2 \right) \mathrm{H}(P)^t \, \mathrm{H}(Q)^t$$
$$\ll |P(\xi)| \cdot Q^2(\xi) \, \mathrm{H}(P)^{t-1} \mathrm{H}(Q)^{t-2},$$

which gives (3.20). In the same way, if q_2 is such that $p_1 \leq q_1 \leq q_2 < p_2$, one ends up with (3.21), and the proof of the lemma is complete. Further, α_1 is real since $p_1 < p_2$.

Let k be sufficiently large such that the polynomials $P_k(X)$ and $Q_k(X)$ have a root in the open disc centered at ξ of radius 1, and apply Lemma 3.2 to the pairs of polynomials (P_k, Q_k). By (3.16), we get

$$\max\{P_k^2(\xi) \cdot \mathrm{H}(P_k)^{n-2} \mathrm{H}(Q_k)^n, \, Q_k^2(\xi) \cdot \mathrm{H}(Q_k)^{n-2} \mathrm{H}(P_k)^n\} \ll \mathrm{H}(P_k)^{-1},$$

since $w_n(\xi) - \varepsilon \geq n - 1/2$. Thus, (3.17) cannot hold for k large enough. If for infinitely many k we are in one of the cases (3.18) or (3.19), we then get $w_n^*(\xi) \geq n$. Consequently, we may assume that for k large enough, we are in one of the cases (3.20) or (3.21). This implies that there exists a real root α_k of the polynomial $P_k Q_k(X)$ such that

$$|\xi - \alpha_k|^2 \ll \mathrm{H}(P_k)^{-2w_n(\xi)+2\varepsilon+n-3} \quad \text{or} \quad |\xi - \alpha_k|^2 \ll \mathrm{H}(P_k)^{-w_n(\xi)+\varepsilon-3}.$$

Since $\mathrm{H}(\alpha_k) \ll \mathrm{H}(P_k)$ and ε can be taken arbitrarily small, we get the lower bound

$$w_n^*(\xi) \geq \min\left\{n, w_n(\xi) - \frac{n-1}{2}, \frac{w_n(\xi)+1}{2}\right\} \geq \min\left\{n, \frac{w_n(\xi)+1}{2}\right\},$$

where we use that $w_n(\xi) \geq n$. Consequently, (3.12) holds when $w_n(\xi) \leq 2n - 1$. If $w_n(\xi) > 2n - 1$, then (3.12) follows from (3.11).

- Beginning of the proofs of (3.14) and (3.13).

Let A and H_A be positive real numbers with $2 < A < n + 1$ and such that, for any algebraic number α of degree at most n and of height at least equal to H_A, we have

$$|\xi - \alpha| \geq \mathrm{H}(\alpha)^{-A}. \tag{3.25}$$

Let $\varepsilon > 0$ and $H \geq 2$ be real numbers. By Minkowski's Theorem B.2, there exists a non-zero integer polynomial $P(X) := a_n X^n + a_{n-1} X^{n-1} + \ldots + a_1 X + a_0$ such that

$$|P(\xi)| \leq H^{-n-\varepsilon}$$
$$|a_1| \leq H^{1+\varepsilon} \qquad (3.26)$$
$$|a_2|, \ldots, |a_n| \leq H.$$

If $|a_1| \leq H$, we then have $H(P) \leq H$, otherwise we get $H(P) = |a_1|$, since our assumption on ξ implies that $|a_0| < H(P)$.

The idea of Bernik and Tishchenko [95] is to consider the quantity $|P'(\xi)|$. When $|P'(\xi)|$ is not too small, Lemma A.5 asserts that $P(X)$ has a root very close to ξ. If $H(P) = |a_1|$, we infer from $0 < \xi < 1/10$ that

$$|P'(\xi)| = |n a_n \xi^{n-1} + \ldots + 2 a_2 \xi + a_1|$$
$$\geq |a_1| - |n a_n \xi^{n-1} + \ldots + 2 a_2 \xi| \geq \frac{|a_1|}{2} = \frac{H(P)}{2}.$$

Recall that ξ is not algebraic of degree at most n. By (3.25), Lemma A.3, and Lemma A.5, if H is large enough, we then get

$$|P(\xi)| \gg |\xi - \alpha| |P'(\xi)| \gg H(P)^{1-A}, \qquad (3.27)$$

where α denotes a root of $P(X)$ with $|\xi - \alpha|$ minimal. Further, by (3.26), we have

$$|P(\xi)| \leq H(P)^{-(n+\varepsilon)/(1+\varepsilon)},$$

which, combined with (3.27), yields

$$\varepsilon \geq \frac{n+1-A}{A-2} - \frac{c_6}{\log 3 H(P)}$$

for some positive constant c_6 depending only on n. Thus, if ε is sufficiently small, then the height of $P(X)$ is strictly less than $|a_1|$ and satisfies $H(P) = \max\{|a_2|, \ldots, |a_n|\}$. In other words, under the hypothesis

$$\varepsilon < \frac{n+1-A}{A-2}, \qquad (3.28)$$

and provided that H is large enough, we have constructed a non-zero integer polynomial $P(X)$ satisfying $H(P) \leq H$ and $|P(\xi)| \leq H^{-n-\varepsilon}$, hence such that

$$|P(\xi)| \leq H(P)^{-n-\varepsilon}. \qquad (3.29)$$

We will exploit this remark in two different ways.

- Proof of the lower bound (3.14).

Let ε be a positive real number satisfying (3.28). Using Theorem B.2 as explained above with arbitrarily large values of H in (3.26), we get that there exist infinitely many integer polynomials $P(X)$ of degree at most n satisfying (3.29). Furthermore, we infer from Lemma A.3 that there are infinitely many integer polynomials $P(X)$, irreducible over $\mathbb{Q}$ and primitive, such that $|P(\xi)| \ll \mathrm{H}(P)^{-n-\varepsilon}$. Let $P(X)$ be such a polynomial and consider the system of inequalities

$$|b_n \xi^n + \ldots + b_0| \leq c_5^{-n} \mathrm{H}(P)^{-n-\varepsilon}$$

$$|b_1| \leq c_5 \mathrm{H}(P)^{1+\varepsilon}$$

$$|b_2|, \ldots, |b_n| \leq c_5 \mathrm{H}(P),$$

where c_5 is the same constant as in the beginning of the proof of (3.12). By Theorem B.2, it has a non-zero integer solution $(b_0, \ldots, b_n)$. Set $Q(X) = b_n X^n + \ldots + b_1 X + b_0$. Arguing as above, we infer from the assumption on ε that $\mathrm{H}(Q) \leq c_5 \mathrm{H}(P)$ if $\mathrm{H}(P)$ is sufficiently large. Consequently, our choice of c_5 implies that the polynomials $P(X)$ and $Q(X)$ have no common factor.

Hence, one can build two sequences $(P_k)_{k \geq 1}$ and $(Q_k)_{k \geq 1}$ of non-zero integer polynomials of degree at most n, such that the height of $P_k(X)$ tends to infinity with k,

$$|P_k(\xi)| \leq \mathrm{H}(P_k)^{-n-\varepsilon}, \quad \mathrm{H}(Q_k) \ll \mathrm{H}(P_k),$$
$$|Q_k(\xi)| \ll \mathrm{H}(P_k)^{-n-\varepsilon} \quad (k \geq 1), \tag{3.30}$$

and

$$P_k(X) \text{ and } Q_k(X) \text{ are coprime } (k \geq 1).$$

As in the proof of (3.12), we apply Lemma 3.2 to the pairs of polynomials (P_k, Q_k). We infer from (3.30) that (3.17) cannot hold for k large enough. If for infinitely many k we are in one of the cases (3.18) or (3.19), we then get $w_n^*(\xi) \geq n$ and (3.14) holds. Otherwise, we may assume that for k large enough, we are in one of the cases (3.20) or (3.21). This means that there exists a real root α_k of the polynomial $P_k Q_k(X)$ such that

$$\mathrm{H}(\alpha_k)^{-2A} \ll |\xi - \alpha_k|^2 \ll \mathrm{H}(\alpha_k)^{-3-n-3\varepsilon}. \tag{3.31}$$

Letting ε tend to the upper bound $(n+1-A)/(A-2)$ given in (3.28), it follows from (3.31) that A satisfies the inequality

$$2 A^2 - (n + 4) A + (3 - n) \geq 0,$$

whence

$$A \geq \frac{n + 4}{4} + \frac{\sqrt{n^2 + 16n - 8}}{4}.$$

The claimed result is proved since $w_n^*(\xi) \geq A - 1$.

• Proof of the lower bound (3.13).

Since the method allows us to construct infinitely many polynomials satisfying (3.29), we get $w_n(\xi) \geq n + \varepsilon$. Letting ε tend to the upper bound $(n + 1 - A)/(A - 2)$ given in (3.28), we obtain

$$w_n(\xi) \geq n + \frac{n + 1 - A}{A - 2},$$

hence,

$$A \geq \frac{2 w_n(\xi) + 1 - n}{w_n(\xi) + 1 - n},$$

and (3.13) follows from Lemma 3.1 and $w_n^{*c}(\xi) \geq A - 1$.

For approximation by real quadratic numbers, Wirsing's statement has been improved upon by Davenport and Schmidt [180], who have obtained the following result, best possible up to the value of the numerical constant.

THEOREM 3.7. *For any real number ξ which is neither rational nor quadratic, and for any real number c_7 greater than $160/9$, there exist infinitely many rational or quadratic real numbers α satisfying*

$$|\xi - \alpha| \leq c_7 \max\{1, |\xi|^2\} H(\alpha)^{-3}.$$

Theorem 3.7 is proved in [180] and also in [512]. It has been extended by Davenport and Schmidt [181] (up to the value of the numerical constant) as follows.

THEOREM 3.8. *Let $n \geq 2$ be an integer and let ξ be a real number which is not algebraic of degree at most n. Then there exists an effectively computable constant c_8, depending only on ξ and on n, an integer d with $1 \leq d \leq n - 1$, and infinitely many integer polynomials $P(X)$ of degree n whose roots $\alpha_1, \ldots, \alpha_n$ can be numbered in such a way that*

$$|(\xi - \alpha_1) \ldots (\xi - \alpha_d)| \leq c_8 H(P)^{-n-1}.$$

Tishchenko [560, 561, 564, 565] obtained slight sharpenings of (3.14) for small values of n. Furthermore, he proved [563] that there exists a sequence $(\gamma_n)_{n \geq 1}$

of positive real numbers, tending to 3, such that, for any real transcendental number ξ and any positive integer n, we have $w_n^*(\xi) \geq n/2 + \gamma_n$. This remains somehow far away from a celebrated conjecture of Wirsing, saying that $w_n^*(\xi) \geq n$ holds for any positive integer n and any real transcendental number ξ. A heuristic support for this conjecture is the fact that there are integer polynomials $P(X)$ with arbitrarily large height such that $|P(\xi)| \ll H(P)^{-n}$. Here and below, the constant implied by $\ll$ only depends on ξ and n. By Lemma A.5, such a polynomial has a root α satisfying $|\xi - \alpha| \ll |P(\xi)/P'(\xi)|$. Furthermore, if $P(X)$ has no other root very close to ξ, the order of magnitude of $|P'(\xi)|$ is $H(P)$ and we get that $|\xi - \alpha| \ll H(P)^{-n-1}$. A difficulty arises since we cannot exclude the case when *all* but finitely many such polynomials have two distinct roots very close to ξ.

The plausibility of Wirsing's Conjecture and (3.11) motivate the following problem.

MAIN PROBLEM. *Let* $(w_n)_{n \geq 1}$ *and* $(w_n^*)_{n \geq 1}$ *be two non-decreasing sequences in* $[1, +\infty]$ *such that*

$$n \leq w_n^* \leq w_n \leq w_n^* + n - 1, \quad \text{for any } n \geq 1.$$

Then there exists a real transcendental number ξ *such that*

$$w_n(\xi) = w_n \quad \text{and} \quad w_n^*(\xi) = w_n^* \quad \text{for any } n \geq 1.$$

There is no evidence against (or for!) the Main Problem, which, in other words, claims that all that is not trivially impossible may occur. It will be the *fil rouge* of the next Chapters. A summary of the results obtained towards its resolution is given in Section 7.8.

3.5 Some examples

In this Section, we discuss where some classical numbers, like π, e, and $\log \alpha$ for a non-zero algebraic number α, are located in Mahler's classification.

Popken [463] proved that for any positive integer n there exists a positive constant $c_9(n)$, depending only on n, such that

$$|P(e)| \geq H^{-n-c_9(n)/(\log\log H)} \tag{3.32}$$

for any integer polynomial of degree n and height H sufficiently large. It immediately follows that we have $w_n(e) = n$ for any positive integer n, thus e is an S-number of type 1 and, by Corollary 3.2, an S^*-number of $*$-type 1. Popken's result was refined by Mahler [376], who showed that we can take $c_9(n) = c_{10}n^2 \log n$ in (3.32), for some suitable positive, absolute, constant

c_{10}. In the same paper, Mahler proved that e^α is an S-number for any non-zero algebraic number α, whose type can be explictly bounded (see, for example, Diaz [187]).

In [384, 385], Mahler showed that π and $\log \alpha$ are not U-numbers, where $\alpha \neq 1$ is a non-zero algebraic number. Subsequently, A. Baker [40] (see also Sorokin [528]) established that for any positive integer n and any positive real number ε, we have $w_n(\log(a/b)) \leq n + \varepsilon$ if a and b are positive integers, sufficiently large in terms of n, ε, and $|a - b|$.

Further results can be found in the doctoral dissertation of Cijsouw [166] or in the books of Feldman and Nesterenko [244] (Chapter 2) and of Chudnovsky [164] (Chapter 2). For example, if α and β are algebraic numbers with $\alpha \neq 0, 1$ and β irrational, then α^β is either an S-number or a T-number of type at most 2.

3.6 Exponents of Diophantine approximation

After having defined the functions w_n and w_n^* in Sections 3.1 and 3.3, we introduce below four other exponents of Diophantine approximation. We survey known results on these and discuss briefly some related questions.

Davenport and Schmidt [182] investigated the approximation of real numbers by algebraic integers of bounded degree. This is an inhomogeneous problem which needs alternative methods, since, for instance, Theorem B.2 is useless in this context. Their approach yields the following result which establishes a link between simultaneous rational approximation and approximation by algebraic numbers of bounded degree.

PROPOSITION 3.3. *Let $n \geq 2$ be an integer and let ξ be a real number. Assume that there exist positive real numbers κ and w and arbitrarily large values of X for which the inequalities*

$$|x_0| \leq X, \quad \max_{1 \leq m \leq n} |x_0 \xi^m - x_m| \leq \kappa X^{-1/w} \tag{3.33}$$

have no solution in integers $x_0, \ldots, x_n$ not all 0. We then have $w_n^*(\xi) \geq w$.

PROOF. We may assume that $\kappa < 1$. Let X be one of the large numbers specified in the statement of the proposition and set $Y = X^{(w+1)/(wn+w)}$. Let $\mathcal{C}(Y)$ be the parallelepiped defined by

$$|x_0| \leq Y^n, \quad |x_0 \xi^m - x_m| \leq Y^{-1}, \quad 1 \leq m \leq n.$$

The first of its successive minima, denoted by $\lambda_1(Y)$, satisfies

$$\lambda_1(Y) \geq \kappa \, Y^{-(n-w)/(w+1)}, \tag{3.34}$$

since otherwise there would be a non-zero solution to the system (3.33).

The polar body $C^*(Y)$ of $C(Y)$ is defined by

$$|x_n\xi^n + \ldots + x_1\xi + x_0| \leq Y^{-n}, \quad |x_m| \leq Y, \quad 1 \leq m \leq n,$$

and it follows from (3.34) and Theorem B.4 that the $(n+1)$-th minimum $\lambda^*_{n+1}(Y)$ of $C^*(Y)$ satisfies

$$\lambda^*_{n+1}(Y) \leq c_{11} Y^{(n-w)/(w+1)},$$

for some positive constant c_{11} depending only on ξ and n. Thus, there exist $n+1$ linearly independent integer polynomials $P_j(X) := x_n^{(j)} X^n + \ldots + x_1^{(j)} X + x_0^{(j)}$, of degree at most n, with

$$|P_j(\xi)| \leq c_{11} Y^{-w(n+1)/(w+1)} \quad \text{and} \quad H(P_j) \leq c_{11} Y^{(n+1)/(w+1)},$$

for any $j = 1, \ldots, n+1$ and, by Theorem B.3, such that $|\det(x_i^{(j)})| \leq (n+1)!$. Arguing then as in the proof of Theorem 2.11, we use the Eisenstein Criterion to construct an irreducible integer polynomial $P(X)$ of degree n by taking a suitable linear combination of $P_1(X), \ldots, P_{n+1}(X)$ in such a way that $P(X)$ has a real root α with

$$|\xi - \alpha| \leq c_{12} H(\alpha)^{-1-w},$$

where c_{12} only depends on ξ and n. This implies that $w_n^*(\xi) \geq w$, as claimed.

Proposition 3.3 invites us to introduce the functions w_n' and $\hat{w}_n'$.

DEFINITION 3.9. *Let n be a positive integer and let ξ be a real number. We denote by $w_n'(\xi)$ the infimum of the real numbers w such that there are arbitrarily large values of X for which the inequalities*

$$|x_0| \leq X, \quad 0 < \max_{1 \leq m \leq n} |x_0\xi^m - x_m| \leq X^{-1/w}$$

have a solution in integers $x_0, \ldots, x_n$. Further, we denote by $\hat{w}_n'(\xi)$ the infimum of the real numbers w such that for all sufficiently large values of X the inequalities

$$|x_0| \leq X, \quad 0 < \max_{1 \leq m \leq n} |x_0\xi^m - x_m| \leq X^{-1/w} \tag{3.35}$$

have a solution in integers $x_0, \ldots, x_n$.

It is readily verified that $\hat{w}'_n(\xi)$ is the supremum of the real numbers w such that there are arbitrarily large values of X for which the inequalities (3.35) have no solution in integers $x_0, \ldots, x_n$.

Any real number ξ satisfies $w'_1(\xi) = 1/w_1(\xi)$ and $w'_n(\xi) \leq \hat{w}'_n(\xi)$. For any positive integer n and any real algebraic number ξ of degree $d \geq 2$, we have $w'_n(\xi) = \hat{w}'_n(\xi) = \min\{n, d-1\}$. Furthermore, it follows from Theorem 1.1 and Dirichlet's *Schubfachprinzip*, that $\max\{1, w'_n(\xi)\} \leq w'_n(\xi) \leq n$ holds for any integer $n \geq 2$ and any irrational real number ξ.

Not surprisingly, $w_n(\xi)$ and $w_n^*(\xi)$ are related to the simultaneous rational approximation of $\xi, \xi^2, \ldots, \xi^n$.

THEOREM 3.9. *For any positive integer n and any real number ξ not algebraic of degree at most n, we have*

$$\frac{n}{w_n(\xi) - n + 1} \leq w'_n(\xi) \leq \frac{(n-1)w_n(\xi) + n}{w_n(\xi)} \leq n$$

and $w_n^*(\xi) \geq \hat{w}'_n(\xi)$. *In particular, we have* $w_n(\xi) > n$ *if, and only if,* $w'_n(\xi) < n$ *holds.*

The first assertion of Theorem 3.9 is a direct consequence of Khintchine's Transference Theorem B.5, while the second one follows from Proposition 3.3.

The method of Davenport and Schmidt [182] offers much more flexibility than the approach Wirsing [598] used to prove Theorem 3.4 above, insofar as we may impose several constraints on the algebraic approximants, for example to be algebraic integers [182, 140] (see also the proof of Theorem 2.11) or algebraic units [549]. Let $n \geq 2$ be an integer and let ξ be a real transcendental number. Applying a result of Davenport and Schmidt [182] which asserts that the assumptions of Proposition 3.3 are satisfied with $w = [n/2]$ and a suitable positive constant κ, Bugeaud and Teulié [140, 549] deduced from Proposition 3.3 the existence of infinitely many real algebraic numbers α of degree n satisfying $|\xi - \alpha| \leq c_{13} H(\alpha)^{-1-[n/2]}$, with a suitable constant c_{13} depending only on ξ and n. Laurent [354] slightly improved upon the result of Davenport and Schmidt for odd values of n. To state a consequence of his result, we introduce the following notation: for any positive real number x, set $\lceil x \rceil = x$ when x is an integer and $\lceil x \rceil = [x] + 1$, otherwise.

THEOREM 3.10. *Let n be a positive integer and let ξ be a real number which is not an algebraic number of degree less than or equal to $\lceil n/2 \rceil$. We then have $\hat{w}'_n(\xi) \geq \lceil n/2 \rceil$ and there exists a constant c_{14}, depending only on ξ and n, infinitely many real algebraic numbers α of degree n, and infinitely many real* *algebraic integers α of degree $n + 1$ such that*

$$|\xi - \alpha| \leq c_{14} \, H(\alpha)^{-1-\lceil n/2 \rceil}.$$

Actually, Laurent's result for $n = 2$ is weaker than the following one, due to Davenport and Schmidt [182].

THEOREM 3.11. *Let ξ be a real number which is neither rational nor a quadratic irrationality. We then have $\hat{w}_2'(\xi) \geq (1 + \sqrt{5})/2$ and there exist a constant c_{15}, depending only on ξ, infinitely many real quadratic numbers α, and infinitely many real algebraic integers α of degree 3 such that*

$$|\xi - \alpha| \leq c_{15} \, H(\alpha)^{-(3+\sqrt{5})/2}.$$

The exponent of $H(\alpha)$ obtained in Theorem 3.10 yields a slightly weaker lower bound for $w_n^*(\xi)$ than (3.14), but we know the exact degree of the approximants. Furthermore, Theorem 3.7 does not imply the first assertion of Theorem 3.11.

After having defined w_n, w_n^*, w_n', and $\hat{w}_n'$, we introduce a fifth function $\hat{w}_n$, which appeared for the first time in [182].

DEFINITION 3.10. *Let n be a positive integer and let ξ be a real number. We denote by $\hat{w}_n(\xi)$ the supremum of the real numbers w such that for all sufficiently large values of X the inequalities*

$$0 < |x_n \xi^n + \ldots + x_1 \xi + x_0| \leq X^{-w}, \quad |x_m| \leq X, \quad 1 \leq m \leq n, \qquad (3.36)$$

have a solution in integers $x_0, \ldots, x_n$.

It is readily verified that $\hat{w}_n(\xi)$ is the infimum of the real numbers w such that there are arbitrarily large values of X for which the inequalities (3.36) have no solution in integers $x_0, \ldots, x_n$.

For any positive integer n and any real algebraic number ξ of degree d, we have $\hat{w}_n(\xi) = \min\{n, d - 1\}$, by Théorème 1 of [123]. Assume now that n is a positive integer and ξ is a real number not algebraic of degree at most n. It follows from Dirichlet's *Schubfachprinzip* that we have $\hat{w}_n(\xi) \geq n$, and Davenport and Schmidt [182] (see also Theorem 4.2 of [484]) proved that $\hat{w}_n(\xi) \leq 2n - 1$ holds. Furthermore, the proof of Khintchine's Transference Theorem B.5 shows that the values $\hat{w}_n'(\xi)$ and $\hat{w}_n(\xi)$ are related by

$$\frac{n}{\hat{w}_n(\xi) - n + 1} \leq \hat{w}_n'(\xi) \leq \frac{(n - 1)\hat{w}_n(\xi) + n}{\hat{w}_n(\xi)} \leq n. \qquad (3.37)$$

Davenport and Schmidt [183] (for $n = 2$), R. C. Baker [49] (for $n = 3$), and

Bugeaud [123, 125] (for general n) proved that $\hat{w}_n(\xi) = n$ and $\hat{w}'_n(\xi) = n$ (this is a direct consequence of (3.37)) hold for almost all real numbers ξ.

If $w < n$, the volume of the convex body defined by (3.33) tends to zero as X tends to infinity. Thus, it is tempting to conjecture that for any integer $n \geq 2$ and any real transcendental number ξ the assumptions of Proposition 3.3 are satisfied with a suitable positive constant κ and $w = n$, or, at least, that $\hat{w}'_n(\xi) = n$ holds. This is however surprisingly not true, as was proved by Roy [479, 480]. Let $\{a, b\}^*$ denote the monoid of words on the alphabet $\{a, b\}$ for the product given by the concatenation. The Fibonacci sequence in $\{a, b\}^*$ is the sequence of words $(f_i)_{i \geq 0}$ defined recursively by

$$f_0 = b, \quad f_1 = a, \quad \text{and} \quad f_i = f_{i-1} f_{i-2} \, (i \geq 2).$$

Since, for every $i \geq 1$, the word f_i is a prefix of f_{i+1}, this sequence converges to an infinite word $f = abaabab \ldots$ called the Fibonacci word on $\{a, b\}$. For two positive distinct integers a and b, let $\xi_{a,b} = [0; a, b, a, a, b, a, \ldots]$ be the real number whose sequence of partial quotients is given by the letters of the Fibonacci word on $\{a, b\}$, and, for any positive integer m, set

$$\xi_m := (m + 1 + \xi_{m,m+2})^{-1} = [0; m + 1, m, m + 2, m, m, m + 2, \ldots].$$

Let γ denote the Golden Section $(1 + \sqrt{5})/2$. Roy showed that ξ_m is transcendental and that we have $\hat{w}'_2(\xi_m) = \gamma$ and $\hat{w}_2(\xi_m) \geq \gamma^2$. More precisely, there exists a positive constant c_{16}, depending only on m, such that, for any $X \geq 1$, the inequalities

$$0 < x_0 \leq X, \quad |x_0 \xi_m - x_1| \leq c_{16} X^{-1/\gamma} \quad \text{and} \quad |x_0 \xi_m^2 - x_2| \leq c_{16} X^{-1/\gamma}$$

have a non-zero integer solution (x_0, x_1, x_2). If c_{16} is small enough, the same conclusion also holds true for the inequalities

$$|x_1| \leq X, \quad |x_2| \leq X, \quad |x_2 \xi_m^2 + x_1 \xi_m + x_0| \leq c_{16} X^{-\gamma^2}.$$

Furthermore, Roy [480] studied the approximation of ξ_m by real quadratic numbers and proved the existence of positive constants c_{17} and c_{18}, depending only on m, such that $|\xi_m - \alpha| \leq c_{17} H(\alpha)^{-2\gamma^2}$ has infinitely many solutions in algebraic numbers α of degree 2, while $|\xi_m - \alpha| \leq c_{18} H(\alpha)^{-2\gamma^2}$ has no solution in quadratic irrationals α. In addition, he established [481, 482] that there exists a positive constant c_{19}, depending only on m, such that

$$|\xi_m - \alpha| \geq c_{19} H(\alpha)^{-\gamma - 1},$$

for any algebraic integer α of degree at most 3, and

$$|P(\xi_m)| \geq c_{19} \, \mathrm{H}(P)^{-\gamma},$$

for any monic, integer polynomial $P(X)$ of degree at most 3. This proves that the last assertion of Theorem 3.11 is the best possible (there is a gap in [567]) and disproves the conjecture claiming that, for any real transcendental number ξ, for any integer $n \geq 2$, and any positive real number ε, there exist infinitely many algebraic integers α of degree at most n such that $|\xi - \alpha| \leq \mathrm{H}(\alpha)^{-n+\varepsilon}$. This conjecture was supported by a metric statement saying that almost all real numbers share this approximation property ([140], Corollaire 1). Moreover, Arbour and Roy [32] proved that $\hat{w}_2(\xi) \leq \gamma^2$ holds for any real number not algebraic of degree at most 2. Actually, they showed that if the real number ξ is such that for any sufficiently large positive real number H there exists a non-zero integer polynomial $P(X)$ of degree at most 2 and height at most H with $|P(\xi)| \leq H^{-\gamma^2}/4$, then ξ is algebraic of degree at most 2. It is worth noting that

$$1 = w_2'(\xi_m) < \frac{1+\sqrt{5}}{2} = \hat{w}_2'(\xi_m) < \frac{3+\sqrt{5}}{2} = \hat{w}_2(\xi_m)$$
$$< w_2^*(\xi_m) = 2 + \sqrt{5},$$

holds, while we know that the functions w_2', $\hat{w}_2'$, $\hat{w}_2$, and w_2^* take the value 2 for almost all real numbers.

To conclude this Section, we define another function, introduced by Bugeaud and Laurent [137] and motivated by Dirichlet's Theorem 1.1.

DEFINITION 3.11. *Let n be a positive integer and let ξ be a real number. We denote by $\hat{w}_n^*(\xi)$ the supremum of the real numbers w such that for all sufficiently large values of X there exists a real algebraic number α of degree at most n with*

$$0 < |\xi - \alpha| \leq \mathrm{H}(\alpha)^{-1} X^{-w} \quad and \quad \mathrm{H}(\alpha) \leq X. \tag{3.38}$$

The discussion at the beginning of Section 3.3 and Dirichlet's Theorem 1.1 invite us to consider (3.38) rather than the system of inequalities $|\xi - \alpha| \leq X^{-w-1}, \mathrm{H}(\alpha) \leq X$.

For any positive integer n and any real algebraic number ξ of degree d, we have $\hat{w}_n(\xi) = \min\{n, d-1\}$, by Théorème 1 of [123]. Dirichlet's Theorem 1.1 obviously implies that $\hat{w}_n^*(\xi) \geq 1$ holds for any irrational number ξ and Corollary A.2 yields the upper bound $\hat{w}_n^*(\xi) \leq 2n-1$. Furthermore, Corollaire 1 of Bugeaud [123] asserts that we have $\hat{w}_n^*(\xi) = n$ for all positive integers n and almost all real numbers ξ. Notice that the numbers ξ_m defined above satisfy

$\hat{w}_2^*(\xi_m) = (3 + \sqrt{5})/2$. Other examples of real numbers ξ with $\hat{w}_2^*(\xi) > 2$ and $\hat{w}_2'(\xi) < 2$ have been given by Bugeaud and Laurent [137]. They further improved (3.13) by showing that

$$\hat{w}_n^*(\xi) \geq \frac{w_n(\xi)}{w_n(\xi) - n + 1} \quad \text{and} \quad w_n^*(\xi) \geq \frac{\hat{w}_n(\xi)}{\hat{w}_n(\xi) - n + 1} \tag{3.39}$$

hold for any positive integer n and any real number ξ not algebraic of degree at most n (see Exercises 3.4 and 3.5).

We conclude by some consideration on inhomogeneous Diophantine approximation. Let n be a positive integer and θ be a real number. For any real number ξ, let $w_n(\xi, \theta)$ denote the supremum of the real numbers w for which there exist infinitely many integer polynomials $P(X)$ of degree at most n satisfying

$$0 < |P(\xi) + \theta| \leq \mathrm{H}(P)^{-w}.$$

Define further the n-th exponent of uniform inhomogeneous approximation w_n^{inh} by

$$w_n^{inh}(\xi) = \inf_{\theta \in \mathbb{R}} w_n(\xi, \theta).$$

Bugeaud and Laurent [138] proved that $\hat{w}_n'(\xi) = w_n^{inh}(\xi)$ for any positive integer n and any irrational real number ξ. It would be of interest to study more closely the exponents $w_n(\xi, \theta)$.

The questions evoked in this Section yield several open problems listed in Chapter 10.

3.7 Exercises

EXERCISE 3.1. Let a, b, c, and d be rational integers with $ad - bc \neq 0$ and let ξ be a real number. Let n be a positive integer. Prove that we have $w_n(\xi) = w_n((a\xi + b)/(c\xi + d))$ and $w_n^*(\xi) = w_n^*((a\xi + b)/(c\xi + d))$. For a fixed real algebraic number α, compare $w_n(\xi + \alpha)$ with $w_n(\xi)$ and $w_n^*(\xi + \alpha)$ with $w_n^*(\xi)$.

EXERCISE 3.2. Let $\xi = [a_0; a_1, a_2, \dots]$ be an irrational real number with convergents $p_0/q_0, p_1/q_1, \dots$ Following Güting [267], use Theorems 1.3 and 1.7 to prove that we have

$$w_1(\xi) = w_1^*(\xi) = 1 + \limsup_{n \to +\infty} \frac{\log a_n}{\log q_{n-1}},$$

if this lim sup is finite, and $w_1(\xi) = w_1^*(\xi) = +\infty$ otherwise.

Deduce that $w_1(\xi) = 1$ if the sequence $(\log \sqrt[n]{a_1 a_2 \ldots a_n})_{n \geq 1}$ converges (Khintchine [321, 323] proved that this sequence tends to $\sum_{j \geq 1} \{\log(j(j + 2) + 1) - \log(j(j + 2))\}(\log j)/\log 2$ for almost all real numbers ξ).

EXERCISE 3.3. Use Bombieri and Mueller's method described in Theorem 2.10 to get the lower bound $w_n^*(\xi) \geq n/(w_n(\xi) - n + 1)$, for any integer $n \geq 1$ and any real number ξ which is not algebraic of degree at most n. Prove the same result by means of the approach of Theorem 2.11.

EXERCISE 3.4. Use the method of Wirsing [598] to prove (3.13) and Theorem 2.10 (up to the numerical constants).

Hint. Let $n \geq 2$ be an integer and let ξ be a real number which is not algebraic of degree at most n. Let $\varepsilon > 0$ and set $w = w_n(\xi) (1 + \varepsilon)^2$. Use Theorem B.2 to prove that there exists a positive constant c_{20}, depending only on ξ and n, and, for any positive H, an integer polynomial $P(X)$ such that

$$|P(\xi)| \leq H^{-w}, \quad |P(1)|, \ldots, |P(n-1)| \leq H \quad \text{and}$$
$$|P(n)| \leq c_{20} H^{w-n+1}.$$

Show that $H(P) \geq c_{21} H^{1+\varepsilon}$ for a suitable constant c_{21}, depending only on ξ, n, and ε, and apply Corollary A.1 to get that, when H is large enough, the polynomial $P(X)$ has a single (real) root in each small disc centered at the points $\xi, 1, \ldots, n - 1$; in particular, there exists a suitable constant c_{22}, depending only on ξ, n, and ε, such that $P(X)$ has a real root α with $|\xi - \alpha| \leq c_{22} H(\alpha)^{-1+\varepsilon-w/(w-n+1)}$. Conclude and prove the first inequality of (3.39).

With the same approach, use Theorem A.1 to get Theorem 2.10 (up to the numerical constants).

EXERCISE 3.5. Show that the proof of (3.13) given in Section 3.4 actually yields the second inequality of (3.39).

EXERCISE 3.6. Let ξ and η be two algebraically dependent transcendental real numbers. Let $F(X, Y)$ be an irreducible polynomial with integer coefficients vanishing at the point (ξ, η). Denote by M (*resp.* by N) the degree of $F(X, Y)$ in X (*resp.* in Y). Following Schmidt [508], prove that the inequality

$$w_n^*(\xi) + 1 \leq M\left(w_{nN}^*(\eta) + 1\right)$$

holds for any positive integer n.

Hint. Use the implicit function theorem to show that there exists a neighbourhood of ξ and a one-to-one differentiable function y defined on this neighbourhood such that $y(\xi) = \eta$ and $F(x, y(x))$ is identically zero. Let $H \geq 1$ be a real number. Assume that α_n is an algebraic number of degree at most n and height at most H such that $w_n^*(\xi, H) = |\xi - \alpha_n|$ and denote by $Q(X) := a_n(X - \alpha_n^{(1)})\ldots(X - \alpha_n^{(n)})$ its minimal polynomial over $\mathbb{Z}$, with $\alpha_n^{(1)} = \alpha_n$. Observe that $|\eta - y(\alpha_n)| \ll w_n^*(\xi, H)$. Here, and in the sequel of the proof, the constant implied in $\ll$ depends only on ξ and on η. Since $y(\alpha_n)$ is a root of the polynomial

$$R(Y) := a_n^M F(\alpha_n^{(1)}, Y)\ldots F(\alpha_n^{(n)}, Y),$$

which is not identically zero, $y(\alpha_n)$ is algebraic of degree less than or equal to nN. Use Exercise A.1 and Lemma A.3 to prove that $H(y(\alpha_n)) \ll H^M$. Conclude.

EXERCISE 3.7. Show that for any non-negative real number δ, any positive integer k, and any real number $w \geq k + 1$, there exist real numbers ξ such that

$$w_1^*(\xi) = k(w + 1) - 1 \quad \text{and} \quad w_1^*(\xi^k) = w + \delta.$$

Hint. Construct inductively two increasing sequences of prime numbers $(q_n)_{n\geq 1}$, $(v_n)_{n\geq 1}$ and two increasing integer sequences $(p_n)_{n\geq 1}$, $(u_n)_{n\geq 1}$ such that the intervals

$$I_{2n} := \left[\frac{p_n^k}{q_n^k} - \frac{1}{q_n^{k(w+1)}}, \frac{p_n^k}{q_n^k} - \frac{1}{2q_n^{k(w+1)}}\right] \quad \text{and}$$

$$I_{2n+1} := \left[\frac{u_n}{v_n} - \frac{1}{v_n^{w+1+\delta}}, \frac{u_n}{v_n} - \frac{1}{2v_n^{w+1+\delta}}\right]$$

satisfy $I_2 \supset I_3 \supset \ldots$. Let ξ be the positive real number such that $\{\xi^k\}$ is the intersection of the intervals I_n, $n \geq 2$. Using triangular inequalities, estimate the difference $|\xi - a/b|$ for any rational number a/b with $q_n^k < b < q_{n+1}^k$ (distinguish three cases, depending on the size of $|a/b - p_n^k/q_n^k|$, $|a/b - u_n/v_n|$, and $|a/b - p_{n+1}^k/q_{n+1}^k|$). Prove that if $(q_n)_{n\geq 1}$ and $(v_n)_{n\geq 1}$ do not increase too rapidly, we then have $w_1^*(\xi^k) = w + \delta$ and $|\xi^k - a/b| \geq b^{-w-1}$ for any integers a and b such that b does not belong to the sequence $(v_n)_{n\geq 1}$.

3.8 Notes

• Another presentation of Koksma's classification, based on equivalence classes, can be found in [508].

• Let $\mathcal{Q}$ be a set of integer polynomials and ξ be a transcendental real number. At the end of [311], Kasch and Volkmann introduced the quantities

$w_n(\xi, H, \mathcal{Q})$ defined as the minimum of $|P(\xi)|$ over the set of integer polynomials $P(X)$ of degrees at most n and heights at most H which belong to $\mathcal{Q}$.

• A link between Theorem B.3 and the approximation of real numbers by algebraic numbers has been discussed by Mahler [401]. Notice that his remark 'In the case of (3.29), (...) this property' on page 159 is incorrect, since the results of Cassels [154] and Davenport [175] quoted therein do not deal with n-tuples of the particular form $(a, a^2, \ldots, a^n)$.

• Khintchine [318] proved that the set of real numbers ξ for which there exist a positive integer n and a positive constant κ such that $|P(\xi)| \geq \kappa \, \mathrm{H}(P)^{-n}$ for any integer polynomial of degree at most n is a null set.

• Galočkin [254] proved that values at some algebraic points of Mahler functions satisfying certain functional equations are S-numbers. Improvements and extensions to more general Mahler functions are due to Becker-Landeck [58], Wass [595], Becker [56], Nishioka [438] (see also [439], pages 137 and 138), Nishioka and Töpfer [440], and Töpfer [569, 570]. Amou [27] gave explicit upper bounds for the type of S-numbers obtained as values at algebraic points of functions satisfying certain functional equations of Mahler's type.

• Lang [348] established that values taken by E-functions at algebraic points are S-numbers (see Shidlovskii [522] for further references and results). Chudnowsky [163] established that if the Weierstrass $\wp$ function has algebraic invariants and complex multiplication, then $\wp(\alpha)$ is an S-number for any nonzero algebraic number α.

• In a very original paper, Duverney [222] showed that the sum of the reciprocals of the Fermat numbers $\sum_{n \geq 0} 1/(2^{2^n} + 1)$ is either an S-number or a T-number of type at most 2 (actually, with a slight modification of the proof, it is possible to show that this sum is an S-number of type at most 4, as kindly communicated to me by Duverney). His proof starts from an irrationality measure and then works by induction.

• A. Baker [37] used Theorem 2.5 and a method of Maillet [403] to construct explicitly families of badly approximable real transcendental numbers. He further gave [38] a condition under which such numbers are U-numbers of type 2. His criterion, however in a less general form, has been improved by Baxa [54] and Mkaouar [425].

• A. Baker [38] established that the Champernowne number is not a U-number. Further, he proved that the set of badly approximable real transcendental numbers contains T-numbers or S-numbers ξ with $w_1(\xi)$ arbitrarily large. His main tool is a refinement of Theorem 2.5, providing a criterion which, when satisfied by a real number ξ, ensures that ξ is transcendental but not a U-number. This has been used by Bundschuh [142] who dealt with a class of real numbers having explicit g-adic and continued fraction expansions. A. Baker's criterion has also been used by several authors, including Oryan [446], Zeren [611], Yilmaz [606], and Gürses [264], to prove that values of specific power series evaluated at algebraic numbers are transcendental numbers but not U-numbers. Furthermore, Baxa [53] applied it to show that Cahen's constant $\sum_{n \geq 0} (-1)^n / (S_n - 1)$, where $S_0 = 2$ and $S_{n+1} = S_n^2 - S_n + 1$ for $n \geq 0$, is transcendental but is not a U-number. Actually, Töpfer (Corollary 8 of [569]) proved that Cahen's constant is an S-number.

• Adhikari, Saradha, Shorey and Tijdeman [3] (see also Saradha and Tijdeman [493]) applied A. Baker's theory of linear forms in logarithms of algebraic numbers with algebraic coefficients to prove that numbers such as $\sum_{n \geq 0}((3n + 1)(3n + 2)(3n + 3))^{-1}$, $\sum_{n \geq 1} \chi(n)/n$, $\sum_{n \geq 1} 2^{-n} F_n/n$, where χ is any non-principal Dirichlet character and $(F_n)_{n \geq 0}$ the Fibonacci sequence, are transcendental but are not U-numbers.

• Dress, Elkies, and Luca [208] established a characterization of U-numbers by simultaneous rational approximation. For any given positive integer n, they prove that a real number ξ is algebraic of degree at most $n + 1$ or satisfies $w_n(\xi) = +\infty$ if, and only if, for each positive integer d, there exist a positive constant M_d and infinitely many integer $(d + 1)$-tuples $(x_0, \ldots x_d)$ such that $|x_0 \xi^k - x_k| < M_d \, x_0^{-1/n}$ for each $k = 1, 2, \ldots, d$.

• Under suitable assumptions on the approximation properties of the T-number (or the U-number) α and the U-number β, Caveny [159] proved that α^β is transcendental. Caveny and Tubbs [160] established a quantitative version of the particular case of Theorem 3.2 saying that if ξ_1 is a U-number and ξ_1 and ξ_2 are algebraically dependent, then ξ_2 is a U-number.

• A. Baker [42] established the existence of pairs of distinct U-numbers which are badly approximable pairs (see the Notes of Chapter 1 for the definition).

• A possible generalization of Wirsing's problem in dimension 2 is the following question: does there exist a function $\Psi : \mathbb{Z}_{\geq 1} \to \mathbb{R}_{>0}$ such that, for any pair (ξ_1, ξ_2) of real transcendental numbers and any integer $n \geq 1$, there are infinitely many integers H such that the equations

$$|\xi_1 - \alpha_1| \leq H^{-\Psi(n)}, \quad |\xi_2 - \alpha_2| \leq H^{-\Psi(n)}$$

have a solution in real algebraic numbers α_1 and α_2 of degree at most n and height at most H? If ξ_1 or ξ_2 is an S-number, it is proved in [131] that the answer is positive with $\Psi(n) = c_{23}n$, for some positive constant c_{23}, depending only on ξ_1 and ξ_2. See also Tishchenko [568]. In the opposite direction, for any integer $n \geq 1$, Roy and Waldschmidt [484] have given explicit examples of pairs (ξ_1, ξ_2) of Liouville numbers such that $\max\{|\xi_1 - \alpha_1|, |\xi_2 - \alpha_2|\} > H^{-3\sqrt{n}}$ holds for all algebraic numbers α_1 and α_2 of degree at most n and height at most H, when H is sufficiently large.

• Kopetzky and Schnitzer [336] applied Theorem 3.7 to the approximation on the unit circle by points whose coordinates are algebraic of degree at most 2.

• Diophantine approximation by conjugate algebraic integers has been studied by Roy and Waldschmidt [484], and by Arbour and Roy [32].

• Yu [609] defined a multidimensional generalization of Mahler's classification: all points in $\mathbb{C}^n$ (or in $\mathbb{R}^n$) are classified into $3n + 1$ disjoint non-empty classes, denoted by A^n, S_t^n, T_t^n and U_t^n with $t = 1, \ldots, n$, in such a way that any two algebraic equivalent (over $\mathbb{Q}$) points in $\mathbb{C}^n$ fall into the same class.

• Mahler and Szekeres [402] investigated approximation of real numbers by roots of integers.

4

Mahler's Conjecture on S-numbers

In Theorem 3.3, we used the Borel–Cantelli Lemma 1.2 to prove that almost all real numbers ξ are S^*-numbers of $*$-type less than or equal to 1. A similar statement is however much more difficult to establish when we consider Mahler's classification. The first important result in this direction is due to Mahler [377], who showed in 1932 that almost all real numbers ξ satisfy $\sup_{n \geq 1}(w_n(\xi)/n) \leq 4$. At the end of [377], he made the conjecture that the upper bound 4 could be replaced by 1.

Until Sprindžuk [536, 537, 538] gave in 1965 a complete affirmative answer to that conjecture, there appeared various improvements of Mahler's result. First, Koksma [333] showed that $\sup_{n \geq 1}(w_n(\xi)/n) \leq 3$ for almost all real numbers ξ. This has been strengthened by LeVeque [362], who replaced the upper bound 3 in Koksma's result by 2. Later on, refining a method introduced by Kasch and Volkmann [311], Schmidt [501] proved the inequality $w_n(\xi) \leq 2n - 7/3$ for almost all real numbers ξ and all positive integers n. Lastly, Volkmann [583, 584] showed that $w_n(\xi) \leq 4n/3$ holds for almost all real numbers ξ and all positive integers n. At the same time, Sprindžuk [533] obtained a slightly stronger result than Volkmann's, shortly before his resolution of Mahler's Conjecture. Apart from these general results, it had been established that $w_n(\xi) = n$ for almost all real numbers ξ and for $1 \leq n \leq 3$. The case $n = 1$ is nothing but Khintchine's Theorem 1.10, while Kubilsus [344] used Vinogradov's method of exponential sums to solve the case $n = 2$. A simpler proof is due to Kasch [310], see Exercise 4.1. Furthermore, Volkmann [582] applied results of Davenport [176] on binary cubic forms for the case $n = 3$.

In the present Chapter, we state Sprindžuk's result, its refinement obtained shortly thereafter by A. Baker [41], and the subsequent improvement due to Bernik [81], which appears to be the best possible. The proof of

Bernik is lengthy and complicated, and we choose not to quote it, since it is included in Bernik and Dodson [86]. Instead, we give a full proof of A. Baker's result, following largely Harman's exposition [273]. The end of Section 4.1 is devoted to results on multiplicative Diophantine approximation obtained by an approach originated in a work of Kleinbock and Margulis [330].

4.1 Statements of the theorems

The purpose of the present Chapter is to give a full proof of the following result.

THEOREM 4.1. *Almost all real numbers ξ satisfy $w_n(\xi) = w_n^*(\xi) = n$ for every positive integer n.*

COROLLARY 4.1. *Almost all real numbers are S-numbers of type 1 and S^*-numbers of $*$-type 1.*

Corollary 4.1 improves upon Theorem 3.3. In Chapter 5, it is proved that the set of S-numbers of type greater than 1 is not 'too small', in the sense that its Hausdorff dimension is equal to 1.

In view of Corollary 3.2, which asserts that $w_n^*(\xi) = n$ holds if $w_n(\xi) = n$, Theorem 4.1 is an easy consequence of the following result of Sprindžuk [538, 539].

THEOREM 4.2. *Let n be a positive integer and ε be a positive real number. Then, for almost all real numbers ξ, the equation*

$$|P(\xi)| < H^{-n-\varepsilon} \tag{4.1}$$

has only a finite number of solutions in integer polynomials $P(X)$ of degree at most n and of height at most H.

Keeping in mind Khintchine's Theorem 1.10, it becomes natural to ask whether the conclusion of Theorem 4.2 remains true if the functions $H \mapsto H^{-n-\varepsilon}$ occurring in (4.1) are replaced by more general (non-increasing) functions. This has been confirmed by A. Baker [41] in 1966.

THEOREM 4.3. *Let n be a positive integer and $\Psi : \mathbb{R}_{\geq 1} \to \mathbb{R}_{>0}$ be a non-increasing continuous function such that the series $\sum_{h \geq 1} \Psi(h)$ converges. Then, for almost all real numbers ξ, the equation*

$$|P(\xi)| < \Psi^n(H) \tag{4.2}$$

has only a finite number of solutions in integer polynomials $P(X)$ of degree at most n and of height at most H.

According to A. Baker [44], 'it seems likely that the function $\Psi^n(H)$ [in (4.2)] can be replaced by $H^{-n+1}\Psi(H)$, and this conjecture is in fact established for $n \leq 3$.' A heuristic support for A. Baker's conjecture is the observation that there are about H^n integer polynomials $P(X)$ of height exactly H and degree at most n. If one of these satisfies $|P(\xi)| < H^{-n+1}\Psi(H)$, then it could have (this is not always true!) a root α close to ξ, say (see the discussion after Theorem 3.8) with $|\xi - \alpha| \ll H^{-n}\Psi(H)$, where, as below, the constant implied by $\ll$ depends only on ξ and n. Hence, ξ would lie in the union of about H^n intervals centered at the real roots of integer polynomials of degree at most n and height H. The Lebesgue measure of this union is approximately equal to $\Psi(H)$. Assuming that the sum of the $\Psi(H)$ converges, it would then follow from the Borel–Cantelli Lemma 1.2 that the set of real numbers ξ for which $|P(\xi)| < H^{-n+1}\Psi(H)$ is satisfied for *infinitely* many integer polynomials. $P(X)$ is a null set.

The main difficulty in this approach is that the estimate $|\xi - \alpha| \ll H^{-n}\Psi(H)$ does not always hold. To overcome this problem, very precise information on the distribution of roots of integer polynomials is needed. Sprindžuk introduced the notion of essential and inessential domains (see Section 14 of Chapter 2 of [540] for the description of the general idea) to solve it when $\Psi(H) = H^{-1-\varepsilon}$ for some positive real number ε. His method was refined by A. Baker [41], and then by Bernik [78, 81], who, in a lengthy and intricate paper, managed to get rid of the induction lying at the heart of the proofs of Theorems 4.2 and 4.3 in order to establish the above-mentioned conjecture of A. Baker.

THEOREM 4.4. *Let n be a positive integer and $\Psi : \mathbb{R}_{\geq 1} \to \mathbb{R}_{>0}$ be a non-increasing continuous function such that the series $\sum_{h \geq 1} \Psi(h)$ converges. Then, for almost all real numbers ξ, the equation*

$$|P(\xi)| < H^{-n+1}\Psi(H)$$

has only a finite number of solutions in integer polynomials $P(X)$ of degree at most n and of height at most H.

Apart from the monotonicity condition imposed on Ψ, which may presumably be somewhat relaxed (this is however a difficult problem, keeping in mind the Duffin–Schaeffer Conjecture mentioned in Section 1.4), Theorem 4.4 is best possible, as shown by Beresnevich [61], whose results yield the following strengthening of Theorem 4.1.

THEOREM 4.5. *Let* $\Psi : \mathbb{R}_{\geq 1} \rightarrow \mathbb{R}_{>0}$ *be a non-increasing continuous function. If the series* $\sum_{h \geq 1} \Psi(h)$ *diverges, then for almost all real numbers* ξ *and any positive integer n the equations*

$$|P(\xi)| < H^{-n+1} \Psi(H) \quad and \quad |\xi - \alpha| < H^{-n} \Psi(H) \qquad (4.3)$$

have infinitely many solutions in integer polynomials $P(X)$ *and real algebraic numbers* α, *of degree at most n and of height at most H, respectively. If the series* $\sum_{h \geq 1} \Psi(h)$ *converges, then, for almost all real numbers* ξ, *the equations (4.3) have only finitely many solutions in integer polynomials* $P(X)$ *and real algebraic numbers* α, *of degree at most n and of height at most H, respectively.*

The first part of Theorem 4.5 is Theorem 3 of [61], quoted in Chapter 6 as Theorem 6.6. The second part derives from Theorem 4.4, using Lemma A.6.

We end this Section by quoting two deep results on multiplicative approximation. Following a suggestion of A. Baker [44], another way to generalize Theorem 4.2 consists in replacing the height $H(P)$ of a polynomial $P(X) = a_n X^n + \ldots + a_1 X + a_0$ by the function Π_+ defined by

$$\Pi_+(P) := \prod_{1 \leq i \leq n} \max\{1, |a_i|\}.$$

A. Baker conjectured that, for almost all real numbers ξ and for any positive real number ε, the equation $|P(\xi)| \leq \Pi_+^{-1-\varepsilon}(P)$ has only finitely many solutions in integer polynomials $P(X)$ of degree at most n. (Actually, he formulated the dual problem, asking whether for any positive real number ε and any positive integer n there exist, for almost all real numbers ξ, only finitely many $(n + 1)$-tuples $(q, p_1, \ldots, p_n)$ of integers such that $q^{1+\varepsilon} |q\xi - p_1| \ldots |q\xi^n - p_n| < 1$. A classical transference theorem [594, 516, 604] shows that both questions are equivalent.) This has been established by Sprindžuk [540] for $n = 2$ (see also Theorem 2 of Yu [608]), by Bernik and Borbat [82] for $n = 3, 4$, and for general n by Kleinbock and Margulis [330], who developed a rather new approach based on the correspondence between approximation properties of numbers and orbit properties of certain flows on homogeneous spaces (see, for example, Kleinbock's survey [326], Margulis survey [405], and Starkov's book [542]).

THEOREM 4.6. *Let n be a positive integer and* ε *be a positive real number. Then, for almost all real numbers* ξ, *the equation*

$$|P(\xi)| < \Pi_+^{-1-\varepsilon}(P)$$

has only a finite number of solutions in integer polynomials $P(X)$ *of degree at most n.*

Shortly thereafter, Bernik, Kleinbock, and Margulis [89] improved upon Theorem 4.6.

THEOREM 4.7. *Let n be a positive integer and $\Psi : \mathbb{R}_{\geq 1} \to \mathbb{R}_{>0}$ be a non-increasing continuous function such that*

$$\sum_{h \geq 1} (\log h)^{n-1} \, \Psi(h) < +\infty.$$

Then, for almost all real numbers ξ, the equation

$$|P(\xi)| < \Psi\big(\Pi_+(P)\big) \tag{4.4}$$

has only a finite number of solutions in integer polynomials $P(X)$ of degree at most n.

Observe that the Borel–Cantelli Lemma 1.2 implies that for a function Ψ as in Theorem 4.7 the set of real n-tuples $(\xi_1, \ldots, \xi_n)$ such that there are infinitely many $(n+1)$-tuples $(a_0, a_1, \ldots, a_n)$ of integers with $|a_0 + a_1 \xi_1 + \ldots + a_n \xi_n| < \Psi(\prod_{i=1}^n \max\{|a_i|, 1\})$ has Lebesgue measure zero. A presumably difficult open question is to prove that Theorem 4.7 is best possible in the sense that (4.4) has infinitely many solutions when the sum $\sum_{h \geq 1} (\log h)^{n-1} \, \Psi(h)$ diverges.

4.2 An auxiliary result

The lemma below deals with integer polynomials having two zeros close to each other. Since it is very specific for the problem considered in the present Chapter, we have not included it in Appendix A.

LEMMA 4.1. *Let $P(X) = a(X - \alpha_1)\ldots(X - \alpha_n)$ and $Q(X) = b(X - \beta_1)\ldots(X - \beta_n)$ be two distinct, irreducible, integer polynomials of degree $n \geq 3$. Set*

$$p^* = |a^n(\alpha_1 - \alpha_3)\ldots(\alpha_1 - \alpha_n)|^{-1/2} \quad and$$
$$q^* = |b^n(\beta_1 - \beta_3)\ldots(\beta_1 - \beta_n)|^{-1/2}.$$

Assume that

$$|\alpha_1 - \alpha_2| \leq |\alpha_1 - \alpha_j| \quad and \quad |\beta_1 - \beta_2| \leq |\beta_1 - \beta_j| \tag{4.5}$$

for $j = 2, \ldots, n$, and also that

$$|\alpha_1 - \alpha_2| < p^* \quad and \quad |\beta_1 - \beta_2| < q^*. \tag{4.6}$$

Then, there exists a positive constant c_1, depending only on n and

$\max_{1 \leq j \leq n} \max\{|\alpha_j|, |\beta_j|\}$, *such that*

$$|\alpha_1 - \beta_1| \geq c_1 \min\{p^*, q^*\}.$$

PROOF. Throughout the present proof, the constants implied by $\ll$ and $\gg$ only depend (at most) on n and $\max_{1 \leq j \leq n} \max\{|\alpha_j|, |\beta_j|\}$. We set

$$\rho := \frac{|\alpha_1 - \beta_1|}{\min\{p^*, q^*\}}$$

and we aim to show that $\rho \gg 1$. The integer polynomial $P(X)$ being without multiple roots, its discriminant (see Definition A.1) is a non-zero integer, and we thus get

$$|a|^{n-1} \prod_{1 \leq i < j \leq n} |\alpha_i - \alpha_j| \geq 1. \tag{4.7}$$

Let j be an integer with $3 \leq j \leq n$. It follows from (4.7) that

$$|\alpha_1 - \alpha_2| \cdot |\alpha_2 - \alpha_j| \gg (p^*)^2 \tag{4.8}$$

holds. By (4.5), we have $|\alpha_2 - \alpha_j| \leq 2|\alpha_1 - \alpha_j|$, which, together with (4.6) and (4.8), implies that $p^* \ll |\alpha_1 - \alpha_j|$. Further, using the definition of ρ, we get

$$|\alpha_j - \beta_1| \leq |\alpha_1 - \beta_1| + |\alpha_1 - \alpha_j| \leq \rho p^* + |\alpha_1 - \alpha_j|$$
$$\ll (1 + \rho)|\alpha_1 - \alpha_j|.$$

Taking the product over all the integers j with $3 \leq j \leq n$, we obtain

$$|a^{n-1} P(\beta_1)| \ll |\alpha_1 - \beta_1| \cdot |\alpha_2 - \beta_2| \cdot |a^n| \cdot \prod_{3 \leq j \leq n} (1 + \rho)|\alpha_1 - \alpha_j|$$
$$\ll (1 + \rho)^{n-2} (p^*)^{-2} |\alpha_1 - \beta_1| \cdot |\alpha_2 - \beta_1|, \tag{4.9}$$

and, reversing the roles of $P(X)$ and $Q(X)$, we get

$$|b^{n-1} Q(\alpha_1)| \ll (1 + \rho)^{n-2} (q^*)^{-2} |\alpha_1 - \beta_1| \cdot |\alpha_1 - \beta_2|. \tag{4.10}$$

Since the integer polynomials $P(X)$ and $Q(X)$ have no common roots, their resultant (see Definition A.1) is a non-zero integer and we have

$$|ab|^n \prod_{1 \leq i, j \leq n} |\alpha_i - \beta_j| \geq 1,$$

hence,

$$|ab|^{n-1} |P(\beta_1)| \cdot |Q(\alpha_1)| \cdot |\alpha_2 - \beta_2| \cdot |\alpha_1 - \beta_1|^{-1} \gg 1. \tag{4.11}$$

Combining (4.9), (4.10), and (4.11) we get

$$|\alpha_1 - \beta_1| \cdot |\alpha_1 - \beta_2| \cdot |\alpha_2 - \beta_1| \cdot |\alpha_2 - \beta_2| \gg (1 + \rho)^{-2n+4} (p^* q^*)^2.$$
(4.12)

Further, (4.6) and the definition of ρ yield that

$$|\alpha_2 - \beta_1| \le |\alpha_1 - \alpha_2| + |\alpha_1 - \beta_1| \ll (1 + \rho) p^*,$$
$$|\alpha_1 - \beta_2| \le |\alpha_1 - \beta_1| + |\beta_1 - \beta_2| \ll (1 + \rho) q^*,$$

and

$$|\alpha_2 - \beta_2| \le |\alpha_1 - \alpha_2| + |\alpha_1 - \beta_2| \ll (1 + \rho) \max\{p^*, q^*\}.$$

It follows from these three inequalities and (4.12) that

$$|\alpha_1 - \beta_1| \gg (1 + \rho)^{-2n+1} \min\{p^*, q^*\},$$

that is,

$$\rho \gg (1 + \rho)^{-2n+1}.$$

Thus, we have proved that $\rho \gg 1$, as wanted.

4.3 Proof of Theorem 4.3

Difficulties encountered and general strategy

Let n be a positive integer and $\Psi : \mathbb{R}_{\ge 1} \to \mathbb{R}_{>0}$ be a non-increasing continuous function such that the series $\sum_{h \ge 1} \Psi(h)$ converges. We begin by explaining the general idea underlying the method of Sprindžuk.

Let $P(X)$ be an integer polynomial of degree at most n and height at most H, and let $E(P)$ denote the set of real numbers ξ satisfying $|P(\xi)| < \Psi^n(H)$. The precise description of $E(P)$ is very difficult, hence, in most cases we are unable to show the convergence of the series $\sum \lambda(E(P))$, where the summation is over all non-zero integer polynomials of degree at most n. This would imply that

$$\lambda \left(\bigcup_{H(P) \ge H_0} E(P) \right) \xrightarrow[H_0 \to +\infty]{} 0,$$
(4.13)

which allows us to conclude by the Borel–Cantelli lemma. However, we may hope to prove (4.13) even if the series $\sum \lambda(E(P))$ diverges: indeed, (4.13) may possibly hold if the sets $E(P)$ intersect 'frequently' and a lot. Thus, we are led to analyze the relative arrangement of the sets $E(P)$ in order to evaluate the multiple intersections.

The basic idea, explained in Section 14 of Chapter 2 of [540] (see also [538], pp. 220–221), consists in considering auxiliary sets $\tilde{E}(P)$ containing $E(P)$, and in dividing them into two classes:

- The essential intervals, which are the sets $\tilde{E}(P)$ such that $\tilde{E}(P)$ has a small intersection with any interval $\tilde{E}(Q)$, where $Q(X)$ is an integer polynomial of same height and same degree as $P(X)$.

- The inessential intervals, which are the sets $\tilde{E}(P)$ such that $\tilde{E}(P)$ has a large intersection with some interval $\tilde{E}(Q)$, where $Q(X)$ is an integer polynomial of same height and same degree as $P(X)$.

Then, the first case is treated by means of Lemma 4.1 above and the Borel–Cantelli Lemma 1.2, while induction and Lemma 4.1 are used for the second one.

We proceed by induction on the integer n. For $n = 1$, the result easily follows from Lemma 1.2, as in the proof of the easy half of Theorem 1.10. Let $n \geq 2$ be a given integer and assume that Theorem 4.3 holds for any positive integer less than or equal to $n - 1$. We introduce in the sequel various sets of real numbers (or of polynomials) which depend on n, but we choose not to indicate this dependence to simplify our notation.

Two reductions of the problem

First, we show that, in order to establish Theorem 4.2, it is plainly enough to prove that the set of real numbers ξ such that (4.2) has infinitely many solutions in integer polynomials $P(X)$ which are irreducible and leading (this terminology means that the leading coefficient of $P(X)$ is equal to its height) has Lebesgue measure zero.

Let $\mathcal{A}_{red}$ be the set of real transcendental numbers ξ for which there exist infinitely many reducible polynomials $P(X)$ of degree n such that $|P(\xi)| < \Psi^n(H(P))$. These polynomials can be written $P(X) = P_1(X) \cdot P_2(X)$, where the degrees of the polynomials $P_1(X)$ and $P_2(X)$ are integers between 1 and $n - 1$. By Lemma A.3, we have $H(P_1) \leq 2^n H(P)$ and, since Ψ is non-increasing, there exist an integer $1 \leq \ell \leq n - 1$ and infinitely many polynomials $P(X)$ of degree ℓ such that $|P(\xi)| < \Psi^\ell(2^{-n} H(P))$. For any positive integer h, set $\tilde{\Psi}(h) = \Psi(2^{-n}h)$. The function $\tilde{\Psi}$ is non-increasing and the sum $\sum_{h \geq 1} \tilde{\Psi}(h)$ converges, hence, by our inductive hypothesis applied to the integer ℓ, the set $\mathcal{A}_{red}$ is a null set, as expected.

We now deal with the second reduction. By Lagrange interpolation formula, there exists a positive constant c_2, depending only on n, such that any integer polynomial $P(X)$ of degree n belongs to a set

$$\mathcal{P}_j := \{P(X) : P(X) \in \mathbb{Z}[X] \text{ of degree } n \text{ such that } |P(j)| \geq c_2 \, H(P)\},$$

for some integer j with $0 \leq j \leq n$. Let $\mathcal{A}$ denote the set of real transcendental numbers ξ for which $|P(\xi)| < \Psi^n(H(P))$ has infinitely many solutions with $P(X)$ irreducible and, for $j = 0, \ldots, n$, set

$$\mathcal{A}_j = \{\xi \in \mathcal{A} : |P(\xi)| < \Psi^n(H(P)) \text{ for infinitely many}$$

$$\text{irreducible polynomials } P(X) \text{ in } \mathcal{P}_j\}.$$

Let $P(X)$ be in $\mathcal{P}_j$ and irreducible. Set $G(X) = P(X + j)$ or $-P(X + j)$, in such a way that $G(0)$ is positive. There exists an integer c_3, that we take greater than c_2^{-1}, depending only on n, such that $H(G) \leq c_3 \, H(P)$. The leading coefficient $G(0)$ of the polynomial $Q(X) := X^n G(X^{-1})$ then satisfies $G(0) > c_3^{-1} H(P)$. Set $c_4 = c_3^2 + 1$ and $R(X) := Q(c_4 X)$. The leading coefficient of $R(X)$ is at least equal to $c_4^n c_3^{-1} H(P)$, while its other coefficients are, in absolute value, bounded by $c_4^{n-1} c_3 \, H(P)$, which is strictly less than $c_4^n c_3^{-1} H(P)$. Hence, the polynomial $R(X)$ is irreducible and leading, and it satisfies $H(R) = c_4^n G(0) \leq c_5 H(P)$, for some positive constant c_5 depending only on n.

Thus, for any integer $j = 0, \ldots, n$ and for any real number ξ in $\mathcal{A}_j$, there exist infinitely many irreducible and leading polynomials $R(X)$ satisfying

$$\left| R\left(\frac{1}{c_4(\xi - j)} \right) \right| = |\xi - j|^{-n} |P(\xi)| < |\xi - j|^{-n} \Psi^n\big(H(P)\big)$$

$$\leq |\xi - j|^{-n} \Psi^n\big(c_5^{-1} \, H(R)\big).$$

Since the function Ψ decreases monotonically, for any given positive real number κ, the sum $\sum_{h \geq 1} \Psi^n(h)$ diverges if, and only if, the sum $\sum_{h \geq 1} \Psi^n(\kappa h)$ diverges. Consequently, if we prove that the set of real transcendental numbers ξ such that (4.2) has infinitely many solutions in leading, irreducible, integer polynomials $P(X)$ has Lebesgue measure zero, we then get that $\mathcal{A}_j$ (and, thus, the set $\mathcal{A}$) is a null set.

Some auxiliary quantities

Since Ψ is non-increasing and the series $\sum_{h \geq 1} \Psi(h)$ converges, $h\Psi(h)$ tends to 0 when h tends to infinity. Thus, there exists a positive integer h_0 such that

$$h\Psi(h) \leq 1/4 \qquad\qquad (4.14)$$

holds for $h \geq h_0$.

Let $\mathcal{P}(h)$ denote the set of leading irreducible integer polynomials $P(X) = a_n X^n + \ldots + a_1 X + a_0$ of height $h = a_n$. Let $P(X)$ be in $\mathcal{P}(h)$ and denote by $\alpha_1, \ldots, \alpha_n$ its roots. Observe that for any integer $j = 1, \ldots, n$ we have

$$h|\alpha_j|^n \leq |a_0| + \ldots + \max\{1, |\alpha_j|^{n-1}\}|a_{n-1}|,$$

hence,

$$|\alpha_j|^n \leq 1 + |\alpha_j| + \ldots + |\alpha_j|^{n-1} \quad \text{if } |\alpha_j| \geq 1.$$

Consequently, we get

$$|\alpha_j| \leq 2, \quad \text{for } j = 1, \ldots, n. \tag{4.15}$$

We number the roots of $P(X)$ in such a way that $|\alpha_1 - \alpha_2|$ is the minimum among the $|\alpha_1 - \alpha_j|$s with $j = 2, \ldots, n$. We then set

$$\tau = |\alpha_1 - \alpha_2|,$$
$$\nu = \frac{2^n \Psi^n(h)}{|P'(\alpha_1)|},$$
$$\mu = \min\{\nu, (\tau\nu)^{1/2}\};$$

these quantities obviously depend on $P(X)$, but, to simplify the notation, we choose not to indicate this dependence. If the real transcendental number ξ satisfies $|P(\xi)| < \Psi^n(h)$ and if α_1 is a root of $P(X)$ with $|\xi - \alpha_1|$ minimal, then we have $|\xi - \alpha_1| < \nu$ and $|\xi - \alpha_1|^2 < \nu\tau$ by Lemma A.5. This implies that $|\xi - \alpha_1| < \mu$. Consequently, in order to prove Theorem 4.3, it is sufficient to show that, for almost all real numbers ξ, there exist only finitely many integer polynomials $P(X)$ in $\cup_{h \geq h_0} \mathcal{P}(h)$ with a root α_1 such that $|\xi - \alpha_1| < \mu$.

Let $h \geq h_0$ be an integer. For an integer polynomial $P(X)$ in $\mathcal{P}(h)$, we set

$$I'(P) = \{\xi \in \mathbb{R} : |\xi - \alpha_1| < \mu\},$$
$$I(P) = \{\xi \in \mathbb{R} : |\xi - \alpha_1| < \mu\Psi^{-1}(h)\}.$$

By (4.14), these intervals satisfy $I'(P) \subset I(P)$ and $|I'(P)| = \Psi(h)|I(P)|$.

We introduce the quantity p^* related to the distribution of the roots of $P(X)$ defined by $p^* = h^{-1}$ if $n = 2$ and by

$$p^* = |h^n(\alpha_1 - \alpha_3) \ldots (\alpha_1 - \alpha_n)|^{-1/2},$$

otherwise. We observe that, for any integer $n \geq 2$, we have

$$(p^*)^{-2}|\alpha_1 - \alpha_2| = h^{n-1}|P'(\alpha_1)|, \tag{4.16}$$

and we divide $\mathcal{P}(h)$ into two classes. The subset $\mathcal{P}_1(h)$ of $\mathcal{P}(h)$ is composed of the polynomials satisfying $\tau \geq p^*$, while the subset $\mathcal{P}_2(h)$ of $\mathcal{P}(h)$ contains exactly those with $\tau < p^*$. Observe that for $n = 2$ the set $\mathcal{P}_2(h)$ is empty, since $\tau \geq p^*$ holds for any quadratic polynomial with integer coefficients. In the first step, we show that, for almost all real numbers ξ, there are only finitely many integer polynomials $P(X)$ in $\cup_{h \geq h_0} \mathcal{P}_1(h)$ having a root α_1 such that $|\xi - \alpha_1| < \mu$. In the second step, we establish a similar result with $\mathcal{P}_1(h)$ replaced by $\mathcal{P}_2(h)$.

The sets $\mathcal{P}_1(h)$, essential and inessential intervals

Let $h \geq h_0$ be an integer. An interval $I(P)$ is said to be inessential if there exists a polynomial $Q(X)$ in $\mathcal{P}_1(h)$ such that the Lebesgue measure of the intersection $I(P) \cap I(Q)$ is greater than $|I(P)|/2$. We observe that the interval $I(Q)$ is not necessarily inessential since $I(P)$ and $I(Q)$ do not have, a priori, the same length. We denote by $\overline{\mathcal{E}}(h)$ the set

$$\{I(P) : \text{there exists } Q(X) \text{ in } \mathcal{P}_1(h) \text{ such that} |I(P) \cap I(Q)| > |I(P)|/2\}$$

of inessential intervals. The set of essential intervals is then

$$\mathcal{E}(h) := \left(\bigcup_{P \in \mathcal{P}_1(h)} I(P) \right) \setminus \overline{\mathcal{E}}(h).$$

Let $P(X)$ be in $\mathcal{P}_1(h)$ and ξ be a real number in $I(P)$. By (4.14), (4.16), and the hypothesis $p^* \leq \tau$, we have

$$|\xi - \alpha_1| < \mu \Psi^{-1}(h) \leq \nu \Psi^{-1}(h) = 2^n \Psi^{n-1}(h) |P'(\alpha_1)|^{-1}$$
$$< h^{1-n} |P'(\alpha_1)|^{-1} = |\alpha_1 - \alpha_2|^{-1}(p^*)^2 \leq |\alpha_1 - \alpha_2|. \quad (4.17)$$

Our choice of α_2 ensures that the assumption of the last statement of Lemma A.5 is fulfilled, hence, we get

$$|P(\xi)| \leq 2^{n-1}|P'(\alpha_1)| \cdot |\xi - \alpha_1|$$
$$< 2^{n-1}\nu|P'(\alpha_1)| \cdot \Psi^{-1}(h) = 2^{2n-1}\Psi^{n-1}(h).$$

In particular, if $Q(X)$ is in $\mathcal{P}_1(h)$ and is such that ξ is in $I(Q)$, we then have $|Q(\xi)| \leq 2^{2n-1} \Psi^{n-1}(h)$ and the polynomial $R(X) := P(X) - Q(X)$, which is of degree less than or equal to $n - 1$ since $P(X)$ and $Q(X)$ have the same leading coefficient, satisfies

$$|R(\xi)| < 2^{2n} \Psi^{n-1}(h) \leq 2^{2n} \Psi^{n-1}\big(H(R)/2\big). \quad (4.18)$$

By our inductive assumption, for almost all ξ, only a finite number of integer polynomials $R(X)$ of degree at most $n-1$ satisfy (4.18). Consequently, almost all ξ belong to only a finite number of intersections $I(P) \cap I(Q)$, where $P(X)$

are in some set $\mathcal{P}_1(h)$ with $h \geq h_0$ and $I(P)$ is an inessential interval. This means that the decreasing intersection

$$\bigcap_{H \geq h_0} \bigcup_{h \geq H} \bigcup_{\substack{P(X) \in \mathcal{P}_1(h), Q(X) \in \mathcal{P}_1(h), \\ P(X) \neq Q(X), |I(P) \cap I(Q)| > |I(P)|/2}} I(P) \cap I(Q)$$

is a null set. Consequently, the Lebesgue measure of

$$\bigcup_{h \geq H} \bigcup_{\substack{P(X) \in \mathcal{P}_1(h), Q(X) \in \mathcal{P}_1(h), \\ P(X) \neq Q(X), |I(P) \cap I(Q)| > |I(P)|/2}} I(P) \cap I(Q)$$

tends to zero when H goes to infinity. However, for polynomials $P(X)$ and $Q(X)$ in $\mathcal{P}_1(h)$ with $|I(P) \cap I(Q)| > |I(P)|/2$, putting $I_{P,Q} = I(P) \cap I(Q)$, the interval $I(P)$ is contained in the union

$$I_{P,Q} \cup (I_{P,Q} + \ell_{P,Q}) \cup (I_{P,Q} - \ell_{P,Q}),$$

where the plus sign and the minus sign mean 'translation', and $\ell_{P,Q}$ denotes the length of $I_{P,Q}$. Thus, the Lebesgue measure of

$$\bigcup_{h \geq H} \bigcup_{\substack{P(X) \in \mathcal{P}_1(h) \\ I(P) \text{ inessential}}} I(P)$$

tends to zero when H goes to infinity, and the set

$$\bigcap_{H \geq h_0} \bigcup_{h \geq H} \bigcup_{\substack{P(X) \in \mathcal{P}_1(h) \\ I(P) \text{ inessential}}} I(P)$$

is a null set, as expected. In other words, for almost all real numbers ξ, there exist only finitely many polynomials $P(X)$ in $\cup_{h \geq h_0} \mathcal{P}_1(h)$ such that $I(P)$ is inessential and ξ belongs to $I(P)$, hence, to $I'(P)$.

We now consider the essential intervals, still under the assumption that $P(X)$ belongs to $\mathcal{P}_1(h)$, and we set

$$\mathcal{A}(h) = \bigcup_{I(P) \in \mathcal{E}(h)} I'(P).$$

Since the interval $I(P)$ is essential, a point in $I(P)$ does not belong to an intersection $I(Q) \cap I(R) \cap I(S)$ of three distinct essential intervals. Indeed, otherwise, one among the intervals $I(P)$, $I(Q)$, $I(R)$, and $I(S)$ would cover strictly more than half of the length of another one, which would then be an inessential interval. Consequently, any point in $I(P)$ belongs to at most three essential intervals and

$$\lambda\big(\mathcal{A}(h)\big) \leq \Psi(h) \sum_{I(P) \in \mathcal{E}(h)} |I(P)| \leq 3\Psi(h)\lambda\left(\bigcup_{I(P) \in \mathcal{E}(h)} I(P)\right) \leq 36\Psi(h),$$

since, by (4.15) and (4.17), any interval $I(P)$ is included in $[-6, 6]$. It then follows from Lemma 1.2 and the convergence of $\sum_{h \geq 1} \Psi(h)$ that almost all ξ belong to only a finite number of intervals $I'(P)$.

Thus, we have proved that, for almost all ξ, there exist only a finite number of polynomials $P(X)$ in $\cup_{h \geq h_0} \mathcal{P}_1(h)$ such that $I(P)$ is essential and ξ belongs to $I'(P)$.

The sets $\mathcal{P}_2(h)$

Since $\mathcal{P}_2(h)$ is empty for $n = 2$, we assume that $n \geq 3$. We aim to show that, for almost all real numbers ξ, there are only finitely many polynomials $P(X)$ in $\cup_{h \geq h_0} \mathcal{P}_2(h)$ with a root α_1 such that $|\xi - \alpha_1| < \mu$. Let $h \geq h_0$ be an integer and let $P(X)$ be a polynomial in $\mathcal{P}_2(h)$. We have $\tau < p^*$, hence, by (4.14) and (4.16), we get

$$\mu^2 \leq \tau \nu = |P'(\alpha)|^{-1} \Psi^n(h) 2^n |\alpha_1 - \alpha_2|$$
$$< h^{1-n} \Psi(h) |P'(\alpha)|^{-1} |\alpha_1 - \alpha_2| \leq \Psi(h)(p^*)^2. \qquad (4.19)$$

Using the discriminant as for the proof of (4.8), we infer from (4.15) that there exists an integer ℓ_0, depending only on n, such that $p^* \leq 2^{-\ell_0+1}$. For integers $j \geq 1$ and $\ell \geq \ell_0$ with $4^{j-1} \geq h_0$, set

$$\mathcal{P}_3(j) = \bigcup_{4^{j-1} \leq h < 4^j} \mathcal{P}_2(h)$$

and

$$\mathcal{P}_4(j, \ell) = \{P(X) \in \mathcal{P}_3(j) : 2^{-\ell} < p^* \leq 2^{-\ell+1}\}.$$

By (4.15), we have $p^* > 2^{-n(j+1)}$ for any polynomial $P(X)$ in $\mathcal{P}_3(j)$. Consequently, the set $\mathcal{P}_4(j, \ell)$ is empty for $\ell > n(j + 1)$.

Let j and ℓ be integers such that $\mathcal{P}_4(j, \ell)$ is non-empty and let $P(X)$ be in $\mathcal{P}_4(j, \ell)$ with $I'(P)$ non-empty. Then we have $|\Im m \, \alpha_1| < \mu$ by the definition of $I'(P)$ and $|\Re e \, \alpha_1| \leq 2$, by (4.15). Assume that, for some real number $c_6 > 1$, there are $c_6 2^\ell$ or more such irreducible polynomials $P(X)$ in $\mathcal{P}_4(j, \ell)$. They furnish $c_6 2^\ell$ distinct complex numbers α_1. Denote by $\delta \leq 1$ the smallest distance between two of these numbers. Then, a simple covering argument gives that

$$c_6 \, 2^\ell \pi \left(\frac{\delta}{2}\right)^2 < 10(\mu + \delta). \qquad (4.20)$$

Further, we infer from (4.14) and (4.19) that $\mu \leq p^* \leq 2^{-\ell+1}$. By (4.15) and $\tau < p^*$, we can apply Lemma 4.1. It yields that there exists a positive constant

c_7, depending only on n, such that $\delta \geq c_7 2^{-\ell}$. Together with (4.20), this gives that there exist positive constants c_8 and c_9, depending on n but not on ℓ and j, such that

$$c_6 \leq c_8 2^{-\ell} \delta^{-2} (\mu + \delta) \leq c_8 2^{-\ell} \delta^{-1} (\mu \delta^{-1} + 1) \leq c_9.$$

Thus, the real number c_6 depends only on n, and we have proved that for at most c_{10} polynomials $P(X)$ in $\mathcal{P}_4(j, \ell)$ the interval $I'(P)$ is non-empty, where c_{10} is a constant depending only on n. Recall that, by (4.14) and (4.19), we have

$$\mu \leq 2^{-\ell+1} \sqrt{\Psi(4^{j-1})} \leq 2^{-\ell-j+1}$$

for these polynomials. Thus, there exists a positive real number c_{11}, depending only on n, such that

$$\sum_{P(X) \in \mathcal{P}_3(j)} |I'(P)| = \sum_{\ell=\ell_0}^{n(j+1)} \sum_{P(X) \in \mathcal{P}_4(j,\ell)} |I'(P)|$$
$$\leq c_{10} \sum_{\ell=\ell_0}^{n(j+1)} 2^{\ell+1} \mu \leq c_{11} j \, 2^{-j}.$$

Since the series $\sum_{j \geq 1} j 2^{-j}$ converges, Lemma 1.2 shows that almost all ξ belong to only a finite number of intervals $I'(P)$ with $P(X)$ in $\mathcal{P}_2(h)$. Consequently, we have proved that for almost all real numbers ξ there exist only a finite number of polynomials $P(X)$ in $\cup_{h \geq h_0} \mathcal{P}_2(h)$ having a root α_1 with $|\xi - \alpha_1| < \mu$, that is, such that ξ belongs to $I'(P)$.

Conclusion

We have proved that for almost all real numbers ξ there exist only a finite number of polynomials $P(X)$ in $\cup_{h \geq h_0} \mathcal{P}(h)$ having a root α_1 with $|\xi - \alpha_1| < \mu$. Thanks to the observations following the definition of μ, this implies that for almost all real numbers ξ there exist only a finite number of integer polynomials $P(X)$ with $|P(\xi)| < \Psi^n(H(P))$. The proof of Theorem 4.3 is complete.

4.4 Exercise

EXERCISE 4.1. Proof of Theorem 4.2 for $n = 2$, following Kasch [310].

1) Let $P(X) = a_2 X^2 + a_1 X + a_0$ be an integer polynomial with distinct roots α_1, α_2 and let $\mathrm{Disc}(P) := a_1^2 - 4a_0 a_2$ denote its discriminant. Prove that

for any complex number ξ we have

$$\min\{|\xi - \alpha_1|, |\xi - \alpha_2|\} \leq \frac{2}{\sqrt{|\mathrm{Disc}(P)|}}\, |P(\xi)|.$$

2) Let s be a real number with $0 < s < 1$. Prove that there exists a positive constant c_{12}, depending only on s, such that, for any positive integer H, the inequality

$$\sum |\mathrm{Disc}(P)|^{-s} \leq c_{12}\, H^{2(1-s)}$$

holds, where the summation is taken over all integer polynomials $P(X)$ of degree at most 2 and of height H with non-zero discriminant.

3) Combine 1) and 2) with $s = 1/2$ to conclude using the Borel–Cantelli Lemma 1.2.

4.5 Notes

• Kubilius' result [344] has been refined by Cassels [153] and extended by Sprindžuk [531] (see also the last Section of [539] and Section 11 of Chapter 2 of [540]).

• Bernik and Dombrovskiĭ [88] established a metric refinement of Theorem 4.2 for general n.

• An effective version of Theorem 4.2 has been obtained by Beresnevich [60] for $n = 2$ and by Kalosha [307] for $n = 3$.

• Harman [275] established asymptotic formulae for the number of solutions to (4.2) when Ψ diverges and satisfies some extra assumptions.

• Using a terminology introduced by Sprindžuk, a curve $(f_1, \ldots, f_n)$ defined on an open subset U of $\mathbb{R}$ is called extremal if, for almost all (in the sense of the Lebesgue measure) points ξ in U, the supremum of the real numbers w for which there exist infinitely many integer $(n + 1)$-tuples $(q_1, \ldots, q_n, p)$ such that $|q_1 f_1(\xi) + \ldots + q_n f_n(\xi) + p| < \max\{|q_1|, \ldots, |q_n|\}^{-w}$ is equal to n. Theorem 4.2 asserts that the curve $\{(\xi, \ldots, \xi^n) : \xi \in \mathbb{R}\}$ is extremal. Many extensions of this result have been obtained, see, for example, [89] and the Notes of Chapter 2 of [86]. As for multiplicative approximation, a curve is called strongly extremal if, for almost all points ξ in U, the supremum of the real numbers w for which there exist infinitely many integer $(n + 1)$-tuples $(q_1, \ldots, q_n, p)$ such that $|q_1 f_1(\xi) + \ldots + q_n f_n(\xi) + p| < \prod_{i=1}^{n} \max\{|q_i|, 1\}^{-w/n}$ is equal to n. The first general results on strong extremality are due to Kleinbock and Margulis [330] and imply Theorem 4.6. They have been superseded by Bernik, Kleinbock, and Margulis [89]. Observe that these results cover the approximation by lacunary polynomials, since they include that for any n-tuple $(a_1, \ldots, a_n)$ of distinct positive integers the curve $\{(\xi^{a_1}, \ldots, \xi^{a_n}) : \xi \in \mathbb{R}\}$ is strongly extremal. Further results are due to Beresnevich [63], Dickinson and Dodson [193], Beresnevich, Bernik, Kleinbock, and Margulis [68], and Kleinbock [327]. Kleinbock, Lindenstrauss, and Weiss [329] identified purely geometrical conditions on measures which are sufficient to guarantee strong extremality, obtaining thus a generalization of the results from [330].

• Theorem 4.3 for $n = 2$ has been extended by Yu [608] and Mashanov [409].

• Bernik, Dickinson, and Dodson [84] proved the inhomogeneous analogue of Theorem 4.4, namely that for any positive integer n and any real number θ the inequality $|P(\xi) + \theta| < H^{-n+1}\Psi(H)$ has, for almost all real numbers ξ, only a finite number of solutions in integer polynomials $P(X)$ of degree at most n and height at most H, where Ψ is as in Theorem 4.4. The case $n = 2$ was already established by Beresnevich, Bernik, and Dodson [66], who proved the inhomogeneous analogue of Theorem 4.5 for $n = 2$.

• The analogue of Theorem 4.4 with the coefficients of the polynomials being prime numbers has been establish by Bernik, Vasiliev, and Dodson [97].

• Bernik [76] confirmed a conjecture of Sprindžuk ([538], Problem C) extending Theorem 4.2. He proved that for any integers k, n with $1 \leq k \leq n$ and for any positive real number ε, the inequality $\prod_{i=1}^{k} |P(\xi_i)| < H^{-n+k-1-\varepsilon}$ has only finitely many solutions in integer polynomials $P(X)$ of degree at most n and height at most H for almost all real k-tuples $(\xi_1, \ldots, \xi_k)$.

• Bernik and Borbat [83] established a two-dimensional analogue of Theorem 4.4.

• Slesoraĭtene ([527] and earlier works) established analogues of Theorem 4.2 for polynomials of small degree in two variables. For arbitrary degrees, a slightly weaker result is due to Vinogradov and Chudnovsky [579] (see also Bernik [74]).

• As for multiplicative approximation, Gallagher [253] proved that for any positive integer n the inequality $\|q\xi_1\| \ldots \|q\xi_n\| < q^{-1}(\log q)^{-n}(\log \log q)^{-1}$ has infinitely many solutions in integers q for almost all real n-tuples $(\xi_1, \ldots, \xi_n)$. Asymptotic formulae have been established by Wang and Yu [593]. Some results from [253] were applied in harmonic analysis by Stokolos [545].

5

Hausdorff dimension of exceptional sets

In the preceding Chapters, we have encountered several sets of real numbers of Lebesgue measure zero, including the set of Liouville numbers, the set of real numbers with bounded partial quotients, the set of very well approximable numbers, and the set of S^*-numbers of $*$-type strictly greater than 1. Some of them are certainly strictly larger than others: indeed, as it may be seen by considering continued fraction expansions (see Exercise 1.5), there are very well approximable numbers other than the Liouville numbers. On the other hand, the set of S^*-numbers of $*$-type at least 2 contains the set of S^*-numbers of $*$-type at least 3, but the results of Chapters 1 to 4 do not enable us to decide whether the inclusion is strict or not.

In the present Chapter, we introduce a powerful tool for discriminating between the sets of Lebesgue measure zero, namely the notion of Hausdorff dimension, developed by Hausdorff in 1919 [276]. Shortly thereafter, Jarník [288, 292] and, independently, Besicovitch [100], applied it to number theoretical problems, and they determined the Hausdorff dimension of sets of real numbers very close to infinitely many rational numbers (Theorem 5.2). Their result has been subsequently generalized in many directions. For instance, A. Baker and Schmidt [45] showed in 1970 that there exist S^*-numbers of arbitrarily large but finite $*$-type (Theorem 5.5). In the present Chapter, we prove both these results and we quote some other extensions of the Jarník–Besicovitch Theorem. Further refinements are stated in Chapter 6.

5.1 Hausdorff measure and Hausdorff dimension

Hausdorff's idea [276] consists in measuring a set by covering it by an infinite, countable family of sets of bounded diameter, and then in looking at what happens when the maximal diameter of these covering sets tends to 0. The reader interested in Hausdorff dimension theory is directed, for example, to Rogers [476], Falconer [234, 236], and Mattila [415].

If U is a non-empty subset of $\mathbb{R}^n$, its *diameter*, denoted by diam(U), is by definition

$$\text{diam}(U) = \sup\{\|\underline{x} - \underline{x}'\|_\infty : \underline{x}, \underline{x}' \in U\},$$

where $\|\underline{x}\|_\infty$ denotes the maximum of the absolute values of the coordinates of a point $\underline{x}$ in $\mathbb{R}^n$. Let J be a finite or infinite set of indices. If for some positive real number δ the sets E and U_j satisfy $E \subset \bigcup_{j \in J} U_j$ and $0 < \text{diam}(U_j) \le \delta$ for any j in J, then $\{U_j\}_{j \in J}$ is called a δ-*covering* of E.

Let f be a *dimension function*, that is, a strictly increasing continuous function defined on $\mathbb{R}_{>0}$ (actually, it is enough to assume that f is defined on some open interval $]0, t[$ with t positive) satisfying $\lim_{x \to 0} f(x) = 0$. For any positive real number δ, set

$$\mathcal{H}_\delta^f(E) := \inf_J \sum_{j \in J} f\big(\text{diam}(U_j)\big),$$

where the infimum is taken over all the countable δ-coverings $\{U_j\}_{j \in J}$ of E. Clearly, the function $\delta \mapsto \mathcal{H}_\delta^f(E)$ is non-increasing. Consequently,

$$\mathcal{H}^f(E) := \lim_{\delta \to 0} \mathcal{H}_\delta^f(E) = \sup_{\delta > 0} \mathcal{H}_\delta^f(E)$$

is well-defined and lies in $[0, +\infty]$. If E_1 and E_2 are two subsets of $\mathbb{R}^n$ with E_1 included in E_2, we then have $\mathcal{H}^f(E_1) \le \mathcal{H}^f(E_2)$. Furthermore, $\mathcal{H}^f$ is subadditive and is a regular outer measure for which the Borelian sets are measurable (see, for example, Rogers [476] or Mattila [415]), called the $\mathcal{H}^f$-measure.

Let f and g be two dimension functions. We say that g corresponds to a 'smaller' generalized dimension than f and we write $g \prec f$ if

$$\lim_{x \to 0} \frac{g(x)}{f(x)} = +\infty.$$

Observe that if $g \prec f$, then g increases faster than f in a neighbourhood of the origin, and we have $\mathcal{H}^g(E) \ge \mathcal{H}^f(E)$ for any subset E of $\mathbb{R}^n$. Further, the ordering induced by $\prec$ is not a total ordering since there exist dimension functions f and g satisfying

$$\liminf_{x \to 0} \frac{g(x)}{f(x)} = 0 \quad \text{and} \quad \limsup_{x \to 0} \frac{g(x)}{f(x)} = +\infty.$$

LEMMA 5.1. *Let f, g, and h be dimension functions satisfying $f \prec g \prec h$. If the set E satisfies $0 \le \mathcal{H}^g(E) < +\infty$ (resp. $0 < \mathcal{H}^g(E) \le +\infty$), we then have $\mathcal{H}^h(E) = 0$ (resp. $\mathcal{H}^f(E) = +\infty$).*

PROOF. Assume that E satisfies $0 \leq \mathcal{H}^g(E) < +\infty$ and let ε be a positive real number. There exists $\delta > 0$ such that $h(x) \leq \varepsilon g(x)/(\mathcal{H}^g(E) + 1)$ for any x with $0 < x < \delta$. For any positive real number δ' less than δ, there exists a countable δ'-covering $\{U_j\}_{j \geq 1}$ of E such that

$$\sum_{j \geq 1} g\big(\mathrm{diam}(U_j)\big) \leq \mathcal{H}^g(E) + 1,$$

whence

$$\sum_{j \geq 1} h\big(\mathrm{diam}(U_j)\big) \leq \varepsilon.$$

This means that $\mathcal{H}^h_{\delta'}(E) \leq \varepsilon$ and yields $\mathcal{H}^h(E) = 0$.

Replacing the functions g and h by f and g, respectively, we get by contraposition the result asserted into brackets.

When f is a power function $x \mapsto x^s$, with s a positive real number, we write $\mathcal{H}^s(E)$ instead of $\mathcal{H}^f(E)$ and we call it the s-dimensional Hausdorff measure of the set E. We define $\mathcal{H}^0$ as the counting measure: $\mathcal{H}^0(E)$ is equal to the cardinality of the set E. Furthermore, for any Borelian subset E of $\mathbb{R}^n$, we have

$$\mathcal{H}^n(E) = \frac{2^n \, (n/2)!}{\pi^{n/2}} \, \mathrm{vol}(E),$$

where vol denotes the Lebesgue measure on $\mathbb{R}^n$. In particular, in the real case, $\mathcal{H}^{\mathrm{Id}} = \mathcal{H}^1$ coincides with the Lebesgue measure on $\mathbb{R}$.

We infer from Lemma 5.1 that $\mathcal{H}^s(E) \geq \mathcal{H}^t(E)$ when $t \geq s \geq 0$. Consequently, the function $s \mapsto \mathcal{H}^s(E)$ is non-increasing on $\mathbb{R}_{\geq 0}$.

COROLLARY 5.1. *Let E be a subset of $\mathbb{R}^n$. If there exists $s \geq 0$ such that $\mathcal{H}^s(E) < +\infty$, then $\mathcal{H}^{s+\varepsilon}(E) = 0$ for any $\varepsilon > 0$. If there exists $s > 0$ such that $\mathcal{H}^s(E) > 0$, then $\mathcal{H}^{s-\varepsilon}(E) = +\infty$ for any ε in $]0, s]$.*

PROOF. This follows from Lemma 5.1 since we have $(x \mapsto x^s) \prec (x \mapsto x^t)$ for any real numbers s and t with $t > s > 0$.

Corollary 5.1 shows that there is a critical value of s at which $\mathcal{H}^s(E)$ 'jumps' from $+\infty$ to 0. This value is called the *Hausdorff dimension* of E.

DEFINITION 5.1. *Let E be a subset of $\mathbb{R}^n$. The Hausdorff dimension of E, denoted by $\dim E$, is the unique non-negative real number s_0 such that*

$$\mathcal{H}^s(E) = 0 \quad \text{if } s > s_0$$

and

$$\mathcal{H}^s(E) = +\infty \quad if \, 0 < s < s_0.$$

In other words, with the notation of Definition 5.1, we have

$$\dim E = \inf\{s : \mathcal{H}^s(E) = 0\} = \sup\{s : \mathcal{H}^s(E) = +\infty\}.$$

The main properties of Hausdorff dimension for subsets $E, E_1, E_2, \ldots$ of $\mathbb{R}^n$ are (see for example [86], p. 65):

(i) $\dim E \leq n$;
(ii) If $\mathrm{vol}(E)$ is positive, then $\dim E = n$;
(iii) If $E_1 \subset E_2$, then $\dim E_1 \leq \dim E_2$;
(iv) $\dim \cup_{j=1}^{+\infty} E_j = \sup\{\dim E_j : j \geq 1\}$;
(v) The Hausdorff dimension of a finite or countable set of points is 0;
(vi) Two sets differing by a countable set of points have the same Hausdorff dimension.

Observe that there exist sets of Hausdorff dimension zero which are uncountable (for example, the set of Liouville numbers, by Corollaries 1.1 and 5.2 below) as well as uncountable sets of Lebesgue measure zero and with Hausdorff dimension one (for example, the set of badly approximable real numbers, see Exercise 5.1).

By property (iv), to determine the Hausdorff dimension of a subset E of $\mathbb{R}^n$, it is enough to know the Hausdorff dimension of intersections of E with products of bounded intervals.

The choice of another norm in the definition of the diameter does not affect the measure when it is either 0 or $+\infty$. In particular, the critical exponent is the same. Borelian sets of Hausdorff dimension s with positive finite s-dimensional Hausdorff measure are called s-sets and enjoy some special properties (see, for example, [234], Chapters 2 to 4, or [236], Chapter 5).

The introduction of dimension functions is a tool for discriminating between sets of same Hausdorff dimension, as will be made clear in Chapter 6.

In the remainder of the present Chapter, except in the notes at the end, we study only subsets of $\mathbb{R}$.

5.2 Upper bound for the Hausdorff dimension

The exact calculation of the Hausdorff dimension (and, *a fortiori*, of the $\mathcal{H}^f$-measure) of a set E is in most cases a difficult problem. It is however often possible to bound it from above by applying the following analogue of the Borel–Cantelli Lemma 1.2.

LEMMA 5.2. *Let E be a Borelian subset of $\mathbb{R}$ and $\{U_j\}_{j \geq 1}$ be a countable family of subsets of $\mathbb{R}$ such that*

$$E \subset \{\xi \in \mathbb{R} : \xi \in U_j \text{ for infinitely many } j \geq 1\}.$$

If f is a dimension function such that

$$\sum_{j \geq 1} f\big(\text{diam}(U_j)\big) < +\infty,$$

then $\mathcal{H}^f(E) = 0$. In particular, if there exists a positive real number s such that

$$\sum_{j \geq 1} \big(\text{diam}(U_j)\big)^s < +\infty,$$

then $\mathcal{H}^s(E) = 0$ and $\dim E \leq s$.

PROOF. By assumption, for any positive integer N, the family $\{U_j\}_{j \geq N}$ is a covering of E. Let δ and ε be positive real numbers. Since the sum $\sum_{j \geq 1} f(\text{diam}(U_j))$ converges, there exists a positive integer N such that $\text{diam}(U_j) < \delta$ for any $j \geq N$ and

$$\sum_{j \geq N} f\big(\text{diam}(U_j)\big) < \varepsilon.$$

This means that $\mathcal{H}^f_\delta(E) < \varepsilon$, hence, we get $\mathcal{H}^f(E) \leq \varepsilon$ which, by letting ε tend to 0, implies $\mathcal{H}^f(E) = 0$.

We give at once an application of Lemma 5.2 to Mahler's classification of numbers. Corollary 5.2 is due to Kasch and Volkmann [311], but their proof, slightly more complicated, does not involve Theorem 3.6. A further result is given in Section 7.5.

COROLLARY 5.2. *The set of T-numbers and the set of U-numbers have Hausdorff dimension zero.*

PROOF. For any real number $w > 1$ and any positive integer H_0, the set of T^*-numbers and the set of U^*-numbers are contained in

$$E(H_0, w) := \bigcup_{n \geq 1} \bigcup_{H \geq H_0} \bigcup_{\alpha} \,]\alpha - H^{-nw}, \alpha + H^{-nw}[,$$

where the last union is taken over all real algebraic numbers α of degree n and height H. Since there are no more than $n(n+1)(2H+1)^n$ such algebraic numbers, we have for any real number s with $1/w < s \leq 2/w$ the estimate

$$\mathcal{H}^s\left(E(H_0, w)\right) \leq \sum_{n \geq 1} \sum_{H \geq H_0} 4n(n+1)\, 2^n \left(\frac{2H_0+1}{2H_0}\right)^n H^{-n(ws-1)}.$$

The above double series converges if H_0 is greater than $4^{1/(ws-1)}$. Thus, by Lemma 5.2, the Hausdorff dimension of the set $E(H_0, w)$ is at most $1/w$. Hence, the set of T^*-numbers and the set of U^*-numbers have Hausdorff dimension zero. We conclude by Theorem 3.6.

Lemma 5.2, called the Hausdorff–Cantelli Lemma by Bernik and Dodson ([86], Lemma 3.10), provides in general, but not always (see Dodson [199]), an upper bound which is the exact value of the dimension. Furthermore, a covering argument as Lemma 5.2 is in some cases unfortunately useless, as it can be seen, for example, in Section 5.7 below. An alternative approach resting on the study of exponential sums is discussed in [86], Section 4.4.2.

5.3 The mass distribution principle

Different methods, more or less sophisticated, allow one to bound from below the Hausdorff dimension of a real set.

Since the sets occurring in Diophantine approximation are often of this form, we consider a set $\mathcal{K}$ included in a bounded interval E, and defined as follows. We set $\mathcal{E}_0 = E$ and we assume that, for any positive integer k, there exists a finite family $\mathcal{E}_k$ of disjoint compact intervals in E such that any interval U belonging to $\mathcal{E}_k$ is contained in exactly one of the intervals of $\mathcal{E}_{k-1}$ and contains at least two intervals belonging to $\mathcal{E}_{k+1}$. We also suppose that the maximum of the lengths of the intervals in $\mathcal{E}_k$ tends to 0 when k tends to infinity. For $k \geq 0$, we denote by E_k the union of the intervals belonging to the family $\mathcal{E}_k$. Then, we define a mass distribution μ on the set

$$\mathcal{K} := \bigcap_{k=1}^{+\infty} E_k$$

by successive subdivisions. We set $\mu(E) = 1$, and we distribute this mass between the intervals $U_1^{(E)}, \ldots, U_{m_1}^{(E)}$ composing $\mathcal{E}_1$ in such a way that $\sum_{i=1}^{m_1} \mu(U_i^{(E)}) = 1$. Likewise, if V belongs to the family $\mathcal{E}_k$, we define, for the intervals $U_1^{(V)}, \ldots, U_{m_V}^{(V)}$ composing $\mathcal{E}_{k+1}$ and included in V, the masses $\mu(U_1^{(V)}), \ldots, \mu(U_{m_V}^{(V)})$ in such a way that $\sum_{i=1}^{m_V} \mu(U_i^{(V)}) = \mu(V)$. Further, for $k \geq 0$, we set $\mu(E \setminus E_k) = 0$.

Let $\mathcal{E}$ be the family consisting of the sets $E \setminus E_k$ for $k \geq 0$, and of all the intervals belonging to the families $\mathcal{E}_k$ with $k \geq 0$. We have defined $\mu(A)$ for any set A belonging to $\mathcal{E}$. As claimed in the following result, this is enough in order to define $\mu(A)$ for any Borelian set A included in E.

PROPOSITION 5.1. *If $\mathcal{K}$ and μ are as above, then μ can be extended to all the subsets of $\mathbb{R}$ in such a way that it is a Borelian measure. Moreover, the support of μ is included in $\mathcal{K}$.*

The proof is involved. We just briefly mention that, if A is a Borelian set in E, we set

$$\mu(A) = \inf\left\{\sum_{j\geq 1} \mu(U_j) : A \subset \bigcup_{j\geq 1} U_j \text{ and } U_j \in \mathcal{E} \text{ for } j \geq 1\right\}.$$

The main difficulties occur in showing that μ defined in this way is indeed a measure. See, for example, [236], Proposition 1.7.

After having defined a measure on the set $\mathcal{K}$, it is often possible to bound its Hausdorff dimension from below by using the following lemma, called the mass distribution principle, or the (easy half of the) Frostman Lemma [251].

LEMMA 5.3. *Let μ be a probability measure with support in a bounded real set $\mathcal{K}$. Let f be a dimension function. Assume that there exist positive real numbers κ and δ such that*

$$\mu(J) \leq \kappa f(|J|)$$

for any interval J with length $|J| \leq \delta$. We then have $\mathcal{H}^f(\mathcal{K}) \geq 1/\kappa$. Furthermore, if $f(x) = x^s$ for some real number s in $]0, 1]$, then $\dim \mathcal{K} \geq s$.

PROOF. If $\{U_j\}_{j\geq 1}$ is a countable δ'-covering of $\mathcal{K}$, with $0 < \delta' \leq \delta$, then, denoting by $\tilde{U}_j$ the smallest interval containing U_j, we have $\operatorname{diam}(U_j) = |\tilde{U}_j|$ and

$$1 = \mu(\mathcal{K}) = \mu\left(\bigcup_{j\geq 1} U_j\right) \leq \sum_{j\geq 1} \mu(U_j) \leq \sum_{j\geq 1} \mu(\tilde{U}_j)$$

$$\leq \kappa \sum_{j\geq 1} f\big(\operatorname{diam}(U_j)\big).$$

We deduce that $\mathcal{H}^f_{\delta'}(\mathcal{K}) \geq \kappa^{-1}$ for any $\delta' \leq \delta$, and the first assertion follows by letting δ' tend to 0. Furthermore, if $f(x) = x^s$ for some $s > 0$, then $\mathcal{H}^s(\mathcal{K})$ is positive, which implies that the Hausdorff dimension of $\mathcal{K}$ is at least equal to s.

An illustration of the strength of Lemma 5.3 is provided by the following result, which is Example 4.6 in [236].

PROPOSITION 5.2. *Keep the same notation as above. Assume further that there exists a positive integer k_0 such that, for any $k \geq k_0$, each interval of E_{k-1} contains at least $m_k \geq 2$ intervals of E_k, these being separated by at least ε_k, where $0 < \varepsilon_{k+1} < \varepsilon_k$. We then have*

$$\dim \mathcal{K} \geq \liminf_{k \to +\infty} \frac{\log(m_1 \ldots m_{k-1})}{-\log(m_k \varepsilon_k)}. \tag{5.1}$$

PROOF. Replacing $\mathcal{K}$ by a smaller set if needed, one can assume that, for $k \geq k_0$, each interval of E_{k-1} exactly contains m_k intervals of E_k, and that there exists a positive integer m such that the set E_k is composed of $mm_{k_0} \ldots m_k$ disjoint intervals $U^{(k)}$, to which we uniformly give the mass $\mu(U^{(k)}) = (mm_{k_0} \ldots m_k)^{-1}$. Let U be an interval of length smaller than ε_{k_0}, and let $k \geq k_0 + 1$ be such that

$$\varepsilon_k \leq |U| < \varepsilon_{k-1}. \tag{5.2}$$

Observe that the number of intervals of E_k which intersect U is, on the one hand, at most equal to m_k (since, by (5.2), U intersects a single interval of E_{k-1}) and, on the other hand, at most equal to $1 + |U|/\varepsilon_k$, since the intervals of E_k are distant by at least ε_k. Consequently, we get

$$\mu(U) \leq \frac{1}{mm_{k_0} \ldots m_k} \min\left\{ m_k, \frac{2|U|}{\varepsilon_k} \right\} \leq \frac{1}{mm_{k_0} \ldots m_k m_k^{s-1}} \left(\frac{2|U|}{\varepsilon_k} \right)^s$$

$$\leq \frac{2|U|^s (m_k \varepsilon_k)^{-s}}{mm_{k_0} \ldots m_{k-1}},$$

for any real number s in $[0, 1]$. This implies that $\mu(U)/|U|^s$ is bounded provided that

$$s < \liminf_{k \to +\infty} \frac{\log(mm_{k_0} \ldots m_{k-1})}{-\log(m_k \varepsilon_k)}.$$

Since $m_k \varepsilon_k$ tends to 0 when k grows to infinity, we conclude by applying Lemma 5.3.

As pointed out by Falconer [236], if for any $k \geq 1$ the components of E_k have the same length δ_k and if each interval of E_{k-1} contains exactly m_k intervals of E_k, which are evenly spaced in the sense that $m_k \varepsilon_k \geq c\delta_{k-1}$ for some positive constant c, then (5.1) becomes

$$\dim \mathcal{K} \geq \liminf_{k \to +\infty} \frac{\log(m_1 \ldots m_{k-1})}{-\log(\delta_{k-1})}. \tag{5.3}$$

Conversely, under the same assumption, the set $\mathcal{K}$ can be covered by $m_1 \ldots m_k$

intervals of length δ_k, thus, by the definition of the Hausdorff dimension, we have equality in (5.3). Such examples occur frequently in number theory, as it will be seen throughout this and the next Chapter.

It follows from Proposition 5.2 that the Hausdorff dimension of the triadic Cantor set is $\log 2 / \log 3$. With some additional effort, it is possible to prove that its measure at the critical exponent is equal to 1 (see [234], Theorem 1.14), hence, it is an s-set.

There are other techniques for establishing lower bounds for the Hausdorff dimension. Potential theoretic and Fourier transform methods are briefly discussed in Chapter 4 of [236], see also Sections 3.5.7 and 4.1.3 of [86]. A full treatment can be found Mattila [415].

5.4 Regular systems

The notion of regular system was introduced in 1970 by A. Baker and Schmidt [45] to compute the Hausdorff dimension of sets $\mathcal{K}_{\mathcal{S}}^*$ of real numbers close to infinitely many points of a given countable set $\mathcal{S}$ (namely, in [45], sets of algebraic real numbers of bounded degree). Here, we slightly change their definition, in order to point out more accurately which properties of $\mathcal{S}$ are really required to get the Hausdorff dimension of $\mathcal{K}_{\mathcal{S}}^*$.

DEFINITION 5.2. *Let E be a bounded open real interval. Let $\mathcal{S} = (\alpha_j)_{j \geq 1}$ be a sequence of distinct real numbers. Let $\Phi : \mathbb{Z}_{\geq 1} \to \mathbb{R}_{\geq 1}$ and $\Xi : \mathbb{Z}_{\geq 1} \to \mathbb{R}_{\geq 1}$ be increasing functions such that $\Phi(n) \geq n$ and $\Xi(n) \geq n$ for any $n \geq 1$. The triple $(\mathcal{S}, \Phi, \Xi)$ is called a regular system of points in E if there exist a positive constant $c_1 = c_1(\mathcal{S}, \Phi, \Xi)$ and, for any interval I in E, a number $K_0 = K_0(\mathcal{S}, \Phi, \Xi, I)$ such that, for any $K \geq K_0$, there exist integers*

$$1 \leq i_1 < \ldots < i_t \leq \Phi(\Xi(K))$$

with α_{i_h} in I for any $h = 1, \ldots, t$,

$$|\alpha_{i_h} - \alpha_{i_\ell}| \geq \frac{1}{\Xi(K)} \quad (1 \leq h \neq \ell \leq t),$$

and

$$c_1 |I| \Xi(K) \leq t \leq |I| \Xi(K).$$

In the original work of A. Baker and Schmidt [45], the set $\mathcal{S}$ is not indexed and the function Ξ is the identity. The introduction of Ξ rests on an idea of Rynne [487], who defined the concept of 'weakly regular system', used, for example, in [130] (see Exercise 5.7). We emphasize that we do not assume that *every* point of S belongs to E. Furthermore, we have supposed that E is bounded, although this was not assumed in [45]. This does not involve any loss of generality as far as calculating Hausdorff dimension is concerned since any unbounded set can be covered by a countable collection of bounded, open sets to which the results may be applied.

In the applications, the real number $K_0(S, \Phi, \Xi, I)$ need not be effectively computable.

Ideally, when the points of S are very well distributed, then we may hope that $(S, x \mapsto \kappa_1 x, x \mapsto \kappa_2 x)$ is a regular system for suitable positive real numbers κ_1 and κ_2. If this occurs, S is said to be *optimal* and we can get sharper results on sets of real numbers close to infinitely many points of S, see Chapter 6.

To illustrate the notion of a regular system, we start with an example.

PROPOSITION 5.3. *We order the rational numbers as follows. First, we divide $\mathbb{Q}$ into classes containing rational numbers of the same height, starting with height 0, 1, 2, and so on. Then within each class, the rationals are ordered in the usual way (as real numbers). Then, the triple $(\mathbb{Q}, x \mapsto 4x, x \mapsto x)$ is a regular system in $]0, 1[$.*

PROOF. For any integer $B \geq 2$, the number of pairs of integers (a, b) with $1 \leq a, b \leq B$ and $\gcd(a, b) = 1$ is at least equal to

$$B^2 - \sum_{d=2}^{B/2} \left(\frac{B}{d}\right)^2 \geq B^2 - B^2\left(\frac{\pi^2}{6} - 1\right) \geq \frac{B^2}{3}.$$

On the other hand, strictly fewer than $2B^2$ rational numbers have their height bounded by B. Consequently, for any positive integer j, we have

$$\frac{\sqrt{j}}{2} \leq H(\alpha_j) \leq \sqrt{3(j+1)} + 1. \tag{5.4}$$

Let I be a bounded real interval in $]0, 1[$ and $Q > 100$ be a real number. By Dirichlet's Theorem 1.1, for any ξ in I, there exists a rational number p/q such that

$$\left|\xi - \frac{p}{q}\right| < \frac{1}{qQ} \quad \text{and} \quad 1 \leq q \leq Q. \tag{5.5}$$

The Lebesgue measure of the set E of points ξ in I for which (5.5) is satisfied by a rational number p/q with $q \leq Q/10$ is less than or equal to

$$\sum_{q=1}^{[Q/10]+1} \frac{2}{qQ} (q|I| + 1),$$

hence, to $|I|/4$, provided that $Q \geq 100|I|^{-1} \log(100|I|^{-1})$. Further, for ξ in $I \setminus E$, there exists a rational number p/q such that

$$\left| \xi - \frac{p}{q} \right| < \frac{10}{Q^2} \quad \text{and} \quad Q/10 \leq q \leq Q.$$

Let $\{p_j/q_j\}_{1 \leq j \leq t}$ be a maximal family of rational numbers in I with $Q/10 \leq q_j \leq Q$ and $|p_j/q_j - p_\ell/q_\ell| \geq 10/Q^2$ for $1 \leq j \neq \ell \leq t$. The union of the intervals

$$\left] \frac{p_j}{q_j} - \frac{20}{Q^2}, \frac{p_j}{q_j} + \frac{20}{Q^2} \right[, \quad 1 \leq j \leq t,$$

covers $I \setminus E$, thus its Lebesgue measure is at least $3|I|/4$. Consequently, we get

$$t \geq \frac{3}{160} Q^2 |I|.$$

Furthermore, it follows from (5.4) that for any $j = 1, \ldots, t$ we have $p_j/q_j = \alpha_{i_j}$ with $Q^2/400 \leq i_j \leq 4Q^2$. We have proved that for any $K \geq K_0$ with

$$K_0 \geq 10^4 |I|^{-2} \log^2\left(100|I|^{-1}\right) \tag{5.6}$$

there exist integers $1 \leq i_1 < \ldots < i_t \leq 4K$ such that $|\alpha_{i_j} - \alpha_{i_\ell}| \geq 1/K$ for $1 \leq j \neq \ell \leq t$ and $t \geq 3K|I|/160$. This shows that the triple $(\mathbb{Q}, x \mapsto 4x, x \mapsto x)$ is a regular system in $]0, 1[$.

Proposition 5.4 shows that the set of algebraic numbers of bounded degree forms a regular system. Further examples are given at the end of this Section and in Section 6.1.

Thanks to Proposition 5.2, we are able to compute a lower bound for the Hausdorff dimension of sets of real numbers close to infinitely many points in a given regular system.

THEOREM 5.1. *Let E be a bounded open real interval. Let $S = (\alpha_j)_{j \geq 1}$ be a sequence of distinct real numbers. Let $\Phi : \mathbb{Z}_{\geq 1} \to \mathbb{R}_{\geq 1}$ and $\Xi : \mathbb{Z}_{\geq 1} \to \mathbb{R}_{\geq 1}$ be increasing and assume that the triple (S, Φ, Ξ) is a regular system of points in E. Let $\Psi : \mathbb{R}_{\geq 1} \to \mathbb{R}_{>0}$ be continuous, non-increasing and such that*

$$\limsup_{x \to \infty} x \Psi(\Phi(x)) < +\infty.$$

Then, the Hausdorff dimension of the set

$$\mathcal{K}_S^*(\Psi) = \limsup_{j \to +\infty} \{ \xi \in E : |\xi - \alpha_j| < \Psi(j) \}$$

satisfies

$$\dim \mathcal{K}_{\mathcal{S}}^*(\Psi) \geq \liminf_{x \to +\infty} \frac{-\log x}{\log \Psi(\Phi(x))}.$$

Obviously, the set $\mathcal{K}_{\mathcal{S}}^*(\Psi)$ also depends on E, but we choose not to indicate this dependence in the notation; this should not cause any trouble in the sequel.

Although the approximation function Ψ needs only to be defined for every sufficiently large integer, we choose for commodity (as in the statement of Theorem 1.10 and in Chapters 4 and 6) to assume that it is defined on $\mathbb{R}_{\geq 1}$ and continuous.

We point out that the lower bound for $\dim \mathcal{K}_{\mathcal{S}}^*(\Psi)$ obtained in Theorem 5.1 does not depend on the function Ξ, as it was observed by Rynne [487].

As will be apparent below, Theorem 5.1 shows that the points in $\mathcal{S}$ do not need to be *very* well distributed in order that we get the exact dimension of $\mathcal{K}_{\mathcal{S}}^*(\Psi)$.

PROOF. We proceed as follows: we construct a Cantor set $\mathcal{K}$ of the form considered in Section 5.3 and contained in $\mathcal{K}_{\mathcal{S}}^*(\Psi)$; we then apply Proposition 5.2 to estimate its Hausdorff dimension from below. Set $E_0 = E$. Let c_2 and c_3 be positive real numbers such that $c_2 < 1/3$ and $x\Psi(\Phi(x)) < (3c_2)^{-1}$ for any $x \geq c_3$. By Definition 5.2, there exist a positive constant c_4, depending only on $\mathcal{S}$, Φ, and Ξ, an integer t_1, a real number $K_1 \geq c_3$ with $c_4|E_0|K_1 \leq t_1 \leq |E_0|K_1$, and a set $A_1' = \{i_1^{(1)}, \ldots, i_{t_1+2}^{(1)}\}$ included in $\{1, \ldots, \Phi(K_1)\}$ such that α_h is in E_0 and $|\alpha_h - \alpha_\ell| \geq 1/K_1$ for any distinct integers h, ℓ in A_1'. We may further assume that K_1 is a value taken by the function Ξ. Define

$$E_1 = \bigcup_{j \in A_1} \left[\alpha_j - c_2 \Psi(\Phi(K_1)), \alpha_j + c_2 \Psi(\Phi(K_1))\right],$$

where A_1 is a subset of A_1' such that E_1 is contained in E_0 and has maximal cardinality. Clearly, E_1 is the union of at least t_1 intervals, which are pairwise disjoint since we have assumed $K_1 \geq c_3$.

We proceed inductively and we let $k \geq 1$ be an integer such that we have constructed a set E_k contained in E of the form

$$E_k = \bigcup_{j \in A_k} \left[\alpha_j - c_2 \Psi(\Phi(K_k)), \alpha_j + c_2 \Psi(\Phi(K_k))\right],$$

where K_k is in $\Xi(\mathbb{Z}_{\geq 1})$ and A_k is a set of t_k integers in $\{1, \ldots, \Phi(K_k)\}$.

For j in A_k, denote by U_j the interval $[\alpha_j - c_2\Psi(\Phi(K_k)), \alpha_j + c_2\Psi(\Phi(K_k))]$ and by $K_{U_j} = K_0(\mathcal{S}, \Phi, \Xi, U_j)$ the constant given by Definition 5.2. We construct a new set E_{U_j} as a finite union of closed intervals contained in U_j. Let K_{k+1} be a value taken by Ξ and greater than $\Xi(K_{U_j})$ for any j in A_k. To simplify the notation, we omit the subscript j and denote by U one of the U_js. By the definition of a regular system, there exist a positive integer t_U with $c_4|U|K_{k+1} \le t_U \le |U|K_{k+1}$ and a set $A'_U = \{i_1^{(U)}, \ldots, i_{t_U+2}^{(U)}\}$ included in $\{1, \ldots, \Phi(K_{k+1})\}$ such that α_h is in U and $|\alpha_h - \alpha_\ell| \ge 1/K_{k+1}$ for any distinct integers h, ℓ in A'_U. Define

$$E_U = \bigcup_{h \in A_U} [\alpha_h - c_2\,\Psi(\Phi(K_{k+1})), \alpha_h + c_2\,\Psi(\Phi(K_{k+1}))],$$

where A_U is a subset of A'_U such that E_U is contained in U and has maximal cardinality. Clearly, E_U is the union of at least t_U intervals, which are pairwise disjoint since $K_{k+1} \ge c_3$.

Set

$$E_{k+1} = \bigcup_{j \in A_k} E_U \quad \text{and} \quad \mathcal{K} := \bigcap_{k \ge 1} E_k.$$

Each interval U_j of E_k contains at least

$$t_{U_j} \ge c_4|U_j|K_{k+1} \ge \left[c_2 c_4 \Psi(\Phi(K_k))\, K_{k+1}\right] =: m_{k+1}$$

intervals of E_{k+1}, these being separated by at least $c_2 K_{k+1}^{-1} =: \varepsilon_{k+1}$, since $c_2 < 1/3$.

All the assumptions of Proposition 5.2 are fulfilled, hence, the Hausdorff dimension of $\mathcal{K}$ satisfies

$$\dim \mathcal{K} \ge \liminf_{k \to +\infty} \frac{\displaystyle\sum_{\ell=1}^{k-1} \log\big(c_2 c_4 \Psi\big(\Phi(K_\ell)\big) K_{\ell+1} - 1\big)}{-\log\big(c_2 c_4 \Psi\big(\Phi(K_k)\big) K_{k+1}\varepsilon_{k+1}\big)}. \tag{5.7}$$

For any positive real number δ and any $k \ge 2$, we may select K_k large enough in such a way that the numerator in (5.7) is larger than $(\log K_k)/(1 + \delta)$. Consequently, we have

$$\dim \mathcal{K} \ge \liminf_{k \to +\infty} \frac{\log K_k}{-(1 + \delta)\log\big(c_2^2 c_4 \Psi(\Phi(K_k))\big)}.$$

Since δ can be taken arbitrarily small, we get

$$\dim \mathcal{K} \ge \liminf_{x \to +\infty} \frac{-\log x}{\log \Psi(\Phi(x))}.$$

Further, we observe that $\mathcal{K}_{\mathcal{S}}^*(\Psi)$ contains $\mathcal{K}$, since Ψ is non-increasing and $\Phi(n) \ge n$ for any $n \ge 1$. Hence, the result follows.

A closer look at the proof of Theorem 5.1 suggests that we are able to consider simultaneously several regular systems, and even a countable set of regular systems. Indeed, at each step k, we use a regular system in order to construct the set E_{k+1} from the set E_k. If we have at our disposal several regular systems, we may use any one of them at each step and, provided that we use all of them infinitely many times, the resulting set, that is, $\mathcal{K} := \cap_{k \geq 1} E_k$, will be composed of points close to infinitely many elements of each regular system. It remains for us to bound the Hausdorff dimension of $\mathcal{K}$ from below. To this end, assume that, for any positive integer n, we have a regular system $(\mathcal{S}_n, \Phi_n, \Xi_n)$ of points in $\mathcal{K}$, and a non-increasing continuous function $\Psi_n : \mathbb{R}_{\geq 1} \to \mathbb{R}_{>0}$ with

$$\liminf_{x \to +\infty} x \Psi_n(\Phi_n(x)) < +\infty.$$

Then, following the proof of Theorem 5.1, we get

$$\dim \mathcal{K} \geq \liminf_{x \to \infty} \frac{-\log x}{\log\left(\min_{n \geq 1} \Psi_n(\Phi_n(x))\right)}. \tag{5.8}$$

This observation is the key point for the proof of Theorem 5.4.

As observed by Bernik and Dodson [86], page 104, points with an asymptotic distribution in $]0, 1[$ with a suitable error term also form a regular system. Indeed, let $\mathcal{S} = (\alpha_j)_{j \geq 1}$ be a sequence of real numbers in $]0, 1[$. Assume that its discrepancy is controlled in the sense that there exists a positive constant $\kappa < 1$ such that, for any interval I in $]0, 1[$ and any real number $Q \geq 1$, the number of points $\alpha_1, \ldots, \alpha_Q$ contained in I is asymptotically equal to $Q|I| + O(Q^\kappa)$, where the numerical constant implied in O is independent on I and Q. Define the function Φ on the set of positive integers by $\Phi(n) = [n^{1/(1-\kappa)}]$. Then (see [86], page 104) the triple $(\mathcal{S}, \Phi, \Xi)$ forms a regular system for some suitable linear function Ξ.

As an application to inhomogeneous approximation, Bernik and Dodson [86] proved that, for almost all real numbers α and for any real number $\tau > 1$, the set of real numbers β such that $|q\alpha - \beta - p| < q^{-\tau}$ holds for infinitely many pairs of integers (p, q) with q positive has Hausdorff dimension $1/\tau$. This result holds in fact for *all* irrational numbers α (see Exercise 5.7 for a proof of this assertion).

5.5 The theorem of Jarník–Besicovitch

It follows from Corollary 1.5 that, for any $\tau > 1$, the set

$$\mathcal{K}_1^*(\tau) := \left\{ \xi \in \mathbb{R} : \left| \xi - \frac{p}{q} \right| < \frac{1}{q^{2\tau}} \text{ for infinitely many rationals } \frac{p}{q} \right\}$$

has Lebesgue measure zero. It is uncountable since it contains the Liouville numbers. Its Hausdorff dimension has been calculated by Jarník [289], and, later and independently, by Besicovitch [100].

THEOREM 5.2. *For any* $\tau \geq 1$, *the Hausdorff dimension of the set* $\mathcal{K}_1^*(\tau)$ *is equal to* $1/\tau$.

COROLLARY 5.3. *For any real number* $\tau \geq 1$, *we have*
$$\dim\{\xi \in \mathbb{R} : w_1^*(\xi) \geq 2\tau - 1\} = \frac{1}{\tau}.$$

PROOF. This is a straigthforward consequence of Theorem 5.2. Indeed, for any real numbers τ and ε with $\tau > 1$ and $0 < \varepsilon < \tau - 1$, the set introduced in the corollary is contained in $\mathcal{K}_1^*(\tau - \varepsilon)$ and contains $\mathcal{K}_1^*(\tau + \varepsilon)$.

Corollary 5.3 does not imply that the Hausdorff dimension of the set
$$\mathcal{W}_1^*(\tau) := \{\xi \in \mathbb{R} : w_1^*(\xi) = 2\tau - 1\}$$

is equal to $1/\tau$. This result, proved in Exercise 5.3, is a straightforward consequence of Satz 4 of Jarník [292]. It was rediscovered by Güting [266], thanks to a refinement of Besicovitch's proof [100] of Theorem 5.2.

The set of very well approximable numbers (Definition 1.4) has Lebesgue measure zero (Corollary 1.5) and Hausdorff dimension 1, by Theorem 5.2. This is also the case for the set of badly approximable numbers (Definition 1.3), by Corollary 1.6 and Exercise 5.1.

We deduce Theorem 5.2 from Proposition 5.3 and Theorem 5.1 above. Another proof, which directly depends on Proposition 5.2, is given in Section 5 of Falconer's book [236].

PROOF OF THEOREM 5.2. By Theorem 1.1, $\mathcal{K}_1^*(1)$ is the set of irrational numbers, hence, its Hausdorff dimension is 1. Let $\tau > 1$ be a real number. Our aim is to prove that $\dim(\mathcal{K}_1^*(\tau) \cap]0, 1[) = 1/\tau$, which implies Theorem 5.2. For any positive integer q, denote by $E(q)$ the set of ξ in $]0, 1[$ for which there exists an integer p such that $|\xi - p/q| < q^{-2\tau}$. Each point of the set $\mathcal{K}_1^*(\tau) \cap]0, 1[$ is contained in infinitely many sets $E(q)$. Since the series
$$\sum_{q \geq 1} (q + 1)(2q^{-2\tau})^s$$

converges for any real number s greater than $1/\tau$, it follows from Lemma 5.2 that the Hausdorff dimension of $\mathcal{K}_1^*(\tau)$ is less than or equal to $1/\tau$.

In order to prove the reverse inequality, denote by $\mathcal{S} = (\alpha_j)_{j \geq 1}$ the set of non-zero rational numbers, ordered as in the statement of Proposition 5.3. For any integer $j \geq 1$, set

$$\Psi(j) = \mathrm{H}(\alpha_j)^{-2\tau}.$$

By Proposition 5.3, the triple $((\alpha_j)_{j\geq 1}, x \mapsto 4x, x \mapsto x)$ is a regular system and we get from (5.4) that $x\Psi(4x) \leq 1$ holds when x is large enough. Since the sets $\mathcal{K}_{\mathcal{S}}^*(\Psi)$ (defined with $E =]0, 1[$) and $\mathcal{K}_1^*(\tau) \cap]0, 1[$ coincide, it follows from Theorem 5.1 and (5.4) that

$$\dim\left(\mathcal{K}_1^*(\tau) \cap]0, 1[\right) \geq \liminf_{x \to \infty} \frac{-\log x}{\log \Psi(4x)} \geq \liminf_{x \to \infty} \frac{\log x}{2\tau \log \mathrm{H}(\alpha_{4x})} = \frac{1}{\tau},$$

as claimed.

5.6 Hausdorff dimension of sets of S^*-numbers

Theorem 3.3 asserts that almost all real numbers are S^*-numbers of $*$-type less than or equal to 1 (see also Corollary 4.1). However, we have not proved up to now that S^*-numbers of $*$-type strictly greater than 1 do exist. We will achieve this in the present Section.

Recall that $\mathbb{A}_n$ denotes the set of real algebraic numbers of degree bounded by n. A. Baker and Schmidt [45] have extended Theorem 5.2 as follows.

THEOREM 5.3. *For any integer $n \geq 1$ and any real number $\tau \geq 1$, the Hausdorff dimension of the set*

$$\mathcal{K}_n^*(\tau) := \{\xi \in \mathbb{R} : |\xi - \alpha| < \mathrm{H}(\alpha)^{-\tau(n+1)} \text{ for infinitely many } \alpha \in \mathbb{A}_n\}$$
is equal to $1/\tau$.

Exactly as we derived Corollary 5.3 from Theorem 5.2, we have the following consequence of Theorem 5.3.

COROLLARY 5.4. *For any integer $n \geq 1$ and any real number $\tau \geq 1$, we have*

$$\dim\{\xi \in \mathbb{R} : w_n^*(\xi) \geq \tau(n+1) - 1\} = \frac{1}{\tau}.$$

In the proof of Theorem 5.3 detailed below we have simplified the approach of A. Baker and Schmidt by using the weaker form of Frostman's Lemma given by Proposition 5.2. However, a slight modification of our argument and the use of Lemma 5.3 instead of Proposition 5.2 yields a sharper result than Corollary 5.4.

THEOREM 5.4. *For any integer $n \geq 1$ and any real number $\tau \geq 1$, the Hausdorff dimension of the set*

$$\mathcal{W}_n^*(\tau) := \{\xi \in \mathbb{R} : w_n^*(\xi) = \tau(n+1) - 1\}$$

is equal to $1/\tau$.

PROOF. See Exercise 5.3.

Theorem 5.4, due to A. Baker and Schmidt [45], obviously implies that for any positive integer n, the set of values taken by the function w_n^* includes the interval $[n, +\infty[$. It should be pointed out that no other method is known for proving this last statement.

We now turn to the proof of Theorem 5.3, which proceeds in several steps. First, we use results from Chapter 4 to show that real algebraic numbers of bounded degree are well distributed (Proposition 5.4). Then we suitably order the sets $\mathbb{A}_n$ (Lemma 5.4), and we conclude by applying Theorem 5.1.

PROPOSITION 5.4. *Let $n \geq 2$ be an integer. For any bounded real interval I there exist a real number $K(I)$ and, for any $K \geq K(I)$, algebraic numbers $\gamma_1, \ldots, \gamma_t$ in $\mathbb{A}_n \cap I$ such that*

$$\mathrm{H}(\gamma_h) \leq K^{1/(n+1)}(\log K)^n, \quad |\gamma_h - \gamma_k| \geq K^{-1} \ (1 \leq h < k \leq t) \qquad (5.9)$$

and

$$t \geq \frac{|I|K}{8}.$$

Using Theorem 4.3, A. Baker and Schmidt established Proposition 5.4 (actually under a slightly weaker form). We could apply Theorem 4.4 to get a sharper statement (that is, with a smaller exponent for $(\log K)$ in (5.9)), as in [86], page 101, but the result obtained would not yield more applications than Proposition 5.4. In particular, it would not allow us to get the results established in Chapter 6.

Proposition 5.4 does not look best possible because of the factor $(\log K)^n$ in (5.9): if the real algebraic numbers of degree at most n are 'evenly' distributed, then Proposition 5.4 should hold with $\mathrm{H}(\gamma_j) \leq cK^{1/(n+1)}$ in (5.9) for some constant c. This is indeed the case, as was proved by Beresnevich [61], see Section 6.1.

PROOF OF PROPOSITION 5.4. For any integer $H \geq 3$, denote by $R_I(H)$ the set of ξ in I for which there exists a real algebraic number α of degree at most n and of height at most H such that

$$|\xi - \alpha| < H^{-n-1}(\log H)^{n(n+1)}. \qquad (5.10)$$

Let ξ be a real transcendental number in I. Denote by $M \geq 1$ a real number such that I is contained in $[-M, M]$ and let $H > 3^{4Mn^2}$ be a real number. By Theorem B.2, there exists a non-zero integer polynomial $P(X) = a_n X^n +$

$\ldots + a_1 X + a_0$, of degree at most n, such that

$$0 < |P(\xi)| \leq (4M)^n \, H^{-n} (\log H)^{n^2-1}$$
$$|a_1| \leq (4M)^{-n} \, H$$
$$|a_j| \leq H (\log H)^{-n-1} \quad (2 \leq j \leq n). \tag{5.11}$$

Since the height of $P(X)$ is bounded by $2^{-n} H$, the height of any root of $P(X)$ is at most equal to H, by Lemma A.3. We have several cases to distinguish. In the sequel, we adopt the convention that the constants implied by $\ll$ depend only on n and M.

Assume first that $|P'(\xi)| \leq n \, (4M)^n \, H (\log H)^{-n-1}$. We infer from (5.11) that we have $|a_1| \ll H (\log H)^{-n-1}$, hence, $\mathrm{H}(P) \ll H (\log H)^{-n-1}$. Consequently, we get

$$|P(\xi)| \ll \mathrm{H}(P)^{-n} (\log \mathrm{H}(P))^{-n-1} < \mathrm{H}(P)^{-n} (\log \mathrm{H}(P))^{-n-1/2}, \tag{5.12}$$

when H is sufficiently large, since $\mathrm{H}(P)$ tends to infinity with H. However, since the series $\sum_{h \geq 2} h^{-1} (\log h)^{-1-1/2n}$ converges, Theorem 4.3 asserts that for almost all real numbers η, the equation

$$|Q(\eta)| < H^{-n} (\log H)^{-n-1/2}$$

has only a finite number of solutions in integer polynomials $Q(X)$ of degree at most n and of height at most H. This shows that the Lebesgue measure of the set of real numbers ξ satisfying (5.12) tends to zero when H tends to infinity.

Assume now that $|P'(\xi)| > n \, (4M)^n \, H (\log H)^{-n-1}$. Lemma A.5 ensures that there exists a zero α of $P(X)$ satisfying

$$|\xi - \alpha| \leq n \, \frac{|P(\xi)|}{|P'(\xi)|} < H^{-n-1} (\log H)^{n(n+1)},$$

whence ξ belongs to $R_I(H)$.

Consequently, the Lebesgue measure of the set $R_I(H)$ tends to $|I|$ when H tends to infinity. In particular, there exists H_0 such that, for any $H \geq H_0$, we have $\lambda(R_I(H)) \geq |I|/2$. Let $H \geq H_0$ be large enough such that $K := H^{n+1} (\log H)^{-n(n+1)}$ is greater than H. Let $\gamma_1, \ldots, \gamma_t$ be a maximal subset of $\mathbb{A}_n \cap I$ composed of algebraic numbers satisfying

$$\mathrm{H}(\gamma_i) \leq H \leq K^{1/(n+1)} (\log K)^n$$

and

$$|\gamma_i - \gamma_j| \geq H^{-n-1} (\log H)^{n(n+1)} = K^{-1}.$$

By maximality, any real algebraic number γ in $\mathbb{A}_n$ such that $H(\gamma) \leq H$ is separated by at most K^{-1} from some γ_i. Thus, by (5.10), any real number ξ belonging to $R_I(H)$ is separated by at most $2K^{-1}$ from some γ_i, and the Lebesgue measure of the union of the intervals centered at the points γ_i for $i = 1, \ldots, t$ and of radius $2K^{-1}$ is greater than or equal to $\lambda(R_I(H))$. Hence, we get $|I|/2 \leq 4t \, K^{-1}$, as claimed.

In order to apply Theorem 5.1, we have to define an ordering on the set $\mathbb{A}_n$ of real algebraic numbers of degree at most n.

LEMMA 5.4. *Let $n \geq 1$ be an integer. We order the set $\mathbb{A}_n := (\alpha_j)_{j \geq 1}$ as follows. First, we put its elements into classes containing algebraic numbers of the same height, starting with height 0, 1, 2, and so on. Then within each class, the elements of $\mathbb{A}_n$ are ordered in the usual way (as real numbers). Then, there exist two positive constants c_5 and c_6, depending only on n, such that, for any $j \geq 1$, we have*

$$c_5 \, j^{1/(n+1)} \leq H(\alpha_j) \leq c_6 \, j^{1/(n+1)}. \tag{5.13}$$

PROOF. The left-hand side inequality in (5.13) is clear, since, for any positive integer H, there are at most $n(2H+1)^{n+1}$ algebraic numbers of height at most H and degree at most n. As for the right-hand side, let $h \geq 5$ be an odd integer. Consider an integer polynomial

$$P(X) := hX^n - a_{n-1}X^{n-1} - \ldots - a_1 X - a_0,$$

where a_0 is congruent to 2 modulo 4 and, for $0 \leq j \leq n - 1$, the integer a_j is even and belongs to $\{0, 2, \ldots, 2[h/2]\}$. By Eisenstein's Criterion, the polynomial $P(X)$ is irreducible. Furthermore, it has (at least) one real root. Consequently, there are at least $c_7 h^n$ real algebraic numbers of height h and degree n, for some positive constant c_7 depending only on n. Hence, there exists a positive constant c_8, depending only on n, such that, for any positive integer H, there are at least $c_8 H^{n+1}$ real algebraic numbers of height at most H and degree at most n. This proves the right-hand side inequality of (5.13). $\quad\square$

We now are able to complete the proof of Theorem 5.3.

PROOF OF THEOREM 5.3. Let $\tau > 1$ be a real number and $n \geq 1$ be an integer. By Theorem 5.2, we assume $n \geq 2$. For any $H \geq 1$, denote by $E(H, n)$ the set of real numbers ξ in $]0, 1[$ for which there exists an algebraic number α in $\mathbb{A}_n$ with $H(\alpha) = H$ and

$$|\xi - \alpha| < H(\alpha)^{-\tau(n+1)}.$$

This is the union of at most $n(n + 1)(2H + 1)^n$ intervals, each of which with length at most $2H^{-\tau(n+1)}$. Furthermore, the set $\cup_{H \geq 1} E(H, n)$ is a countable covering of $\mathcal{K}_n^*(\tau)$. Since for any real number $s > 1/\tau$ the series

$$\sum_{H \geq 1} n(n + 1)(2H + 1)^n \left(2H^{-\tau(n+1)}\right)^s$$

converges, it follows from Lemma 5.2 that the Hausdorff dimension of $\mathcal{K}_n^*(\tau)$ is less than or equal to $1/\tau$.

In order to prove the reverse inequality, denote by $(\alpha_j)_{j \geq 1}$ the set $\mathbb{A}_n$ ordered as in Lemma 5.4. For any integer $j \geq 1$, set

$$\Psi(j) = H(\alpha_j)^{-\tau(n+1)}.$$

By Proposition 5.4 and Lemma 5.4, the triple $(\mathbb{A}_n, \Phi_n, x \mapsto x)$ is a regular system, where

$$\Phi_n(x) = \left[c_9 \, x \, (\log 3x)^{n(n+1)}\right] \quad (x \geq 1), \tag{5.14}$$

for a suitable positive constant c_9, depending only on n. Further, by (5.13), we have $x \Psi(\Phi_n(x)) \leq 1$ when x is large enough. Consequently, we infer from Theorem 5.1 that

$$\dim \mathcal{K}_n^*(\tau) \geq \liminf_{x \to \infty} \frac{-\log x}{\log \Psi(\Phi_n(x))}$$

$$\geq \liminf_{x \to \infty} \frac{-\log x}{-\tau \left(\log x - c_{10} \log \log x\right)} = \frac{1}{\tau},$$

where c_{10} is a positive constant depending only on n. This completes the proof that $\dim \mathcal{K}_n^*(\tau) = 1/\tau$, and this implies that $\dim \mathcal{K}_n^*(1) = 1$.

As announced at the end of Section 5.4, we may use simultaneously infinitely many regular systems.

THEOREM 5.5. *Let $(\tau_n)_{n \geq 1}$ be a sequence of real numbers greater than or equal to 1. We then have*

$$\dim \bigcap_{n \geq 1} \mathcal{K}_n^*(\tau_n) = \frac{1}{\sup_{n \geq 1} \tau_n}.$$

In particular, for any real number $\tau \geq 1$, the Hausdorff dimension of the set

$$\bigcap_{n \geq 1} \mathcal{K}_n^*(\tau)$$

is equal to $1/\tau$. Thus, there exist S*-numbers of arbitrarily large *-type.

PROOF. The upper bound immediately follows from Theorem 5.3. For the lower bound, we consider the regular systems $(\mathbb{A}_n, \Phi_n, x \mapsto x)$, with Φ_n as in (5.14), and the functions

$$\Psi_n : j \mapsto \mathrm{H}(\alpha_j)^{-\tau(n+1)}.$$

We conclude by (5.8).

Theorem 5.5 asserts that the Hausdorff dimension of a countable intersection of sets $\mathcal{K}_n^*(\tau_n)$ is the infimum of the dimensions of the sets $\mathcal{K}_n^*(\tau_n)$. In the next Chapter, we refine this result by using the notion of *intersective sets*, introduced by Falconer.

With a slight modification of the proof of Theorem 5.5, similar to that suggested for getting Theorem 5.4, it is possible to prove that there exist S^*-numbers with any prescribed *-type greater than 1.

THEOREM 5.6. *For any real number* $\tau \geq 1$, *the Hausdorff dimension of the set of real S^*-numbers of *-type τ is equal to $1/\tau$.*

PROOF. See Exercise 5.4.

Theorems 5.4 and 5.6 also follow from deeper results obtained in Chapter 6 (Theorem 6.3). It seems that Theorem 5.6 appeared for the first time in [132]. However, it can be relatively easily deduced from the results of A. Baker and Schmidt [45]. Theorems 5.4 and 5.6 remain true if the *-type of an S^*-number ξ is defined as the supremum of the sequence $((w_n^*(\xi) + 1)/(n+1))_{n \geq 1}$.

5.7 Hausdorff dimension of sets of S-numbers

In the previous Section, we have calculated the Hausdorff dimension of sets of real numbers related to Koksma's classification. A parallel with the classification of Mahler leads us to consider the sets

$$\mathcal{K}_n(\tau) = \{\xi \in \mathbb{R} : |P(\xi)| < \mathrm{H}(P)^{-\tau(n+1)+1} \quad \text{for infinitely many}$$
$$\text{integer polynomials } P(X) \text{ of degree } \leq n\},$$

for any integer $n \geq 1$ and any real number $\tau \geq 1$. Obviously, the sets $\mathcal{K}_1(\tau)$ and $\mathcal{K}_1^*(\tau)$ coincide. By Theorem 4.1, the sets $\mathcal{K}_n(\tau)$ have Lebesgue measure zero when $\tau > 1$, thus we would like to discriminate them using the Hausdorff dimension. It follows from Lemma A.6 that the Hausdorff dimension of $\mathcal{K}_n(\tau)$ is at least equal to that of $\mathcal{K}_n^*(\tau)$; unfortunately, we cannot apply the Hausdorff–Cantelli Lemma 5.2 to bound it from above, for the same reasons as explained in Section 4.1.

A. Baker and Schmidt [45] combined Theorem 3.4 and Lemma 5.2 to show that

$$\frac{1}{\tau} \le \dim \mathcal{K}_n(\tau) < \frac{2}{\tau}, \tag{5.15}$$

which, together with the upper bound $\dim \mathcal{K}_2(\tau) \le 1/\tau$ obtained by Kasch and Volkmann [311], yield that $\dim \mathcal{K}_2(\tau) = 1/\tau$ (see Exercise 5.5). They conjectured that the left inequality in (5.15) is indeed an equality. For n and τ large enough, this was proved by R. C. Baker [48], and, a few years later, a complete proof of the conjecture was established by Bernik [75, 77] (see also [90]).

THEOREM 5.7. *For any $n \ge 1$ and any $\tau \ge 1$, the Hausdorff dimension of the set $\mathcal{K}_n(\tau)$ is equal to $1/\tau$.*

The proof of the upper bound (unlike for Theorem 5.3, this is the most difficult half) is involved and uses the method of essential and inessential domains, at a technical level comparable with (and maybe slightly superior to) the proof of Theorem 4.4. Without much effort, we deduce from Theorem 5.7 the analogues of Theorems 5.4 and 5.6.

THEOREM 5.8. *For any integer $n \ge 1$ and any real number $\tau \ge 1$, the Hausdorff dimension of the set*

$$\mathcal{W}_n(\tau) := \{\xi \in \mathbb{R} : w_n(\xi) = \tau(n+1) - 1\}$$

is equal to $1/\tau$. Furthermore, for any real number $\tau \ge 1$, the Hausdorff dimension of the set of real S-numbers of type τ is equal to $1/\tau$.

PROOF. It follows from Theorem 5.7 that $\dim \mathcal{W}_n(\tau) \le 1/\tau$. Write

$$\mathcal{W}_n^{\ge}(\tau) := \{\xi \in \mathbb{R} : w_n(\xi) \ge \tau(n+1) - 1\},$$

and observe that

$$\mathcal{W}_n^{\ge}(\tau) = \mathcal{W}_n(\tau) \cup \bigcup_{\ell=1}^{+\infty} \mathcal{W}_n^{\ge}(\tau + 1/\ell). \tag{5.16}$$

For any real numbers $\tau' > 1$ and ε with $0 < \varepsilon < \tau' - 1$, the set $\mathcal{W}_n^{\ge}(\tau' + \varepsilon)$ is included in $\mathcal{K}_n(\tau')$, hence, its Hausdorff dimension is at most equal to $1/\tau'$, by Theorem 5.7. Consequently, we have $\mathcal{H}^f(\mathcal{W}_n^{\ge}(\tau + 1/\ell)) = 0$ for any integer $n \ge 1$, where f is the dimension function $f : x \mapsto x^{1/\tau} \exp(\sqrt{\log 1/x})$. Furthermore, we infer from Proposition 3.2 and Exercise 5.3 that $\mathcal{H}^f(\mathcal{W}_n^{\ge}(\tau)) > 0$, and we get $\mathcal{H}^f(\mathcal{W}_n(\tau)) > 0$ by (5.16). Thus the Hausdorff dimension of $\mathcal{W}_n(\tau)$ is at least equal to $1/\tau$, and the first assertion of Theorem 5.8 is proved.

As for the second assertion, we observe that, for any real number $\tau \geq 1$, we have

$$\bigcap_{n=1}^{+\infty} \mathcal{W}_n^{\geq}(\tau) \subset \bigcap_{n=1}^{+\infty} \mathcal{W}_n(\tau) \cup \bigcup_{n=1}^{+\infty} \bigcup_{\ell=1}^{+\infty} \mathcal{W}_n^{\geq}(\tau + 1/\ell).$$

Arguing as above with the same dimension function f, we infer from Proposition 3.2 and Exercise 5.4 that $\mathcal{H}^f(\cap_{n \geq 1} \mathcal{W}_n(\tau)) > 0$. Thus, the Hausdorff dimension of $\cap_{n \geq 1} \mathcal{W}_n(\tau)$ is at least equal to $1/\tau$ and the proof of Theorem 5.8 is complete, since any element of $\cap_{n \geq 1} \mathcal{W}_n(\tau)$ is an S-number of type τ.

The last assertion of Theorem 5.8 remains true if the type of an S-number ξ is defined as the supremum of the sequence $((w_n(\xi) + 1)/(n + 1))_{n \geq 1}$.

We end this Section by introducing a new exponent of Diophantine approximation, related to multiplicative approximation. Let $\tau > 1$ be a real number and let $n \geq 1$ be an integer. For an integer polynomial $P(X) = a_n X^n + \ldots + a_1 X + a_0$ we introduced in Chapter 4 the quantity

$$\Pi_+(P) := \prod_{1 \leq i \leq n} \max\{1, |a_i|\}.$$

For a given real number ξ, let $w_n^+(\xi)$ be the supremum of the real numbers w for which there exist infinitely many integer polynomials $P(X)$ of degree at most n satisfying

$$0 < |P(\xi)| \leq \Pi_+^{-w}(P).$$

Observe that we have $w_n^+(\xi) \geq w_n(\xi)/n$. By Proposition 3.1, this implies that $w_n^+(\xi) \geq 1$ if ξ is transcendental. Furthermore, we set

$$\mathcal{K}_n^+(\tau) = \{\xi \in \mathbb{R} : |P(\xi)| < \Pi_+^{-2\tau+1}(P) \text{ for infinitely many}$$
$$\text{integer polynomials } P(X) \text{ of degree } \leq n\},$$

and

$$\mathcal{W}_n^+(\tau) := \{\xi \in \mathbb{R} : w_n^+(\xi) = 2\tau - 1\}.$$

By Proposition 3.1 and Theorem 4.6, almost all real numbers ξ satisfy $w_n^+(\xi) = 1$, thus the Lebesgue measure of $\mathcal{K}_n^+(\tau)$ is zero. Clearly, $\mathcal{K}_n^+(\tau)$ contains the sets $\mathcal{K}_1(\tau), \ldots, \mathcal{K}_n(\tau)$, therefore its Hausdorff dimension is greater than or equal to $1/\tau$, by Theorem 5.2. Yu [608] conjectured that this is in fact the exact value. The conjecture for $n = 1$ is nothing but Theorem 5.2 and it has been established by Yu [608] for $n = 2$ (see Exercise 5.6). This result has been rediscovered by Beresnevich and Bernik [64], but it remains an open problem for $n \geq 3$. Furthermore, the Hausdorff dimension of $\mathcal{W}_2^+(\tau)$ is equal to $1/\tau$, as proved in Exercise 5.6.

5.8 Restricted Diophantine approximation

The sets $\mathcal{K}_1^*(\tau)$ are defined without any restriction on the rational approximants p/q. We could however impose some conditions on them, for instance, we could ask that q belongs to a given infinite subset of the integers, like the prime numbers or the powers of 2, etc. Such questions have been considered by Borosh and Fraenkel [112], who extended earlier results of Eggleston [225], and established the following statement.

THEOREM 5.9. *Let $\mathcal{N}$ be an infinite set of distinct positive integers, and let v in $[0, 1]$ be the real defined by the conditions*

$$\sum_{q \in \mathcal{N}} q^{-v} \text{ diverges and } \sum_{q \in \mathcal{N}} q^{-v-\varepsilon} \text{ converges for any } \varepsilon > 0.$$

For any real number $\tau \geq (1 + v)/2$, the Hausdorff dimension of the set

$$\left\{ \xi \in \mathbb{R} : \left| \xi - \frac{p}{q} \right| < \frac{1}{q^{2\tau}} \text{ for infinitely many rationals } \frac{p}{q} \text{ with } q \in \mathcal{N} \right\}$$

is equal to $(1 + v)/(2\tau)$.

The proof of Theorem 5.9 given in [112] is quite technical and is inspired by the proof of Theorem 5.3 due to A. Baker and Schmidt [45]. It proceeds by contradiction and does not use the mass distribution principle (see also Chapter 10 of Harman [273]).

In particular, Theorem 5.9 asserts that, for any real number $\tau > 3/2$, the Hausdorff dimension of the set of real numbers ξ such that infinitely many rational numbers p/q satisfy $|\xi - p/q^2| < q^{-\tau}$ is equal to $3/(2\tau)$. Furthermore, it allows us to answer a conjecture of Erdös related to the convergents of the continued fraction expansion of real numbers.

COROLLARY 5.5. *Let $\mathcal{N}$ be an infinite set of distinct positive integers. The Hausdorff dimension of the set of real numbers having infinitely many convergents whose denominators belong to $\mathcal{N}$ is greater than or equal to $1/2$.*

The proof is immediate since any convergent p_n/q_n of the continued fraction expansion of the real number ξ satisfies $|\xi - p_n/q_n| < q_n^{-2}$. Actually, Eggleston's results [225] are sufficient to get Corollary 5.5.

We end this Section by stating a corollary to Theorem 5.9 and a result, due to Harman ([273], Theorem 10.8), on approximation by rational numbers whose denominators and numerators are restricted.

THEOREM 5.10. *Let $\tau \geq 1$ be a real number. The Hausdorff dimensions of the sets*

$$\left\{ \xi \in \mathbb{R} : \left| \xi - \frac{p}{q} \right| < \frac{1}{q^{2\tau}} \text{ for infinitely many rationals } \frac{p}{q} \text{ with } q \text{ prime} \right\}$$

and

$$\left\{ \xi \in \mathbb{R} : \left| \xi - \frac{p}{q} \right| < \frac{1}{q^{2\tau}} \text{ for infinitely many rationals } \frac{p}{q} \text{ with } p \right.$$
$$\left. \text{and } q \text{ prime} \right\}$$

are equal to $1/\tau$.

Restrictions on numerators introduce new difficulties which do not seem to be easy to overcome.

5.9 Exercises

EXERCISE 5.1. The Hausdorff dimension of the set $\mathcal{B}$ of badly approximable real numbers, following Jarník [288].

We keep the notation of Chapter 1. Let $M \geq 2$ be an integer and denote by $\mathcal{B}_M$ the set of real numbers in $[0, 1]$ having their partial quotients at most equal to M. Let s be a real number with $0 < s < 1$.

1) To determine $\mathcal{H}^s(\mathcal{B}_M)$, prove that we need only to consider *finite* coverings $\mathcal{U}$ of $\mathcal{B}_M$ composed of closed intervals having their endpoints in $\mathcal{B}_M$.

2) Let U be an interval belonging to $\mathcal{U}$. Prove that there exist a m-tuple $\underline{k} = (k_1, \dots, k_m)$ of positive integers not greater than M and two integers h and ℓ with $1 \leq h < \ell \leq M$ such that

$$U \cap F_{(\underline{k},h)} \neq \emptyset \text{ and } U \cap F_{(\underline{k},\ell)} \neq \emptyset.$$

Deduce that $\lambda(F_{\underline{k}}) \leq 4 M^3 |U|$.

3) In order to determine whether $\mathcal{H}^s(\mathcal{B}_M)$ is positive or not, prove that we need only consider *finite* coverings $\mathcal{U}$ of $\mathcal{B}_M$ composed of closed intervals $F_{\underline{k}}$, where $\underline{k}$ is an m-tuple of positive integers not greater than M for some positive integer m.

4) Assume that s is such that for any integer $m \geq 0$ and any m-tuple $\underline{k}$ of positive integers not greater than M we have

$$\lambda(F_{\underline{k}})^s \leq \sum_{h=1}^{M} \lambda(F_{(\underline{k},h)})^s.$$

Prove that this implies that the Hausdorff dimension of $\mathcal{B}_M$ is at least equal to s.

5) Use 4) to prove that $\dim \mathcal{B}_2 > 1/4$ and that, for any integer $M \geq 9$, we have $\dim \mathcal{B}_M > 1 - 4/(M \log 2)$.

6) Prove that we have $\dim \mathcal{B}_M < 1 - 1/(8M \log M)$ for any integer $M \geq 9$.

EXERCISE 5.2. Denote by $[0; a_1, a_2, \dots]$ the continued fraction expansion of a real number ξ in $[0, 1[$. Let $b > 1$ and $c > 1$ be real numbers and set

$$\Xi(b, c) := \{\xi : a_n \geq c^{b^n} \quad \text{for infinitely many } n\}$$

and

$$\tilde{\Xi}(b, c) := \{\xi : a_n \geq c^{b^n} \quad \text{for all positive integers } n\}.$$

1) Following Feng, Wu, Liang and Tsen [245], prove that we have

$$\dim \Xi(b, c) = \dim \tilde{\Xi}(b, c) \geq 1/(b + 1).$$

2) Following Ganesa Moorthy [255], prove that $\dim \Xi(b, c) \leq 2/(b + 1)$ and deduce that the set

$$\mathcal{U} = \{\xi : \text{the sequence } (\log a_n)^{1/n} \text{ is unbounded}\}$$

has Hausdorff dimension zero.

Actually, both sets $\Xi(b, c)$ and $\tilde{\Xi}(b, c)$ turn out to have Hausdorff dimension $1/(b + 1)$, as proved by Łuczak [372].

Hint. For 1), apply Proposition 5.2 with the sets

$$E_k := \{\xi \in [0, 1] : c^{b^n} \leq a_n \leq 3c^{b^n} \quad \text{for } 1 \leq n \leq k + 1\}$$

and use Theorems 1.3 and 1.7.

For 2), let $k < b$ be fixed. Arguing by contradiction, prove that for every $\xi = [0; a_1, a_2, \dots]$ with $a_n \geq c^{b^n}$ for infinitely many n, there are infinitely many n such that the n-th convergent p_n/q_n to ξ satisfies $q_{n+1} > q_n^k$. Deduce that for every positive integer m, the collection of intervals $]r/q - q^{-k-1}, r/q + q^{-k-1}[$, with $1 \leq r \leq q - 1$ and $q \geq m$, forms a cover for the set $\tilde{\Xi}(b, c)$. Conclude. Prove that $\dim \mathcal{U} = 0$.

EXERCISE 5.3. Prove that $\dim \mathcal{W}_n^*(\tau) = 1/\tau$ holds for any positive integer n and any real number $\tau > 1$.

Hint. Use the regular system $(\mathbb{A}_n, \Phi_n, x \mapsto x)$ with Φ_n as in (5.14) to construct inductively a Cantor set contained in $\mathcal{K}_n^*(\tau)$ and of the same form as in the proof of Theorem 5.1. Instead of applying Proposition 5.2, argue as in the proof of Proposition 5.2 to define a measure μ on $\mathcal{K}_n^*(\tau)$ and show that $\mu(U) \leq c|U|^{1/\tau} \exp(\sqrt{\log 1/|U|})$ for some absolute constant c and any sufficiently small interval U. Use Lemma 5.3 to prove that for $f : x \mapsto x^{1/\tau} \exp(\sqrt{\log 1/x})$, we have $\mathcal{H}^f(\mathcal{K}_n^*(\tau)) > 0$. Observe that $\mathcal{H}^f(\mathcal{K}_n^*(\tau+1/k)) = 0$ for any positive integer k and prove that $\mathcal{H}^f(\mathcal{W}_n^*(\tau)) > 0$. Conclude.

EXERCISE 5.4. Prove that $\dim \cap_{n\geq 1} \mathcal{W}_n^*(\tau) = 1/\tau$ holds for any positive integer n and any real number $\tau > 1$.

Hint. Use the observation at the end of Section 5.4 and the same general idea as in Exercise 5.3 to show that, with f as in Exercise 5.3, we have $\mathcal{H}^f(\cap_{n\geq 1} \mathcal{W}_n^*(\tau)) > 0$.

EXERCISE 5.5. Following Kasch and Volkmann [311], prove that $\dim \mathcal{K}_2(\tau) = 1/\tau$.

Hint. Use 1) of Exercise 4.1 to find a suitable countable covering of $\mathcal{K}_2(\tau)$ and then use 2) of Exercise 4.1 to prove that $\dim \mathcal{K}_2(\tau) \leq 1/\tau$.

EXERCISE 5.6. Following Beresnevich and Bernik [64], prove that $\dim \mathcal{K}_2^+(\tau) = 1/\tau$. Proceed as in Exercise 5.3 to establish that $\dim \mathcal{W}_2^+(\tau) = 1/\tau$.

Hint. It is sufficient to show that $\dim(\mathcal{K}_2^+(\tau) \cap [1,2]) = 1/\tau$. Let ξ be in $[1,2]$ and $P(X) = a_2 X^2 + a_1 X + a_0$ be an integer polynomial with $|P(\xi)| < \Pi_+^{-2\tau+1}(P)$.

Prove that $H(P) \leq 7 \max\{|a_1|, |a_2|\}$.

Let $J(\tau)$ denote the subset of $\mathcal{K}_2^+(\tau)$ composed of the real numbers ξ in $[1,2]$ for which there are infinitely many integer polynomials $Q(X) = a_2 X^2 + a_1 X + a_0$ with

$$|Q(\xi)| < \Pi_+(Q)^{-2\tau+1} \quad \text{and} \quad |Q'(\xi)| < \frac{H(Q)}{14}.$$

Prove that each of these polynomials satisfies $|a_2| \leq |a_1| \leq 8|a_2|$, and derive from Exercise 5.5 that $\dim J(\tau) \leq 3/(4\tau - 1)$.

Assume now that there are infinitely many integer polynomials $P(X) = a_2 X^2 + a_1 X + a_0$ with

$$|P(\xi)| < \Pi_+(P)^{-2\tau+1} \quad \text{and} \quad |P'(\xi)| \geq \frac{H(P)}{14}. \tag{5.17}$$

For a fixed polynomial $P(X)$, denote by $\sigma(P)$ the set of ξ in $[1,2]$ satisfying (5.17). Prove that $\sigma(P)$ is the union of at most three intervals, whose lengths are at most $28 \, \Pi_+(P)^{-2\tau+1} \, H(P)^{-1}$ (use the Mean Value Theorem). For any ρ in $]1/\tau, 1[$, show that the sum

$$\sum_{P(X)} \sum_{1 \le i \le 3} |\sigma^i(P)|^\rho$$

converges, where the first sum is taken over all the integer polynomials $P(X) = a_2 X^2 + a_1 X + a_0$ (sum up over k and ℓ, and then over $|a_1|$ in $[2^k, 2^{k+1} - 1]$ and $|a_2|$ in $[2^\ell, 2^{\ell+1} - 1]$). Conclude.

EXERCISE 5.7. Prove that, for any real irrational number α and any $\tau > 1$, the Hausdorff dimension of the set

$$\left\{ \xi \in \mathbb{R} : \|n\alpha - \xi\| < \frac{1}{n^\tau} \text{ holds for infinitely many integers } n \right\}$$

is equal to $1/\tau$, a result due to Bugeaud [127] and, independently, to Schmeling and Troubetzkoy [497].

Hint. The *Three Distance Theorem* asserts (see, for example, [529], [525] or [5] [1]) that, for any integer $N \ge 2$, the points $\{\alpha\}, \{2\alpha\}, \ldots, \{N\alpha\}$ divide the interval $[0, 1]$ in $N + 1$ intervals, whose lengths take at most three distinct values, one of these being the sum of the two others. Deduce that there exists a strictly increasing infinite sequence of positive integers $(N_r)_{r \ge 1}$ with the following property: For any interval I in $[0, 1]$ and any integer r large enough (in terms of $|I|$), at least $|I| K_r / 3$ numbers among $\{\alpha\}, \ldots, \{N_r \alpha\}$ belong to I and are mutually distant by at least $1/K_r$, where K_r denotes the greatest even integer smaller than $(N_r + 1)/3$. Conclude by applying Theorem 5.1 with a suitably chosen function Ξ. As observed by Drmota, it is also possible to use Theorem 1.7 instead of the Three Distance Theorem to solve this exercise.

5.10 Notes

• Besides the Hausdorff dimension, there exist various other 'fractal dimensions', like the lower and upper box-counting dimensions, the Fourier dimension and the packing dimension introduced by Tricot [571, 572], see Chapter 3 of [236] and Chapters 5 and 12 of [415]. Falconer [237] proved that, for any $\tau \ge 1$, the set $\mathcal{K}_1^*(\tau)$ has packing dimension equal to 1.

  • Theorem 5.2 has been extended to systems of linear forms by Bovey and Dodson [116].

[1] There is a misprint in the statement of the *Three Distance Theorem* in [5]: η_k is the smallest length, not the largest one.

• An inhomogeneous analogue of Theorem 5.2 together with an extension to systems of linear forms has been obtained by Levesley [365] (see also Exercise 6.3 and [497]). His main theorem implies earlier results of Dickinson [192] and Dodson [202].

• Kaufman [315] (see also Bluhm [102]) gave a new proof of Theorem 5.2. He established the stronger property that, for any real number $\tau > 1$, there exists a positive measure μ_τ with support in a compact subset of $\mathcal{K}_1^*(\tau)$ and whose Fourier-Stieltjes transform satisfies

$$\hat{\mu}_\tau(x) = \int_{\mathbb{R}} \exp(-2i\pi xy) \, \mathrm{d}\mu_\tau(y)$$

$$= o(\log|x|) |x|^{-1/(2\tau)}, \quad |x| \to +\infty.$$

Furthermore, he showed that the Fourier dimension (see [415], page 168, for definition) of $\mathcal{K}_1^*(\tau)$ is equal to its Hausdorff dimension. Sets with this property are called *Salem sets* and are very rare as deterministic sets; the first constructions of such sets, due to Salem, were random.

• Another extension of Theorem 5.2 has been obtained by Falconer [238].

• Let $k \geq 1$ be an integer, $\tau > (k+1)/2$ be a real number and set

$$\mathcal{K}^{(k)}(\tau) := \left\{ (\xi_1, \dots, \xi_k) \in \mathbb{R}^k : \left| \xi - \frac{p_1}{q} \right| \times \dots \times \left| \xi - \frac{p_k}{q} \right| < \frac{1}{q^{2\tau}} \right.$$

$$\left. \text{for infinitely many integers } p_1, \dots, p_k, q \right\}.$$

Bovey and Dodson [115] established that $\dim \mathcal{K}^{(k)}(\tau) = k-1+2/(2\tau-k+1)$. This provides a generalization of Theorem 5.2, and has been extended to systems of linear form by Yu [607]. A function $F : \mathbb{R}^k \to \mathbb{R}$ is a distance function if, by definition, F is continuous, non-negative and satisfies $F(tx) = tF(x)$ for all $t \geq 0$. Examples of distance functions are provided by the 'sup' norm, the usual Euclidean norm, and by $(x_1, \dots, x_k) \mapsto \prod_{j=1}^k |x_j|^{1/k}$. Sets of the form $\{(\xi_1, \dots, \xi_k) \in \mathbb{R}^k : F(\langle qx_1 \rangle, \dots, \langle qx_k \rangle) < \Psi(q)$ for infinitely many integers $q\}$, where F is a distance function and Ψ some approximation function, have been studied by Dodson [198, 199], who obtained generalizations of the above quoted result from [115]. Extensions to systems of linear forms have been worked out by Dodson [198] and Dodson and Hasan [204].

• In order to obtain lower bounds for the Hausdorff dimension of sets of 'very well approximable' points in a multidimensional setting, Dodson, Rynne, and Vickers [206], inspired by geometrical ideas of Besicovitch [100], introduced the notion of ubiquitous systems. Actually, in the one dimension case, ubiquitous and regular systems are almost equivalent, see Rynne [487] for an interesting discussion. Using this powerful tool, Theorem 5.2 has been extended to 'very well approximable' linear forms and to 'very well approximable' systems of linear forms in [206] and by Dickinson [190, 191]. We refer the reader to the survey of Dodson [201], to Chapter 5 of Bernik and Dodson [86], and to the survey of Beresnevich, Bernik and Dodson [67]. A new technique involving a multidimensional analogue of regular systems has been developed by Beresnevich, Bernik, Kleinbock, and Margulis [68].

• Weiss [596, 597] proved that, for any real number $\tau > 1$, almost no points on a Cantor set (with respect to the standard measure supported on it) belong to $\mathcal{K}_1^*(\tau)$. This has been extended by Kleinbock, Lindenstrauss, and Weiss [329].

• Pollington [458] established that the Hausdorff dimension of the set of real numbers in $\mathcal{K}_n^*(\tau)$ which are simply normal (see the Notes of Chapter 1 for the definition) in no base is equal to $1/\tau$.

• Hinokuma and Shiga [281] slightly improved Theorem 5.2. Furthermore, they [282] asserted an extension of Theorem 5.9, but there is apparently a problem in their proof: the term $(\log N_{j-1})^{-3}$ occurring in the denominator of the fraction at the bottom of page 371 should be $(\log N_j)^{-3}$, and this only gives that the Hausdorff dimension is non-negative. Presumably, Proposition 5.2 cannot be used to get Theorem 5.9, since the points of the regular system are not sufficiently spaced from each other. However, a proof of the assertions claimed in [282] can be found in [488].

• Theorem 5.9 has been extended to simultaneous approximation and then generalized in several directions by Rynne [486, 488, 489] and by Rynne and Dickinson [491]. Simultaneous approximation with differing exponents in the approximation functions for each coordinate is studied in [489, 491].

• As a corollary to Theorem 5.9, R. C. Baker and Harman (see Theorem 10.7 in [273]) proved that, given a non-negative function Ψ such that the sum $\sum_{q\geq 1} q\Psi(q)$ diverges, the set of real numbers ξ for which $|\xi - p/q| < \Psi(q)$ has infinitely many solutions in coprime positive integers p and q has Hausdorff dimension 1. This result means that the Duffin–Schaeffer Conjecture (see Section 1.4) cannot 'fail badly'.

• Vilchinskiĭ [577] and Vilchinskiĭ and Dombrovskiĭ [578] obtained various metric results in restricted Diophantine approximation.

• It follows from Exercise 5.1 and the fact that $\mathcal{H}^1$ coincides with the Lebesgue measure that the set $\mathcal{B}$ of badly approximable real numbers is not an s-set.

• The results obtained in Exercise 5.1 have been refined by several authors, including Bumby, Cusick, and Hensley. Among other statements, Hensley [277] established that

$$\dim \mathcal{B}_M = 1 - \frac{6}{\pi^2 M} - \frac{72 \log M}{\pi^4 M^2} + O(M^{-2}),$$

as M tends to infinity. Further, he proved [278] that $\dim \mathcal{B}_2 = 0.5312805\ldots$ and gave the first six significant digits for the Hausdorff dimensions of some other sets of real numbers with finitely many different partial quotients. A more efficient approach to calculating these dimensions has been developed by Jenkinson and Pollicott [301]. Vallée [576] determined the Hausdorff dimension of sets of real numbers whose continued fraction expansion obeys a not necessarily finite but periodic set of constraints.

• Schmidt [503, 505] extended some results of Jarník [288] by showing that, for any positive integers m and n, the Hausdorff dimension of the set of badly approximable $m \times n$ real matrices is equal to mn. Continued fractions are useless for this problem; Schmidt's proof rests on the method of α-β games (see also Chapter III of [512]). An inhomogeneous analogue of Schmidt's results has been obtained by Kleinbock [325], using the methods from [330].

• The Ganesa Moorthy theorem, proved in Exercise 5.2, contains as an immediate corollary a result of Cusick [172] asserting that $\dim\{\xi \in \mathbb{R} : \log_2 \log_2 \log_2 a_n \geq n$ for all $n \geq 1\} = 0$. In the same vein, Hirst [283] showed that $\dim\{\xi \in \mathbb{R} : a_n \geq a^n$ for all $n \geq 1\} = 1/2$. Good [258] claimed that the Hausdorff dimension of $\mathcal{U}$ is zero, but, as noticed, for example, by Hirst [283], there is a gap in his proof. For related results, see Jarník [298] and Ramharter [467, 468].

• Extensions of the result proved in Exercise 5.7 to simultaneous approximation and to approximation of linear forms have been studied by Bugeaud and Chevallier [134].

• As for restricted Diophantine approximation, denoting by p_n/q_n the convergents of a real number, Erdös and Mahler [229] asserted without proof that there exist real numbers for which the greatest prime factor of both p_n and q_n is bounded for infinitely many n. They also conjectured that these numbers, which are necessarily transcendental (by Ridout's Theorem 2.3), are Liouville numbers. The claim of Erdös and Mahler has been proved by Fraenkel [249]. However, their conjecture remains open. Fraenkel and Borosh [250] have determined the Hausdorff dimension of sets of real numbers very close to infinitely many rational numbers, whose numerators and denominators are mainly composed of fixed prime numbers. Furthermore, they have generalized their result to simultaneous Diophantine approximation [113].

• For Hausdorff dimension of exceptional sets of real n-tuples related to n-dimensional extensions of Dirichlet's Theorem 1.1 established by Davenport and Schmidt [184], see R. C. Baker [51] and the references given therein.

• Let $n \geq 2$ be an integer and w_1, w_2 be real numbers with $w_1 \geq w_2 \geq 1$ and $w_1 + w_2 > n + 1$. Using planar regular systems, Dombrovskiĭ [207] computed the Hausdorff dimension of the set of vectors (ξ_1, ξ_2) in the plane for which the inequalities $|\xi_1 - \alpha_1| < \mathrm{H}(P)^{-w_1}$, $|\xi_2 - \alpha_2| < \mathrm{H}(P)^{-w_2}$, hold for infinitely many integer polynomials $P(X)$ of degree at most n, where α_1 and α_2 are two zeros of $P(X)$. Further results are due to Kovalevskaya [338].

• The Hausdorff dimension of sets of points (ξ_1, ξ_2) in $\mathbb{R}^2$ for which the inequality $|P(\xi_1)| \cdot |P(\xi_2)| < \mathrm{H}(P)^{-w}$ has infinitely many solutions in integer polynomials $P(X)$ of bounded degree has been studied by Pereverzeva [451, 452].

• There are many papers concerned with Diophantine approximation on classical curves and Hausdorff dimension. The reader is directed to Bernik and Dodson [86] and to the survey by Beresnevich and Bernik [65]. Recent results are due to Beresnevich [63], Dickinson and Dodson [193], and Rynne [490].

• For applications of results connected with this Chapter to the problem of small denominators or to the Schrödinger equation, see Chapter 7 of Bernik and Dodson [86], Section 7 of Dodson and Kristensen [205], the survey by Dodson [203], and Kristensen [343].

6

Deeper results on the measure
of exceptional sets

Theorem 5.4, due to A. Baker and Schmidt [45], asserts that for any integer $n \geq$ 1 and any real number $\tau \geq 1$ the Hausdorff dimension of the set $\mathcal{W}_n^*(\tau)$ of real numbers ξ with $w_n^*(\xi) = \tau(n+1) - 1$ is equal to $1/\tau$. In the present Chapter, we are concerned with various refinements, including the determination of the Hausdorff measure of $\mathcal{W}_n^*(\tau)$ at the critical exponent (Corollary 6.3 below).

There are essentially two new ingredients. On the one hand, we need an improvement of Proposition 5.4, which is due to Beresnevich [61] and asserts that real algebraic numbers of bounded degree are distributed 'as evenly as they could be'. On the other hand, we present a refined analysis of the Hausdorff measure of sets of real numbers close to infinitely many points in a given real sequence.

One essential tool, introduced in Section 6.1, is the notion of 'optimal regular systems' (also termed 'best possible regular systems' by Beresnevich, Bernik, and Dodson [67]). We state four general results on sets of real numbers close to infinitely many points in an optimal regular system. We establish the first one in Section 6.2, which allows us to give an alternative proof of (a slightly stronger form of) Khintchine's Theorem 1.10. The second one, stated in Section 6.3, provides the Hausdorff dimension of general exceptional sets. The third one, given in Section 6.4, refines Theorem 5.1 inasmuch as it yields the Hausdorff *measure* instead of the Hausdorff dimension. Finally, in Section 6.5, the fourth one shows that the sets of real numbers considered here share a stability property. Namely, the Hausdorff dimension of a countable intersection of a family $(\mathcal{K}_j)_{j \geq 1}$ of such sets is equal to the minimum of the Hausdorff dimensions of the $\mathcal{K}_j$s. Section 6.6 is devoted to applications of the general statements obtained in the previous Sections to Diophantine approximation.

We stress that (apart from Theorems 6.1 and 6.2) none of the results stated in this Chapter are fully proved in the present book. Theorems 6.6 to 6.10 rest on Proposition 6.1, a difficult result on the distribution of algebraic numbers of bounded degree. Furthermore, proofs of Theorems 6.3 to 6.5 are quite technical and lie beyond the scope of this book: the reader is directed to [132, 130] and to the impressive work of Beresnevich, Dickinson, and Velani [71].

6.1 Optimal regular systems

Our aim is to give more accurate information on sets $\mathcal{K}_{\mathcal{S}}^*$ of real numbers close to infinitely many points belonging to a given countable set $\mathcal{S}$. In Chapter 5, we have seen that if the points in $\mathcal{S}$ are evenly distributed in some sense, then we are able to compute the Hausdorff dimension of $\mathcal{K}_{\mathcal{S}}^*$. However, we would like to get more, in particular to obtain Khintchine-type results and to determine the Hausdorff measure of $\mathcal{K}_{\mathcal{S}}^*$ at the critical exponent. It appears that we then need refined information on the distribution of the points in $\mathcal{S}$, whence the introduction of the notion of *optimal* regular system.

DEFINITION 6.1. *Let E be a bounded open real interval. Let $\mathcal{S} = (\alpha_j)_{j \geq 1}$ be a sequence of distinct real numbers. Then $\mathcal{S}$ is called an optimal regular system of points in E if there exist positive constants c_1, c_2, and c_3, depending only on $\mathcal{S}$, and, for any bounded interval I in E, a number $K_0 = K_0(\mathcal{S}, I)$ such that, for any $K \geq K_0$, there exist integers*

$$c_1 K \leq i_1 < \ldots < i_t \leq K$$

with α_{i_h} in I for $h = 1, \ldots, t$,

$$|\alpha_{i_h} - \alpha_{i_\ell}| \geq \frac{c_2}{K} \quad (1 \leq h \neq \ell \leq t)$$

and

$$c_3 |I| K \leq t \leq |I| K.$$

As in the definition of a *regular system*, we index the set $\mathcal{S}$ and we do not assume that all points of $\mathcal{S}$ belong to E.

Ideally, when the points of the set $\mathcal{S}$ are 'very well' distributed, then $\mathcal{S}$ is likely to be an optimal regular system. This is the case for the set of rational numbers, as follows from Proposition 5.3, and for every set $\mathbb{A}_n$ of real algebraic numbers of degree at most n, as shown by Beresnevich ([60] for $n \geq 2$ and [61] for general n). Observe that Proposition 5.4 is not sharp enough to yield this assertion. We do not give the proof of the result from [61], since it lies beyond the scope of the present book. It rests, among others, on a difficult lemma due to Bernik [79] (extended by Borbat [109]) and auxiliary results (see, for example, [86], Section 2.4) used in the proofs of Theorems 4.4 and 5.7. One of these lemmas asserts that, for any given positive real number ε and for almost all real numbers ξ, all but finitely many integer polynomials $P(X)$ of degree n and with $|P(\xi)| \leq H(P)^{-n}$ satisfy $|P'(\xi)| > H(P)^{1-\varepsilon}$; a statement which may be compared with Theorem 4.2. We quote Theorem 3 of [61].

PROPOSITION 6.1. *Let $n \geq 1$ and $M \geq 2$ be integers and let I be an interval contained in $]-M+1, M-1[$. There exist positive constants c_4, c_5, depending only on n, a number $K_0 = K_0(n, I)$ and, for any $K \geq K_0$, there are $\alpha_1, \ldots, \alpha_t$ in $\mathbb{A}_n \cap I$ such that*

$$c_4 M^n K \leq H(\alpha_h) \leq M^n K, \quad (1 \leq h \leq t),$$
$$|\alpha_h - \alpha_\ell| \geq K^{-n-1} \quad (1 \leq h < \ell \leq t),$$
$$t \geq c_5 |I| K^{n+1}.$$

Actually, the existence of c_4 is not proved in [61]. However, it is not difficult to deduce it by following the proof of Beresnevich (see, for example, [124], Théorème G). The constants c_4 and c_5, but not $K_0(n, I)$, are explicitly given in [124]. It follows from (5.6) of Chapter 5 that we can take $K_0(1, I) = 100|I|^{-1} \log(100|I|^{-1})$ for any interval I in $[0, 1]$. Furthermore, Beresnevich [60] showed that Proposition 6.1 holds true with $K_0(2, I) = 48|I|^{-1} \log(24|I|^{-1})$ for any interval I in $[0, 1]$. When n is greater than or equal to 3, an effective value for $K_0(n, I)$ has not been calculated at present.

COROLLARY 6.1. *We number the elements of $\mathbb{A}_n := (\alpha_j)_{j \geq 1}$ by increasing order of their height, and, when the heights are equal, by increasing numerical order. Then, the set $\mathbb{A}_n$ is an optimal regular system in any bounded open real interval.*

Corollary 6.1 immediately follows from Proposition 6.1 and Lemma 5.6. We draw the reader's attention to the assumption that E has to be bounded. Indeed, otherwise the constants c_2 and c_3 occurring in Definition 6.1 would have to depend on I.

Actually, Beresnevich [61] proved that the set of real algebraic numbers of *fixed* degree n is an optimal regular system. Consequently, all the following results also hold with the set $\mathbb{A}_n$ replaced by the set of real algebraic numbers of degree exactly n.

Other examples of optimal regular systems include the set of real algebraic integers of any fixed degree $n \geq 2$, the set of real algebraic units of any fixed degree $n \geq 3$, the set of real algebraic integers in a given (non-totally complex) number field, and the sequence $(\{n\alpha\})_{n \geq 1}$ for any badly approximable real number α (see Exercise 6.4).

6.2 A Khintchine-type result

The aim of this Section is to show that sets of real numbers close to infinitely many points in an optimal regular system satisfy a Khintchine-type statement. This result is due to Beresnevich [61, 62].

THEOREM 6.1. *Let E be a bounded open real interval. Let $\mathcal{S} = (\alpha_j)_{j \geq 1}$ be a sequence of real numbers which is an optimal regular system in E. Let $\Psi : \mathbb{R}_{\geq 1} \to \mathbb{R}_{>0}$ be a non-increasing continuous function and set*

$$\mathcal{K}_{\mathcal{S}}^*(\Psi) = \limsup_{j \to +\infty} \{\xi \in E : |\xi - \alpha_j| < \Psi(j)\}.$$

Then the set $\mathcal{K}_{\mathcal{S}}^*(\Psi)$ is a null set if the sum $\sum_{j \geq 1} \Psi(j)$ converges, and it has full measure if this sum diverges.

We need several auxiliary lemmas to prove the divergence half of Theorem 6.1. Here, we closely follow Beresnevich [61].

LEMMA 6.1. *Let E be a measurable real set. If there exists an absolute positive constant κ such that $\lambda(E \cap I) \geq \kappa \lambda(I)$ for any bounded real interval I, then E has full measure.*

PROOF. If the set $\mathbb{R} \setminus E$ has positive measure, then, by the Lebesgue Density Theorem (see, for example, Riesz and Sz.-Nagy [472], page 13, or Mattila [415], Corollary 2.14), there exists a real number x_0 such that, for any $\varepsilon > 0$, there is $\delta > 0$ with

$$\lambda\big((\mathbb{R} \setminus E) \cap [x_0 - \delta, x_0 + \delta]\big) \geq 2\delta(1 - \varepsilon).$$

Taking $\varepsilon = \kappa/2$, this yields $\lambda(E \cap [x_0 - \delta, x_0 + \delta]) \leq \delta\kappa$, a contradiction with our assumption. Consequently, E has full measure.

Lemma 6.2 provides a converse of the Borel–Cantelli Lemma 1.2. It originates in a work of Chung and Erdős [165].

LEMMA 6.2. *Let E_i, $i \geq 1$, be measurable real sets contained in a bounded interval I. If the sum $\sum_{i=1}^{\infty} \lambda(E_i)$ diverges, then we have*

$$\lambda\big(\limsup_{i \to +\infty} E_i\big) \geq \limsup_{n \to +\infty} \frac{\left(\sum_{i=1}^{n} \lambda(E_i)\right)^2}{\sum_{i=1}^{n} \sum_{j=1}^{n} \lambda(E_i \cap E_j)}.$$

PROOF. We follow Sprindžuk [539], p. 18, and Harman [274]. Let m, n be positive integers with $m \leq n$ and set

$$E_m^n = \bigcup_{j=m}^n E_j, \quad M(m,n) = \sum_{j=m}^n \lambda(E_j),$$

$$\text{and} \quad V(m,n) = \sum_{j=m}^n \sum_{k=m}^n \lambda(E_j \cap E_k).$$

Let $\chi_{m,n}$ denote the characteristic function of E_m^n and χ_j denote that of E_j for any positive integer j. Since

$$M(m,n) = \int_I \sum_{j=m}^n \chi_j(x)\, dx = \int_I \chi_{m,n}(x) \sum_{j=m}^n \chi_j(x)\, dx,$$

the Cauchy–Schwarz inequality yields that

$$M(m,n)^2 \leq \left(\int_I \chi_{m,n}(x)\, dx \right) \left(\int_I \sum_{j=m}^n \sum_{k=m}^n \chi_j(x)\, \chi_k(x)\, dx \right) \tag{6.1}$$

$$= \lambda(E_m^n)\, V(m,n).$$

Since

$$\frac{M(1,n)^2}{V(1,n)} \leq \frac{M(m,n)\big(M(m,n) + 2m|I|\big) + m^2|I|^2}{V(m,n)}$$

and the sum $\sum_{i=1}^\infty \lambda(E_i)$ diverges, we let n tend to infinity and keep m fixed to obtain

$$\limsup_{n \to \infty} \frac{M(m,n)^2}{V(m,n)} \geq \limsup_{n \to \infty} \frac{M(1,n)^2}{V(1,n)}.$$

Combined with (6.1), this gives

$$\lim_{n \to \infty} \lambda(E_m^n) \geq \limsup_{n \to \infty} \frac{M(1,n)^2}{V(1,n)},$$

and we now let m tend to infinity to get the lemma.

LEMMA 6.3. *Let $(a_n)_{n \geq 1}$ be a non-increasing sequence of positive numbers such that the sum $\sum_{n \geq 1} a_n$ diverges. Let κ be a positive real number and, for $n \geq 1$, set $b_n := \min\{a_n, \kappa/n\}$. Then the sequence $(b_n)_{n \geq 1}$ is also non-increasing, and the sum $\sum_{n \geq 1} b_n$ diverges.*

PROOF. Since $(a_n)_{n \geq 1}$ and $(\kappa/n)_{n \geq 1}$ are non-increasing, so is $(b_n)_{n \geq 1}$. Thus, for any integer $\ell \geq 2$, we have

$$\ell b_\ell \leq 2 \sum_{n=[\ell/2]}^\ell b_n. \tag{6.2}$$

Assume that the sum $\sum_{n \geq 1} b_n$ converges. It follows from (6.2) that ℓb_ℓ tends to 0 as ℓ tends to infinity. However, since $\sum_{n \geq 1} a_n$ diverges, the definition of $(b_n)_{n \geq 1}$ yields that there exist infinitely many integers ℓ such that $b_\ell = \kappa / \ell$, which is a contradiction. Hence, the sum $\sum_{n \geq 1} b_n$ diverges.

LEMMA 6.4. *Let* $\Psi : \mathbb{R}_{>0} \to \mathbb{R}_{>0}$ *be a given non-increasing, continuous function such that the sum* $\sum_{h \geq 1} \Psi(h)$ *diverges (resp. converges). Let* κ *and* M *be positive real numbers with* $M > 1$. *Then, the sum* $\sum_{h \geq 1} M^h \, \Psi(\kappa M^h)$ *also diverges (resp. converges).*

PROOF. This follows from the equality

$$\int_1^\infty M^x \Psi(\kappa M^x) dx = \frac{1}{\kappa \log M} \int_{\kappa M}^\infty \Psi(y) dy,$$

which is nothing but the change of variable $y := \kappa M^x$.

We now turn our attention to the proof of Theorem 6.1.

PROOF OF THEOREM 6.1. The convergence half is a straightforward application of Lemma 1.2, thus we omit the details. We assume that the sum $\sum_{j \geq 1} \Psi(j)$ diverges. Let c_6 and c_7 denote the second and the third positive constants given by Definition 6.1 applied with $\mathcal{S}$. For any $j \geq 1$, set $\tilde{\Psi}(j) := \min\{\Psi(j), c_6/(3j)\}$. By Lemmas 6.3 and 6.4, the sums $\sum_{j \geq 1} \tilde{\Psi}(j)$ and $\sum_{k \geq 1} 2^k \tilde{\Psi}(2^k)$ also diverge. Let I be a bounded interval in E and let $K_0 = K_0(\mathcal{S}, I)$ be given by Definition 6.1. For any integer $k \geq K_0$, there is a collection $A_k(I) := \{i_1, \ldots, i_t\}$ of distinct integers such that

$$1 \leq i_1 < \cdots < i_t \leq 2^k,$$

$$|\alpha_{i_h} - \alpha_{i_\ell}| \geq \frac{c_6}{2^k} \quad (1 \leq h < \ell \leq t),$$

$$c_7 |I| 2^k \leq t \leq |I| 2^k.$$

For any j in $A_k(I)$, set

$$E_k(\alpha_j) := \{\xi \in I : |\xi - \alpha_j| < \tilde{\Psi}(2^k)\}$$

and define

$$E_k = \bigcup_{j \in A_k(I)} E_k(\alpha_j) \quad \text{and} \quad E(I) = \limsup_{k \to +\infty} E_k.$$

Since $\tilde{\Psi}(2^k) \leq \Psi(2^k)$ for any positive integer k and since (by Lemma 6.3) $\tilde{\Psi}$ is non-increasing, we get

$$\lambda(E(I)) \leq \lambda\big(\mathcal{K}_\mathcal{S}^*(\Psi) \cap I\big), \tag{6.3}$$

and we now establish a lower bound for $\lambda(E(I))$.

To this end, we observe that for distinct integers j and h in $A_k(I)$, the intervals $E_k(\alpha_j)$ and $E_k(\alpha_h)$ are disjoint, since $|\alpha_j - \alpha_h| \geq c_6 2^{-k}$ and $c_6 \geq 2^{k+1} \tilde{\Psi}(2^k)$, by the definition of $\tilde{\Psi}$. It follows that

$$2^{k+1} c_7 |I| \tilde{\Psi}(2^k) \leq \lambda(E_k) \leq 2^{k+1} |I| \tilde{\Psi}(2^k), \tag{6.4}$$

whence we get that

$$\sum_{k \geq K_0} \lambda(E_k) = +\infty. \tag{6.5}$$

Let N_0 be such that we have

$$\sum_{k=K_0}^{N_0} 2^k \tilde{\Psi}(2^k) > 1, \tag{6.6}$$

and let k, ℓ, and N with $K_0 \leq k < \ell \leq N$ and $N \geq N_0$.

For an integer j in $A_k(I)$, the number of distinct integers h in $A_\ell(I)$ such that $E_\ell(\alpha_h)$ and $E_k(\alpha_j)$ have non-empty intersection is at most

$$2 + \frac{2 \tilde{\Psi}(2^k)}{c_6 2^{-\ell}} = 2 + 2^{\ell+1} c_6^{-1} \tilde{\Psi}(2^k),$$

whence

$$\lambda(E_\ell \cap E_k(\alpha_j)) \leq 4 \tilde{\Psi}(2^\ell)(1 + 2^\ell c_6^{-1} \tilde{\Psi}(2^k)). \tag{6.7}$$

Since the cardinality of $A_k(I)$ is less than or equal to $2^k |I|$, we infer from (6.7) that

$$\lambda(E_\ell \cap E_k) \leq 4|I| \big(2^k \tilde{\Psi}(2^\ell) + c_6^{-1} 2^{k+\ell} \tilde{\Psi}(2^k) \tilde{\Psi}(2^\ell) \big). \tag{6.8}$$

We are now ready to estimate from above the double summation

$$\sum_{\ell=K_0}^{N} \sum_{k=K_0}^{N} \lambda(E_\ell \cap E_k) = \sum_{k=K_0}^{N} \lambda(E_k) + 2 \sum_{\ell=K_0+1}^{N} \sum_{k=K_0}^{\ell-1} \lambda(E_\ell \cap E_k). \tag{6.9}$$

We first observe that

$$\sum_{k=K_0}^{N} \lambda(E_k) \leq 2|I| \sum_{k=K_0}^{N} 2^k \tilde{\Psi}(2^k) \leq 2|I| \left(\sum_{k=K_0}^{N} 2^k \tilde{\Psi}(2^k) \right)^2, \tag{6.10}$$

by (6.6) and $N \geq N_0$. Furthermore, we notice that

$$\sum_{\ell=K_0+1}^{N} \sum_{k=K_0}^{\ell-1} 2^k \tilde{\Psi}(2^\ell) \leq \sum_{\ell=K_0+1}^{N} 2^\ell \tilde{\Psi}(2^\ell) \leq \left(\sum_{k=K_0}^{N} 2^k \tilde{\Psi}(2^k) \right)^2, \tag{6.11}$$

again by (6.6). Finally, we check that

$$\sum_{\ell=K_0+1}^{N} \sum_{k=K_0}^{\ell-1} 2^{k+\ell} \tilde{\Psi}(2^k)\tilde{\Psi}(2^\ell) \leq \left(\sum_{k=K_0}^{N} 2^k \tilde{\Psi}(2^k) \right)^2, \tag{6.12}$$

and we deduce from (6.4) and (6.8)–(6.12) that

$$\frac{(\sum_{k=K_0}^{N} \lambda(E_k))^2}{\sum_{\ell=K_0}^{N} \sum_{k=K_0}^{N} \lambda(E_\ell \cap E_k)} \geq \frac{(2c_7|I|)^2}{(10 + 8c_6^{-1})|I|} =: c_8|I|. \tag{6.13}$$

By (6.3), (6.5), (6.13), and Lemma 6.2, we get

$$\lambda\big(\mathcal{K}_\mathcal{S}^*(\Psi) \cap I\big) \geq \lambda(E(I)) \geq c_8|I|,$$

for any bounded interval I in E. Thus, Lemma 6.1 asserts that $\mathcal{K}_\mathcal{S}^*(\Psi)$ has full measure, as claimed.

We observe that the method of the proof of Theorem 6.1 works as well with a slightly weaker definition of regular system: indeed, the existence of the constant c_1 in Definition 6.1 is not needed. This remark also applies for Theorem 6.2, but it does not seem to hold for Theorem 6.3.

6.3 Hausdorff dimension of exceptional sets

Clearly, with the notation of Theorem 6.1, if Ψ_1 decreases more rapidly than Ψ_2, then the set $\mathcal{K}_\mathcal{S}^*(\Psi_1)$ is smaller than the set $\mathcal{K}_\mathcal{S}^*(\Psi_2)$. The aim of this Section is to determine their Hausdorff dimensions. An indication of how a function $g : \mathbb{R}_{\geq 1} \to \mathbb{R}_{>0}$ grows near infinity is provided by its *lower order at infinity* $\lambda(g)$.

DEFINITION 6.2. *The lower order at infinity* $\lambda(g)$ *of a function* $g : \mathbb{R}_{\geq 1} \to \mathbb{R}_{>0}$ *is defined by*

$$\lambda(g) = \liminf_{x \to +\infty} \frac{\log g(x)}{\log x}.$$

This notion arises naturally in the theory of Hausdorff dimension, see, for example, Dodson [200] and Dickinson [192]. Clearly, $\lambda(g)$ is non-negative when $\lim_{x \to +\infty} g(x) = +\infty$.

THEOREM 6.2. *Keep the notation of Theorem 6.1, and assume that the sum* $\sum_{j \geq 1} \Psi(j)$ *converges. Let* λ *denote the lower order at infinity of the function* $1/\Psi$. *Then the Hausdorff dimension of the set* $\mathcal{K}_\mathcal{S}^*(\Psi)$ *is equal to* $1/\lambda$.

Observe that the assumption of Theorem 6.2, whose proof is left as Exercise 6.1, implies that λ is at least equal to 1.

6.4 Hausdorff measure of exceptional sets

In the previous Section, we determined the Hausdorff dimensions of the sets $\mathcal{K}_{\mathcal{S}}^*(\Psi)$; we are now interested in calculating their Hausdorff measures.

THEOREM 6.3. *Keep the notation of Theorem 6.1. Let f be a dimension function such that $\lim_{x \to 0} f(x)/x = +\infty$ and $x \mapsto f(x)/x$ decreases in a neighbourhood of the origin. Assume that $x \mapsto x f(2\Psi(x))$ is non-increasing and tends to 0 when x goes to infinity. Then we have*

$$\mathcal{H}^f(\mathcal{K}_{\mathcal{S}}^*(\Psi)) = +\infty \quad \text{if} \quad \sum_{j \geq 1} f(2\Psi(j)) \quad \text{diverges}$$

and

$$\mathcal{H}^f(\mathcal{K}_{\mathcal{S}}^*(\Psi)) = 0 \quad \text{if} \quad \sum_{j \geq 1} f\big(2\Psi(j)\big) \quad \text{converges.}$$

We point out that Theorem 6.3 does not imply Theorem 6.2.

Since $x \mapsto f(x)/x$ decreases in a neighbourhood of the origin, the sum $\sum_{j \geq 1} f\big(\Psi(j)\big)$ converges if, and only if, the sum $\sum_{j \geq 1} f\big(2\Psi(j)\big)$ converges. The latter sum arises naturally in the proof of Theorem 6.3, that is the reason why it occurs in its statement, as in Satz 4 of Jarník [292].

Theorem 6.3, proved in [71] and in [130], refines Theorem 6.1. A straightforward application of the Hausdorff–Cantelli Lemma 5.5 yields the convergence half, while the divergence half is quite technical and much more difficult to establish. The general idea is however easy to explain and originates in Jarník's work [292]: we construct inductively a Cantor set 'as large as possible' contained in $\mathcal{K}_{\mathcal{S}}^*(\Psi)$ and to which we apply the full power of the Frostman Lemma 5.3.

6.5 Sets with large intersection properties

Several authors have introduced large classes of sets of Hausdorff dimension at most s which turn out to have the property that countable intersections of such sets also have dimension at most s. Such sets share a *large intersection* property in an s-dimensional sense. Examples include the *regular* sets of A. Baker and Schmidt (see Theorem 5.4), the $\mathcal{M}_\infty^s$-sequences of Rynne [487], constructions using the 'ubiquitous systems' of Dodson, Rynne, and Vickers [206] and the '$\mathcal{M}_\infty^s$-dense' construction of Falconer [235]. These constructions have been unified by Falconer [237], who gave a more direct definition of classes $\mathcal{G}^s$ of subsets of $\mathbb{R}^n$ with a large intersection property. His theory can be easily extended to general dimension functions [132].

DEFINITION 6.3. *Let f be a dimension function concave on a neighbourhood of the origin. We define* $\mathcal{G}^f(\mathbb{R})$ *to be the class of real G_δ-sets F such that*

$$\mathcal{H}^g\left(\cap_{i=1}^{+\infty} f_i(F)\right) = +\infty$$

for any dimension function g with $g \prec f$ and any sequence of similarity transformation $\{f_i\}_{i=1}^{+\infty}$.

THEOREM 6.4. *The class $\mathcal{G}^f(\mathbb{R})$ is closed under countable intersections and under bi-Lipschitz transformations on $\mathbb{R}$.*

THEOREM 6.5. *We keep the notation and hypotheses of Theorem 6.3. We furthermore assume that $\mathcal{S}$ is an optimal regular system in any bounded open real interval. If the sum*

$$\sum_{j \geq 1} f\left(2\Psi(j)\right)$$

diverges, then the set

$$\limsup_{j \to +\infty} \{\xi \in \mathbb{R} : |\xi - \alpha_j| < \Psi(j)\}$$

belongs to the class $\mathcal{G}^f(\mathbb{R})$.

Theorem 6.5, proved in [132], does not imply Theorem 6.3, and vice-versa. It yields some results which seem to be out of reach with the methods used in the proof of Theorem 6.3. Both the stability by countable intersection and the stability by bi-Lipschitz transform are important properties of $\mathcal{G}^f(\mathbb{R})$. We refer the reader to Falconer [235, 236, 237] for more details, and to the next Section for applications in the context of Diophantine approximation; most of which have been pointed out in [235, 236, 237].

6.6 Application to the approximation by algebraic numbers

Since the set of real algebraic numbers of bounded degree is an optimal regular system (Proposition 6.1), many important results derive in a straightforward way from Theorems 6.1 to 6.5. Throughout this Section, n denotes a positive integer and $\Psi : \mathbb{R}_{\geq 1} \to \mathbb{R}_{>0}$ is a continuous, non-increasing function. We set

$$\mathcal{K}_n^*(\Psi) := \left\{\xi \in \mathbb{R} : |\xi - \alpha| < \Psi(H(\alpha)) \text{ for infinitely many } \alpha \in \mathbb{A}_n\right\},$$

and we aim to give accurate metric results on the set $\mathcal{K}_n^*(\Psi)$. Observe that, for any real number $\tau \geq 1$, the set $\mathcal{K}_n^*(x \mapsto x^{-\tau(n+1)})$ coincides with the set $\mathcal{K}_n^*(\tau)$ defined in Chapter 5.

The convergence half of Theorem 6.6 is due to Koksma [333], Satz 12. Its divergence half, established by Beresnevich [61], implies that Theorem 4.4 from Chapter 4 is best possible.

THEOREM 6.6. *The set $\mathcal{K}_n^*(\Psi)$ has Lebesgue measure zero if the sum $\sum_{x=1}^{\infty} x^n \Psi(x)$ converges, and it has full measure if the sum $\sum_{x=1}^{\infty} x^n \Psi(x)$ diverges.*

PROOF. Let E be a bounded real open interval. By Corollary 6.1, the set $\mathbb{A}_n = (\alpha_j)_{j \geq 1}$ is an optimal regular system in E. We apply Theorem 6.1 with the non-increasing function $\tilde{\Psi}$ defined by $\tilde{\Psi}(j) := \Psi(\mathrm{H}(\alpha_j))$ and affine on any interval $[j, j+1]$, with j a positive integer. We only have to check that the sums $\sum_{j \geq 1} \tilde{\Psi}(j)$ and $\sum_{j \geq 1} j^n \Psi(j)$ have the same behaviour. To this end, we infer from Lemma 5.4 that there exist positive constants c_9 and c_{10}, depending only on n, such that, for any $j \geq 1$, we have

$$\Psi\left(c_9 \, j^{1/(n+1)}\right) \leq \tilde{\Psi}(j) \leq \Psi\left(c_{10} \, j^{1/(n+1)}\right). \tag{6.14}$$

It follows from (6.14) and the fact that Ψ is non-increasing that there exist positive constants $c_{11}, \ldots, c_{14}$, depending only on n, such that

$$\sum_{k=1}^{[c_{11} M^{1/(n+1)}]} c_{12} \, k^n \, \Psi(k) \leq \sum_{j=1}^{M} \tilde{\Psi}(j) \leq \sum_{k=1}^{[c_{13} M^{1/(n+1)}]} c_{14} \, k^n \, \Psi(k)$$

holds for any sufficiently large integer M. Consequently, the sum $\sum_{j \geq 1} \tilde{\Psi}(j)$ converges if, and only if, the sum $\sum_{j \geq 1} j^n \Psi(j)$ converges. Hence, we infer from Theorem 6.1 that $\mathcal{K}_n^*(\Psi) \cap E$ has Lebesgue measure zero if the sum $\sum_{x=1}^{\infty} x^n \Psi(x)$ converges, and has full measure otherwise. Theorem 6.6 follows immediately.

We emphasize that we have given a complete proof of Theorem 6.6 for $n = 1$, which depends on Proposition 5.3 and Theorem 6.1. This slightly improves upon Theorem 1.10, established by Khintchine [317] by means of the theory of continued fractions. Indeed, we get that if $\Psi : \mathbb{R}_{\geq 1} \to \mathbb{R}_{>0}$ is a non-increasing continuous function, then the set

$$\mathcal{K}_1^*(\Psi) = \left\{ \xi \in \mathbb{R} : \left| \xi - \frac{p}{q} \right| < \Psi(q) \right.$$

$$\left. \text{for infinitely many rational numbers } \frac{p}{q} \right\}$$

has Lebesgue measure zero if the sum $\sum_{q=1}^{\infty} q \Psi(q)$ converges and full Lebesgue measure otherwise. Theorem 1.10 requires a stronger assumption on Ψ, namely that the function $x \mapsto x^2 \Psi(x)$ is non-increasing. A slightly different proof of Theorem 1.10 can also be found in Cassels [155], Chapter VII.

It follows from Theorem 5.3 that, for any real numbers τ and τ' with $\tau' > \tau \geq 1$, the set $\mathcal{K}_n^*(\tau')$ is strictly contained in $\mathcal{K}_n^*(\tau)$. This motivates the study of the following question, asked in [124].

PROBLEM. *Let Ψ_1 and Ψ_2 be given non-increasing, continuous, positive functions defined on $\mathbb{R}_{\geq 1}$ and such that*

$$\Psi_1(x) < \Psi_2(x) \quad \text{for } x \text{ sufficiently large.}$$

Do there exist real numbers approximable by elements of $\mathbb{A}_n$ at the order Ψ_2, but not at the order Ψ_1? In other words, is the set $\mathcal{K}_n^*(\Psi_2) \setminus \mathcal{K}_n^*(\Psi_1)$ empty or not?

This has been solved by Jarník ([292], Satz 5 and Satz 6; see also Exercise 1.5) under some restriction when $n = 1$, and by A. Baker and Schmidt [45] for general n and functions Ψ_i, $i = 1, 2$, of the form $x \mapsto x^{-2\tau_i}$ (see Theorem 5.3).

To obtain further results on the above question, we determine the Hausdorff dimensions and the generalized Hausdorff measures of the null sets occurring in Theorem 6.6. For the proofs of Theorems 6.7 and 6.8 below, we argue as in the proof of Theorem 6.6, except that we apply Theorems 6.2 to 6.5 instead of Theorem 6.1.

THEOREM 6.7. *Assume that the sum $\sum_{x=1}^{\infty} x^n \Psi(x)$ converges. Denote by λ the lower order at infinity of the function $1/\Psi$. We then have*

$$\dim \mathcal{K}_n^*(\Psi) = \frac{n+1}{\lambda}.$$

THEOREM 6.8. *Let f be a dimension function such that $x \mapsto f(x)/x$ is decreasing in a neighbourhood of the origin and $\lim_{x \to 0} f(x)/x = +\infty$. Assume that $x \mapsto x^{n+1} f(2\Psi(x))$ is non-increasing. Then, we have $\mathcal{H}^f(\mathcal{K}_n^*(\Psi)) = 0$ if the sum*

$$\sum_{x=1}^{+\infty} x^n f(2\Psi(x))$$

converges. Otherwise, we have $\mathcal{H}^f(\mathcal{K}_n^*(\Psi)) = +\infty$ and the set $\mathcal{K}_n^*(\Psi)$ belongs to the class $\mathcal{G}^f(\mathbb{R})$.

Theorem 6.8 allows us to give a rather satisfactory partial answer to the above Problem. Namely, given the approximation functions Ψ_1 and Ψ_2, we construct a suitable dimension function f for discriminating between $\mathcal{K}_n^*(\Psi_1)$ and

$\mathcal{K}_n^*(\Psi_2)$, when these sets have Lebesgue measure zero. The idea goes back to Jarník [292], Satz 5, and yields Corollary 6.2.

COROLLARY 6.2. *Let Ψ_1 and Ψ_2 be continuous, positive, non-increasing functions defined on $\mathbb{R}_{\geq 1}$. For $i = 1, 2$, assume that the sum $\sum_{i=1}^{\infty} x^n \Psi_i(x)$ converges and that the function $x \mapsto x^n \Psi_i(x)$ is non-increasing. If the function*

$$x \mapsto \frac{\Psi_2^{-1} \circ \Psi_1(x)}{x}$$

is non-decreasing and tends to infinity with x, then the set

$$\mathcal{K}_n^*(\Psi_2) \setminus \mathcal{K}_n^*(\Psi_1)$$

is non-empty and has the same Hausdorff dimension as the set $\mathcal{K}_n^*(\Psi_2)$.

For $n = 1$, Bugeaud [129] proved a sharper result, namely that, for any positive real number $c < 1$, any continuous function $\Psi : \mathbb{R}_{\geq 1} \to \mathbb{R}_{>0}$ such that $x \mapsto x^2 \Psi(x)$ is non-increasing and the sum $\sum_{x \geq 1} x \Psi(x)$ converges, the set $\mathcal{K}_1^*(\Psi) \setminus \mathcal{K}_1^*(c\Psi)$ has same Hausdorff dimension as the set $\mathcal{K}_1^*(\Psi)$. This does not follow from Theorem 6.8, which does not help to get an analogous result for general n.

Let $c_1 < c_2$ and $\tau > 1$ be given positive real numbers. For a general $n \geq 2$, we do not even know whether there exist real numbers ξ such that infinitely many algebraic numbers α of degree n satisfy $|\xi - \alpha| < c_2 H(\alpha)^{-(n+1)\tau}$, but only finitely many of them satisfy $|\xi - \alpha| < c_1 H(\alpha)^{-(n+1)\tau}$. However, for $n = 2$, a partial positive answer follows from a nested interval construction based on Beresnevich's effective proof [60] of Proposition 6.1 for $n = 2$.

For a given positive integer n, the exact order of approximation of the real number ξ by algebraic numbers of degree at most n is

$$\sup\{\tau : \xi \in \mathcal{K}_n^*(\tau)\} = \frac{w_n^*(\xi) + 1}{n + 1}.$$

Inspired by Beresnevich, Dickinson, and Velani, who gave in [70] sharp consequences of Theorem 1 of [195], Bugeaud [124] introduced a refined notion of exact order: the exact 'logarithmic' order. Recall that for any positive integer i we denote by $\log_i$ the i-fold iterated logarithm. For $n \geq 1$ and $t \geq 1$, let $v_0 \geq 1, v_1, \ldots, v_{t-1}$ be real numbers. For a non-zero real number τ, denote by $\mathcal{K}_n^*(v_0, \ldots, v_{t-1}, \tau)$ the set of real numbers ξ for which the inequality

$$|\xi - \alpha| < H(\alpha)^{-(n+1)v_0} \left(\log H(\alpha)\right)^{-v_1} \cdots$$
$$\left(\log_{t-1} H(\alpha)\right)^{-v_{t-1}} \left(\log_t H(\alpha)\right)^{-\tau}$$

is satisfied by infinitely many algebraic numbers α of degree at most n. For any real number v_t, we then set $\underline{v} = (v_0, \ldots, v_{t-1}, v_t)$ and we consider the set

$$\mathcal{W}_n^*(\underline{v}) := \{\xi \in \mathbb{R} : \sup\{\tau : \xi \in \mathcal{K}_n^*(v_0, \ldots, v_{t-1}, \tau)\} = v_t\}.$$

In particular, when $t = 0$, we have $\mathcal{W}_n^*((v_0)) = \mathcal{W}_n^*(v_0)$. Theorem 6.8 allows us to describe very precisely the metric structure of the sets $\mathcal{W}_n^*(\underline{v})$: their Hausdorff dimension is equal to $1/v_0$, independently of $v_1, \ldots, v_t$, but they have different sizes, as it can be seen by considering dimension functions.

Let $t \geq 0$ and $\underline{v} = (v_0, \ldots, v_t)$ with $v_0 \geq 1$. For any non-negative real number δ, define the dimension function $f_{\underline{v},\delta}$ for $u > 0$ (and small) by

$$f_{\underline{v},\delta}(u) := u^{\delta+1/v_0} \quad \text{if } t = 0,$$

and

$$f_{\underline{v},\delta}(u) := u^{1/v_0} \prod_{i=1}^{t-1} \left(\log_i \frac{1}{u}\right)^{-1+v_i/v_0} \left(\log_t \frac{1}{u}\right)^{-1-\delta+v_t/v_0} \quad \text{if } t \geq 1.$$

With the above notation, Theorem 6.9 is an almost straigthforward consequence of Theorem 6.8.

THEOREM 6.9. *Let $\underline{v} = (v_0, \ldots, v_t)$ be a real $(t+1)$-tuple with $v_0 > 1$. Then, the Hausdorff dimension of $\mathcal{W}_n^*(\underline{v})$ is equal to $1/v_0$ and, more precisely,*

$$\mathcal{H}^{f_{\underline{v},\delta}}(\mathcal{W}_n^*(\underline{v})) = \begin{cases} +\infty & \text{if } \delta = 0, \\ 0 & \text{if } \delta > 0. \end{cases}$$

COROLLARY 6.3. *For any positive integer n and any real number $\tau > 1$, we have*

$$\mathcal{H}^{1/\tau}(\mathcal{K}_n^*(\tau)) = \mathcal{H}^{1/\tau}(\mathcal{W}_n^*(\tau)) = +\infty.$$

Corollary 6.3 implies that the sets $\mathcal{K}_n^*(\tau)$ and $\mathcal{W}_n^*(\tau)$ are not s-sets. For $\mathcal{K}_1^*(\tau)$, this result goes back to Jarník [290, 291] (see also [205]) and is actually a particular case of a more general statement due to Bugeaud, Dodson, and Kristensen [135]: for any given dimension function f, the $\mathcal{H}^f$-measure of any real set invariant by rational translation is either 0 or $+\infty$ (see Exercise 6.5).

In the case $n = 1$, Theorem 6.9 is contained in Theorem 1 of Beresnevich, Dickinson, and Velani [70], whose Theorem 3 deals with an even more general formalism and is deeper.

A further metric statement is given in [124], where it is pointed out that we do not have any information on the Hausdorff dimension of the sets $\mathcal{W}_n^*((1, 1/2))$. For $n \geq 2$, we even do not know whether these sets are empty or not.

Theorem 6.10, established in [132], provides a nice application of the theory of intersective sets.

THEOREM 6.10. *Let* $\underline{v} = (v_0, \ldots, v_t)$ *with* $v_0 > 1$. *Then, the Hausdorff dimension of the set*

$$\bigcap_{n \geq 1} \bigcap_{k \geq 1} \{\xi \in \mathbb{R} : \xi^k \in \mathcal{W}_n^*(\underline{v})\}$$

is equal to $1/v_0$.

COROLLARY 6.4. *For any real number* $\tau \geq 1$, *the Hausdorff dimension of the set*

$$\bigcap_{n \geq 1} \bigcap_{k \geq 1} \{\xi \in \mathbb{R} : w_n^*(\xi^k) = \tau(n+1) - 1\}$$

is equal to $1/\tau$. *Thus, the set of real numbers* ξ *whose integer powers are all* S^*-*numbers of* $*$-*type* τ *has Hausdorff dimension* $1/\tau$.

Corollary 6.4 remains true if the $*$-type of an S^*-number ξ is defined as the supremum of the sequence $((w_n^*(\xi) + 1)/(n + 1))_{n \geq 1}$. Unlike Theorem 5.6, it can presumably not be proved with the tools developed in Chapter 5. It heavily depends on the notion of intersective sets.

We point out that there exist real transcendental numbers ξ for which $w_n^*(\xi) \neq w_n^*(\xi^2)$, see Theorem 7.7 for explicit examples. However, it is not known whether $t^*(\xi) = t^*(\xi^2)$ holds for any S^*-number ξ.

For distinct positive integers n and n' and given real numbers τ and τ' with $\tau' > \tau \geq 1$, the methods presented here are unable to decide whether the intersection $\mathcal{W}_n^*(\tau) \cap \mathcal{W}_{n'}^*(\tau')$ is empty or not. This is a challenging open problem.

6.7 Exercises

EXERCISE 6.1. Prove Theorem 6.2.

Hint. Use Lemma 5.2 and a similar argument as in the proof of Theorem 5.1.

EXERCISE 6.2. Use Theorem 6.8 to prove Theorems 6.9 and 6.10.

EXERCISE 6.3. Let α be a given real number. The aim of this exercise is to establish an inhomogeneous extension of Theorem 5.2: For any $\tau > 1$, the

Hausdorff dimension of the set

$$\left\{ \xi \in \mathbb{R} : \|q\xi - \alpha\| < \frac{1}{q^{2\tau-1}} \quad \text{holds for infinitely many} \right.$$
$$\left. \text{positive integers } q \right\}$$

is equal to $1/\tau$. To this end, prove that the set $\mathcal{S}_\alpha$ composed of the points

$$\left\{ \frac{p+\alpha}{q} \right\}, \quad p, q \in \mathbb{Z}, \quad q \geq 1, \quad 0 \leq p \leq q - 1,$$

form an optimal regular system in $]0, 1[$ and apply Theorem 5.1.

Hint. Let $Y > 1$ be a real number and I be an interval in $]0, 1[$. Set

$$J(Y) = I \cap \bigcup_{\substack{q=Y/2 \\ (p,q)=1}}^{Y} \left] \frac{p}{q} - \frac{1}{4q^2}, \frac{p}{q} + \frac{1}{4q^2} \right[.$$

Observe that any ξ in $J(Y)$ has a convergent p_h/q_h such that $Y/2 \leq q_h \leq Y < q_{h+1}$. Write $\xi = p_h/q_h + \delta/(Yq_h)$ and $\alpha = (t/q_h) + \delta'/(2q_h)$, with $|\delta| < 1/2$ and $|\delta'| \leq 1$. Prove that there exist integers x and y such that $q_h \leq x < 2q_h$ and $xp_h - yq_h = t$. Show that we have $|\xi - (y+\alpha)/x| \leq 3/Y^2$ and argue as at the end of the proof of Proposition 5.3 to conclude.

EXERCISE 6.4. Let α be a badly approximable real irrational number. Use the *Three Distance Theorem* (or use Theorem 1.7) as in Exercise 5.7 to prove that the sequence $(\{n\alpha\})_{n\geq 1}$ is an optimal regular system.

EXERCISE 6.5. Let f be a dimension function, and let E be a real set, invariant by rational translation. Prove that $\mathcal{H}^f(E) = 0$ or $+\infty$ (see [135]). Taking $f = \text{Id}$, prove that either E or its complementary set is a null set.

Hint. Observe that any real open interval $]a, b[$ can be represented as a countable union of intervals $[u_j/v_j, (u_j + 1)/v_j]$ such that $b - a = \sum_j 1/v_j$. Deduce that $\mathcal{H}^f(]a, b[\cap E \cap [0, 1]) \leq (b-a)\mathcal{H}^f(E \cap [0, 1])$. Observe that if $E \cap [0, 1]$ does not have full Lebesgue measure, then it can be covered by a countable union of open intervals $]a_m, b_m[$ such that $\sum_m (b_m - a_m) < 1$. Conclude.

6.8 Notes

• Hinokuma and Shiga [282] (but see notes in Chapter 5) and Rynne [488] obtained the Hausdorff dimension of $\mathcal{K}_1^*(\Psi)$ for any arbitrary non-negative approximation function Ψ without any monotonicity assumption. See also Dickinson [192].

• Theorem 6.8 for $n = 1$ is due to Jarník [292] and has been extended to systems of linear forms by Dickinson and Velani [195]. Beresnevich, Dickinson, and Velani [70] derived from Theorem 1 of [195] sharp results on the exact order of approximation of systems of linear forms.

• Under suitable assumptions, the notion of ubiquitous system may be used to prove that the Hausdorff measure is infinite at the critical exponent, see Theorem 1 of Dodson, Rynne, and Vickers [206].

• A (slightly) weaker version of Theorem 6.8 (apart from the last assertion) appeared in [124]. Measure theoretic laws for lim sup sets in a very general framework have been established by Beresnevich, Dickinson, and Velani [71]. Applications of their results include Theorems 6.1, 6.2, 6.3, 6.6, 6.7, and 6.8 (apart from the last assertion).

• An inhomogeneous analogue of Theorem 6.8 for $n = 1$ can be found in [130]. See also [71] for generalizations to systems of linear forms.

• For extensions of the case $n = 1$ of Corollary 6.2 to simultaneous approximation and to systems of linear forms, see Černy [161] and Jarník [300].

• It is interesting to compare Theorem 5.2 with the results proved in Exercises 5.7 and 6.3.

7

On T-numbers and U-numbers

As Mahler [376] proposed his classification of real numbers, he observed that
the set of U-numbers is non-empty, for it contains the Liouville numbers (Definition 1.1). He further showed that the set of T-numbers and the set of U-
numbers have Lebesgue measure zero, a result later improved upon by Kasch
and Volkmann [311], who established that both have Hausdorff dimension zero
(Corollary 5.2). However, Mahler was not able to decide whether the set of T-
numbers is non-empty.

The first proof of the existence of T-numbers goes back to 1968, thirty-six
years after Mahler's paper, and is due Schmidt [507], who, shortly thereafter,
gave a simpler proof [508], together with a slightly stronger statement. The
main ideas of Schmidt's proof lie at the heart of every paper providing new results on the set of T-numbers. In the present Chapter, we establish the existence
of T-numbers following R. C. Baker [47]. Sections 7.2 and 7.3 are devoted
to the proof, which incidentally allows us to construct real numbers ξ with
$w_n(\xi) \neq w_n^*(\xi)$ for every $n \geq 2$, a question further discussed in Section 7.4.
We emphasize that the construction of T-numbers is fundamentally a nested
interval construction. This observation is used in Section 7.5 to get one metric
result of R. C. Baker [47], who proved that the set of T-numbers is not 'too
small', although of Hausdorff dimension zero.

Schmidt's construction also establishes the existence of U-numbers of any
given type (Definition 3.4). The problem, however, had already been solved by
LeVeque [361] who, in 1953, produced explicit examples of Liouville numbers
whose positive real m-th root is a U-number of type exactly m. Section 7.6
is devoted to LeVeque's result and includes an alternative construction of U-
numbers of given type. In Section 7.7, we present an approach due to Güting
[269], who, thanks to explicit constructions of real numbers as roots of sums of
convergent series, produced examples of real numbers ξ with prescribed values
for $w_n(\xi)$ and $w_n^*(\xi)$ for finitely many values of n. Finally, in Section 7.8,

we summarize the results obtained in the previous Chapters towards the Main Problem.

7.1 *T*-numbers do exist

R. C. Baker's theorem asserts the existence of T-numbers with specific properties. This provides an important step towards the resolution of the Main Problem stated at the end of Section 3.4.

THEOREM 7.1. *Let $(w_n)_{n\geq 1}$ and $(w_n^*)_{n\geq 1}$ be two non-decreasing sequences in $[1, +\infty]$ such that*

$$w_n^* \leq w_n \leq w_n^* + (n-1)/n, \quad w_n > n^3 + 2n^2 + 4n + 3,$$
$$\text{for any } n \geq 1. \tag{7.1}$$

Then there exists a real transcendental number ξ such that

$$w_n^*(\xi) = w_n^* \quad and \quad w_n(\xi) = w_n, \quad \text{for any } n \geq 1.$$

In particular, T-numbers do exist.

The fact that the function $f : n \mapsto n^3 + 2n^2 + 4n + 3$ occurring in the statement of Theorem 7.1 is of order of magnitude n^3 is due to technical constraints. Presumably, the conclusion of Theorem 7.1 still holds if f is replaced by any non-decreasing function g such that $g(n)/n$ tends to infinity with n or, even, such that $\lim \sup_{n \to +\infty} g(n)/n$ is infinite. Theorem 7.1 implies that there exist T-numbers of any type τ greater than or equal to 3. However, the existence of T-numbers of type strictly smaller than 3 is an open problem. No particular significance should be attached to this value 3, which appears for technical reasons.

Theorem 3.4 asserts that any real transcendental number ξ satisfies $w_n(\xi) \leq w_n^*(\xi) + n - 1$ for every positive integer n. This leads to conjecture that the term $(n-1)/n$ occurring in (7.1) could perhaps be replaced by $n-1$, a question discussed in Section 7.4.

It should be pointed out that Theorem 7.1 does not imply the existence of real numbers ξ with $w(\xi) \neq w^*(\xi)$. This remains an open problem.

Theorem 7.1 is (very slightly) better than the main result of [47], since we use (7.24) below instead of setting $\chi_n = w_n - n + 2$, as R. C. Baker did. Furthermore, it is possible to slightly decrease the lower bound for w_n in (7.1) if we seek for numbers ξ for which $w_n(\xi)$ and $w_n^*(\xi)$ are equal for every positive integer n (see Exercise 7.1).

Schmidt's construction is rather tedious and complicated. The T-numbers are obtained as limits of sequences of algebraic numbers in a 'semi-explicit' way. Indeed, at the heart of his proof, Schmidt used the fact that algebraic numbers are not 'too well' approximable by algebraic numbers of bounded degree, which is a result of an ineffective nature (Theorems 2.6 and 2.7). For this reason, Schmidt's method does not allow us to construct explicit examples of T-numbers. This remains an open problem.

Incidentally, Theorem 7.1 shows the existence of U-numbers of arbitrary given type, a result established by LeVeque [361] in 1953 with a constructive method (Theorem 7.4). Following Schmidt [508], there are, however, alternative ways to split the U-numbers into infinitely many classes, including the following one. Write $\beta(\xi)$ for the infimum of the numbers β such that for every positive integer n we have

$$w_n(\xi, H) \leq c_n \, e^{-H^\beta}, \quad \text{as } H \to +\infty,$$

for some constant c_n depending only on n. Observe that if ξ and ξ' are two algebraically dependent U-numbers, we then have $\beta(\xi) = \beta(\xi')$. Consequently, we may subdivide the set of U-numbers ξ according to the value of $\beta(\xi)$. Using the same ideas as in his proof of the existence of T-numbers, Schmidt [508] established that, for any β in $[0, +\infty]$, there exist uncountably many U-numbers ξ with $\beta(\xi) = \beta$.

Schmidt [508] also proposed a possible generalization of the notion of Liouville numbers. A real number field $\mathbb{K}$ being given, a real number ξ is called $\mathbb{K}$-Liouville if for every positive real number w there is a real number α in $\mathbb{K}$ with $|\xi - \alpha| < \mathrm{H}(\alpha)^{-w}$. According to this definition, the usual Liouville numbers are termed $\mathbb{Q}$-Liouville. Assuming that $\mathcal{F}$ is a collection of real algebraic number fields such that $\mathbb{K}_1 \subset \mathbb{K}_2$ and $\mathbb{K}_1$ in $\mathcal{F}$ implies $\mathbb{K}_2$ in $\mathcal{F}$, Schmidt proved (again with the same kind of construction) that there exist real numbers ξ which are $\mathbb{K}$-Liouville precisely for the fields $\mathbb{K}$ in $\mathcal{F}$.

7.2 The inductive construction

Before proceeding with the construction of sequences of real numbers satisfying various conditions, we recall (part of) Schmidt's Theorem 2.7, which follows from his Subspace Theorem.

THEOREM 2.7. *Let ξ be a real algebraic number and n be a positive integer. Then, for any positive real number ε, there exists a positive (ineffective) constant $\kappa(\xi, n, \varepsilon)$ such that*

$$|\xi - \alpha| > \kappa(\xi, n, \varepsilon) \, \mathrm{H}(\alpha)^{-n-1-\varepsilon} \tag{7.2}$$

for any algebraic number α of degree at most n.

The exponent of H(α) in (7.2) does not depend on ξ: this fact is crucial for establishing the existence of T-numbers. Actually, Schmidt applied Theorem 2.6 (that is, the generalization due to Wirsing of Roth's Theorem 2.1) in [507] and in [508], thus getting a lower bound for w_n of order of magnitude $2n^3$ instead of n^3 in (7.1).

We follow step by step the arguments of R. C. Baker [47], who deduced Theorem 7.1 from Proposition 7.1 below, as will be shown in Section 7.3. In fact, we build a T^*-number; by Theorem 3.6, this is equivalent to constructing a T-number.

PROPOSITION 7.1. *Let* v_1, v_2, ... *be real numbers* > 1 *and* μ_1, μ_2, ... *be real numbers in* [0, 1]. *Let* m_1, m_2, ... *be positive integers and* χ_1, χ_2, ... *be real numbers satisfying* $\chi_n > n^3 + 2n^2 + 4n + 3$ *for any* $n \geq 1$. *Then, there exist positive real numbers* λ_1, λ_2, ..., *prime numbers* g_1, g_2, ..., *and integers* c_1, c_2, ... *such that the following conditions are satisfied.*

(I_j) $g_j \nmid c_j^{m_j} + 2(-1)^{m_j+1}v_j^{m_j}$, *where* $v_j = [g_j^{\mu_j}]$ ($j \geq 1$).

(II_1) $\xi_1 = (c_1 + 2^{1/m_1}v_1)/g_1 \in I_0 :=]1, 2[$.

(II_j) $\xi_j = (c_j + 2^{1/m_j}v_j)/g_j$ *belongs to the interval* I_{j-1} *defined by*

$$\xi_{j-1} + \frac{1}{2}g_{j-1}^{-v_{j-1}} < x < \xi_{j-1} + \frac{3}{4}g_{j-1}^{-v_{j-1}} \quad (j \geq 2).$$

(III_j) $|\xi_j - \alpha_n| \geq \lambda_n \, \mathrm{H}(\alpha_n)^{-\chi_n}$

for any algebraic number α_n *of degree* $n \leq j$ *which is distinct from* $\xi_1, \ldots, \xi_j$ ($j \geq 1$).

Let us see how one should interpret the different parameters introduced in Proposition 7.1. Observe that the sequence $(\xi_j)_{j \geq 1}$ we construct is strictly increasing and bounded, hence, it converges to a limit ξ. Since $c_j \leq 2g_j$, Lemma A.4 implies that for any $j \geq 1$ the height of ξ_j is at most $4^{m_j+1} g_j^{m_j}$. Thus, conditions (II_{j+1}) show that the order of approximation of ξ by the algebraic number ξ_j depends only on v_j and m_j, namely that we have $|\xi - \xi_j| \leq 2^{4v_j} \mathrm{H}(\xi_j)^{-v_j/m_j}$. Consequently, if for some positive integer n and some real number $v > 1$ there are infinitely many indices j such that $m_j = n$ and $v_j = v$, then conditions (II_j) give us that $w_n^*(\xi) \geq v/n - 1$. Furthermore, conditions (III_j) show that the algebraic numbers other than the ξ_js are not too close to ξ and they yield that $w_n^*(\xi) \leq \max\{\chi_n, v/n\} - 1$. With a suitable choice of the sequences $(m_j)_{j \geq 1}$, $(v_j)_{j \geq 1}$, and $(\chi_n)_{n \geq 1}$, the real number ξ obtained is a

T^*-number. However, some technical conditions impose that χ_n has to grow at least as fast as n^3, thus, with this method, it does not seem to be possible to construct T-numbers of type smaller than 3.

The role of the parameter μ is to measure the gap between $w_n(\xi)$ and $w_n^*(\xi)$, as it will be clear in Section 7.3. Taking μ_j equal to 1 for any j, we have $v_j = g_j$ and

$$\xi_j = 2^{1/m_j} + \frac{c_j}{g_j}.$$

The condition (I_j) is then satisfied if g_j does not divide c_j, and the T-numbers ξ obtained in this way satisfy $w_n(\xi) = w_n^*(\xi)$ for any n.

In Schmidt's original construction, the algebraic numbers $2^{1/m_j}$ are replaced by arbitrary real algebraic numbers of degree m_j. These play a pivotal role. The particular choice of the numbers $2^{1/m_j}$ allows us to build real numbers ξ for which $w_n(\xi)$ and $w_n^*(\xi)$ are different.

PROOF OF PROPOSITION 7.1. In all what follows, we denote by α_n a real algebraic number of degree exactly n. We fix a sequence $(\varepsilon_n)_{n \geq 1}$ in $]0, 1[$ such that, for any $n \geq 1$, we have

$$\chi_n > n^3 + 2n^2 + 4n + 3 + 20n^2 \varepsilon_n. \tag{7.3}$$

We begin by adding four extra conditions $(IV_j), \ldots, (VII_j)$ to be satisfied by the numbers ξ_j.

Set $J_0 = I_0$. For any positive integer j, let J_j denote the subset of I_j consisting of the real numbers x in I_j such that

$$|x - \alpha_n| \geq 2\lambda_n H(\alpha_n)^{-\chi_n}$$

for any algebraic number α_n of degree $n \leq j$, distinct from $\xi_1, \ldots, \xi_j, x$ and of height $H(\alpha_n)$ sufficiently large, that is, satisfying

$$H(\alpha_n) \geq (\lambda_n g_j^{v_j})^{1/\chi_n}.$$

The supplementary conditions are the following.

(IV_j) $\qquad\qquad\qquad \xi_j \in J_{j-1} \quad (j \geq 2).$

(V_j) $\qquad |\xi_j - \alpha_j| \geq 2\lambda_j H(\alpha_j)^{-\chi_j}$ for any $\alpha_j \neq \xi_j \quad (j \geq 1).$

(VI_j)
$\qquad$ If $n \leq j$ and $H(\alpha_n) \leq g_j^{1/(n+1+\varepsilon_n)}$, then $|\xi_j - \alpha_n| \geq 1/g_j \quad (j \geq 1)$

(VII_j) $\qquad\qquad\qquad \lambda(J_j) \geq \lambda(I_j)/2 \quad (j \geq 1).$

We construct the numbers $\xi_1, \lambda_1, \xi_2, \lambda_2, \ldots$ by induction. At the j-th stage, there are two distinct steps. Step (A_j) consists in building an algebraic number

$$\xi_j = \frac{c_j + 2^{1/m_j} v_j}{g_j}$$

satisfying conditions (I_j) to (VI_j). In step (B_j), we show that the number ξ_j constructed in (A_j) satisfies (VII_j) as well, provided that g_j is chosen large enough in terms of

$$v_1, \ldots, v_j, \mu_1, \ldots, \mu_j, m_1, \ldots, m_j, \chi_1, \ldots, \chi_j, \varepsilon_1, \ldots, \varepsilon_j,$$
$$\xi_1, \ldots, \xi_{j-1}, \lambda_1, \ldots, \lambda_{j-1}. \tag{7.4}$$

The symbols $o, \gg$ and $\ll$ used throughout steps (A_j) and (B_j) mean that the numerical implicit constants depend (at most) on the quantities (7.4). Furthermore, the symbol o implies 'as g_j tends to infinity'.

Step (A_1) is rather easy. Let $P(X) := X^{m_1} - 2 v_1^{m_1}$ denote the minimal polynomial of $2^{1/m_1} v_1$ over $\mathbb{Z}$ and observe that (I_1) is satisfied if, and only if, g_1 does not divide $P(-c_1)$. Let g be a prime number and assume that g divides $P(d), P(d+1), \ldots, P(d+m_1)$ for some integer d. Then g also divides the difference $P(d+j+1) - P(d+j) = m_1(d+j)^{m_1-1} + \ldots$ for any integer $j = 0, \ldots, m_1 - 1$. If $m_1 \geq 2$, we similarly get that g divides $m_1(m_1-1)(d+j)^{m_1-2} + \ldots$ for any integer $j = 0, \ldots, m_1 - 2$. Continuing in this way, it follows that g divides $m_1!$. Consequently, if the prime number g_1 is larger than m_1, then there are $\gg g_1$ numbers $\xi_1 = (c_1 + 2^{1/m_1} v_1)/g_1$ in the interval $]1, 2[$ satisfying condition (I_1). These $\gg g_1$ numbers have mutual distances at least g_1^{-1} and, since there are only $o(g_1)$ rational numbers α_1 satisfying $H(\alpha_1) \leq g_1^{1/(2+\varepsilon_1)}$, we are able to choose ξ_1 such that (VI_1) is verified. Moreover, by Theorem 2.7 with $n = 1$ (this is Roth's Theorem 2.1), there exists λ_1 in $]0, 1[$ such that both (III_1) and (V_1) hold.

Let $j \geq 2$ be an integer and assume that $\xi_1, \ldots, \xi_{j-1}$ have been constructed. Step (A_j) is much harder to verify, since we have no control on the set J_{j-1}. Thus, it seems difficult to check that the condition (IV_j) holds. To overcome this problem, we introduce a new set J'_{j-1} which contains J_{j-1}.

Set $\xi_j = (c_j + 2^{1/m_j} v_j)/g_j$ for some positive integers c_j and g_j with

$$g_j^{v_j} > 8 g_{j-1}^{v_{j-1}}, \tag{7.5}$$

and denote by J'_{j-1} the set formed by the real numbers x in I_{j-1} satisfying

$$|x - \alpha_n| \geq 2\lambda_n \, H(\alpha_n)^{-\chi_n}$$

for any algebraic number α_n of degree $n \le j$, distinct from $\xi_1, \ldots, \xi_j, x$, and whose height $H(\alpha_n)$ satisfies the inequalities

$$(\lambda_n \, g_{j-1}^{v_{j-1}})^{1/\chi_n} \le H(\alpha_n) \le (2\lambda_n \, g_j^{n^2+n+1+2n\varepsilon_n})^{1/(\chi_n-n-1-\varepsilon_n)}. \quad (7.6)$$

Since

$$\begin{aligned}
\chi_n - n - 1 - \varepsilon_n &> n^3 + 2n^2 + 2n + 1 + 5n^2\varepsilon_n \\
&> (n+1)(n^2 + n + 1 + 2n\varepsilon_n),
\end{aligned} \quad (7.7)$$

the exponent of g_j in the right member of (7.6) is strictly less than $1/(n+1)$. Thus, there are $o(g_j)$ algebraic numbers α_n satisfying (7.6), and we observe that, unlike J_{j-1}, the set J'_{j-1} is a finite union of intervals, and, more precisely, a union of $o(g_j)$ intervals. We will prove that for g_j large enough we have $\gg g_j$ suitable choices for c_j such that conditions (I_j) to (V_j) are fulfilled.

Let α_n be an algebraic number of degree n. By Theorem 2.7, there exists a positive constant $\kappa(m_j, n, \varepsilon_n)$ such that

$$\begin{aligned}
|\xi_j - \alpha_n| &= \frac{v_j}{g_j}\left|2^{1/m_j} - \left(\frac{g_j\alpha_n - c_j}{v_j}\right)\right| \\
&\ge \kappa(m_j, n, \varepsilon_n) \frac{v_j}{g_j} H(v_j^{-1}(g_j\alpha_n - c_j))^{-n-1-\varepsilon_n} \\
&\ge g_j^{-(n^2+n+1+2n\varepsilon_n)} H(\alpha_n)^{-n-1-\varepsilon_n},
\end{aligned} \quad (7.8)$$

if g_j satisfies

$$g_j \ge \kappa(m_j, n, \varepsilon_n)^{-1/(n\varepsilon_n)} 2^{(2n+1)(n+1+\varepsilon_n)/(n\varepsilon_n)}.$$

Here, we have used $v_j \ge 1$, $\max\{c_j, g_j, v_j\} \le 2g_j$, and Lemma A.4. In particular, if g_j is large enough, we have

$$|\xi_j - \alpha_n| \ge 2\lambda_n H(\alpha_n)^{-\chi_n} \quad (7.9)$$

as soon as

$$H(\alpha_n)^{\chi_n-n-1-\varepsilon_n} \ge 2\lambda_n g_j^{n^2+n+1+2n\varepsilon_n}. \quad (7.10)$$

By (VII_{j-1}) and $J'_{j-1} \supset J_{j-1}$, we have $\lambda(J'_{j-1}) \gg 1$. Since the set J'_{j-1} is the union of $o(g_j)$ intervals, if g_j is a sufficiently large prime number, then, arguing as in step (A_1), we get that there exist $\gg g_j$ numbers $\xi_j = (c_j + 2^{1/m_j}v_j)/g_j$ in J'_{j-1} such that (I_j) is satisfied. Such ξ_js also belong to J_{j-1}, since (7.10) implies (7.9), and condition (IV_j) is verified.

Thus, we are left with $\gg g_j$ suitable algebraic numbers ξ_j, mutually distant by at least g_j^{-1}. Only $o(g_j)$ algebraic numbers α_n satisfy

$$H(\alpha_n) \le g_j^{1/(n+1+\varepsilon)}, \qquad (7.11)$$

thus there are $\gg g_j$ algebraic numbers ξ_j such that $|\xi_j - \alpha_n| \ge 1/g_j$ for the numbers α_n verifying (7.11). Further, Theorem 2.7 ensures that there exists λ_j in $]0, 1[$ such that (V_j) is satisfied. Consequently, there are $\gg g_j$ algebraic numbers ξ_j satisfying (I_j), (II_j), (IV_j), (V_j) and (VI_j).

It remains for us to show that such a ξ_j also satisfies (III_j). To this end, it suffices to prove that

$$|\xi_j - \alpha_n| \ge \lambda_n H(\alpha_n)^{-\chi_n}$$

holds for any algebraic number α_n of degree $n < j$, which is different from $\xi_1, \ldots, \xi_j$ and whose height $H(\alpha_n)$ satisfies

$$H(\alpha_n) < (\lambda_n g_{j-1}^{\nu_{j-1}})^{1/\chi_n}.$$

Since the sequence $(g_t^{\nu_t})_{t \ge 1}$ is increasing, we either have

$$g_n^{-\nu_n} < \lambda_n H(\alpha_n)^{-\chi_n}, \qquad (7.12)$$

or there exists an integer t with $n < t < j$ such that

$$g_t^{-\nu_t} < \lambda_n H(\alpha_n)^{-\chi_n} \le g_{t-1}^{-\nu_{t-1}}. \qquad (7.13)$$

In the former case, we infer from (V_n), (7.5) and (7.12) that

$$|\xi_j - \alpha_n| \ge |\xi_n - \alpha_n| - |\xi_j - \xi_n| \ge 2\lambda_n H(\alpha_n)^{-\chi_n} - g_n^{-\nu_n} > \lambda_n H(\alpha_n)^{-\chi_n}.$$

In the latter case, (IV_t), (7.5) and (7.13) yield that

$$|\xi_j - \alpha_n| \ge |\xi_t - \alpha_n| - |\xi_j - \xi_t| \ge 2\lambda_n H(\alpha_n)^{-\chi_n} - g_t^{-\nu_t} > \lambda_n H(\alpha_n)^{-\chi_n}.$$

Thus, condition (III_j) holds and the proof of step (A_j) is completed.

Before going on with step (B_j), we observe that the integer c_j is far from being uniquely determined. Indeed, it follows from the argument used in step (A_1) that there exist at least

$$\frac{g_j \lambda(J_{j-1})}{m_j + 1}$$

candidates ξ_j having the property (I_j). In the course of step (A_j), we excluded only $o(g_j)$ of them. Thus, if g_j is sufficiently large, we have at least

$$\frac{g_j g_{j-1}^{-\nu_{j-1}}}{32 m_j} \qquad (7.14)$$

suitable choices for ξ_j. This observation is valid at any step j and shows that the construction actually gives an uncountable set of T-numbers. It will be used in Section 7.5.

Let $j \geq 1$ be an integer. For the proof of step (B_j), we first establish that if g_j is large enough and if x lies in I_j, then we have

$$|x - \alpha_n| \geq 2\lambda_n H(\alpha_n)^{-\chi_n} \tag{7.15}$$

for any algebraic number $\alpha_n \neq \xi_j$ of degree $n \leq j$ such that

$$(\lambda_n g_j^{\nu_j})^{1/\chi_n} \leq H(\alpha_n) \leq g_j^{\nu_j/(\chi_n - n - 1 - \varepsilon_n)}. \tag{7.16}$$

Let, then, $\alpha_n \neq \xi_j$ be an algebraic number satisfying (7.16) and let x be in I_j, that is, such that

$$\frac{1}{2} g_j^{-\nu_j} < x - \xi_j < \frac{3}{4} g_j^{-\nu_j}. \tag{7.17}$$

If $\nu_j(n + 1 + \varepsilon_n) \leq \chi_n - n - 1 - \varepsilon_n$, then $H(\alpha_n) \leq g_j^{1/(n+1+\varepsilon_n)}$ and it follows from (VI_j), (7.16), (7.17), and the assumption $\nu_j > 1$ that

$$|x - \alpha_n| \geq |\xi_j - \alpha_n| - |\xi_j - x|$$
$$\geq g_j^{-1} - g_j^{-\nu_j} \geq 2g_j^{-\nu_j} \geq 2\lambda_n H(\alpha_n)^{-\chi_n},$$

provided that g_j is large enough.

Otherwise, we have

$$\nu_j(n + 1 + \varepsilon_n) > \chi_n - n - 1 - \varepsilon_n, \tag{7.18}$$

and, by (7.8), we get

$$|x - \alpha_n| \geq |\xi_j - \alpha_n| - |\xi_j - x|$$
$$\geq g_j^{-(n^2 + n + 1 + 2n\varepsilon_n)} H(\alpha_n)^{-n-1-\varepsilon_n} - g_j^{-\nu_j}$$
$$\geq g_j^{-(n^2 + n + 1 + 2n\varepsilon_n)} H(\alpha_n)^{-n-1-\varepsilon_n}/2. \tag{7.19}$$

To check the last inequality, we have to verify that

$$2g_j^{-\nu_j} \leq g_j^{-(n^2 + n + 1 + 2n\varepsilon_n)} H(\alpha_n)^{-n-1-\varepsilon_n}. \tag{7.20}$$

In view of (7.16), inequality (7.20) is true as soon as

$$2g_j^{\nu_j(n+1+\varepsilon_n)/(\chi_n - n - 1 - \varepsilon_n)} \leq g_j^{\nu_j} g_j^{-(n^2 + n + 1 + 2n\varepsilon_n)},$$

which, by (7.18), holds for g_j large enough when

$$\frac{n+1+\varepsilon_n}{\chi_n - n - 1 - \varepsilon_n} < 1 - (n^2 + n + 1 + 2n\varepsilon_n)\frac{n+1+\varepsilon_n}{\chi_n - n - 1 - \varepsilon_n}, \qquad (7.21)$$

and in particular when χ_n satisfies (7.3). Furthermore, we have

$$g_j^{-(n^2+n+1+2n\varepsilon_n)} H(\alpha_n)^{-n-1-\varepsilon_n} \geq 4\lambda_n H(\alpha_n)^{-\chi_n}. \qquad (7.22)$$

Indeed, by (7.16), $\lambda_n < 1$, and (7.18), we get

$$\begin{aligned}
H(\alpha_n)^{\chi_n - n - 1 - \varepsilon_n} &\geq (\lambda_n g_j^{\nu_j})^{(\chi_n - n - 1 - \varepsilon_n)/\chi_n} \\
&\geq \lambda_n g_j^{\nu_j(\chi_n - n - 1 - \varepsilon_n)/\chi_n} \\
&> \lambda_n g_j^{(\chi_n - n - 1 - \varepsilon_n)^2/(\chi_n(n+1+\varepsilon_n))} \geq 4\lambda_n g_j^{n^2+n+1+2n\varepsilon_n},
\end{aligned}$$

since we infer from (7.3) that

$$(\chi_n - n - 1 - \varepsilon_n)^2 > \chi_n(n + 1 + \varepsilon_n)(n^2 + n + 1 + 2n\varepsilon_n). \qquad (7.23)$$

Combining (7.19) and (7.22), we have checked that

$$|x - \alpha_n| \geq 2\lambda_n H(\alpha_n)^{-\chi_n}$$

holds under the assumption (7.18). By (7.18), this implies that (7.15) is true if α_n satisfies (7.16) and is not equal to ξ_j. Consequently, for g_j large enough, the complement J_j^c of J_j in I_j is contained in the union of the intervals

$$]\alpha_n - 2\lambda_n H(\alpha_n)^{-\chi_n}, \alpha_n + 2\lambda_n H(\alpha_n)^{-\chi_n}[,$$

where α_n runs over the real algebraic numbers of degree $n \leq j$ and height greater than $g_j^{\nu_j/(\chi_n - n - 1 - \varepsilon)}$. The Lebesgue measure of J_j^c is then

$$\ll \sum_{n=1}^{j} \sum_{H > g_j^{\nu_j/(\chi_n - n - 1 - \varepsilon_n)}} H^{n - \chi_n} = o(g_j^{-\nu_j}) = o(\lambda(I_j)),$$

since for any positive integers H and n there are at most $(8H)^n$ algebraic numbers of height H and degree n (see (8.5) in Chapter 8). Thus, we conclude that we can find g_j large enough such that $\lambda(J_j) \geq \lambda(I_j)/2$. This completes step (B_j) as well as the proof of Proposition 7.1.

At this point, we may summarize where the condition $\chi_n > n^3 + 2n^2 + 4n + 3$ appears. There are three steps where it is needed, namely (7.7), (7.21) and (7.23). Asymptotically, these three inequalities reduce, respectively, to

$\chi_n > (n+1)(n^2+n+2)$, $\chi_n > (n+1)(n^2+n+3)$, and $(\chi_n - n - 1)^2 > \chi_n(n+1)(n^2+n+1)$. The most restricting condition is given by (7.21), hence, our assumption on χ_n.

7.3 Completion of the proof of Theorem 7.1

The sequence $(\xi_j)_{j\geq 1}$ obtained in Proposition 7.1 is strictly increasing and bounded, thus it converges to a real number which we denote by ξ. It is easily seen that if the sequence $(m_j)_{j\geq 1}$ is chosen in such a way that all the integers $1, 2 \ldots$ appear in it infinitely many times, then ξ is a T-number. Now, we explain how to choose the sequence $(\mu_j)_{j\geq 1}$ such that $w_n(\xi)$ and $w_n^*(\xi)$ are different.

Let $(m_j)_{j\geq 1}$ be a sequence of positive integers taking infinitely many times each value $1, 2, \ldots$. For $j \geq 1$, we set $v_j = m_j(w_{m_j}^* + 1)$ and we define μ_j in $[0, 1]$ by

$$w_{m_j}^* + (m_j - 1)(1 - \mu_j)/m_j = w_{m_j}.$$

Moreover, for any integer $n \geq 1$, we set

$$\chi_n = w_n, \tag{7.24}$$

in such a way that $\chi_n > n^3 + 2n^2 + 4n + 3$. Let $\lambda_1, \lambda_2, \ldots, \xi_1, \xi_2, \ldots$ be as in Proposition 7.1 and denote by ξ the limit of the strictly increasing sequence $(\xi_j)_{j\geq 1}$.

We fix an integer $n \geq 1$. Observe that the minimal polynomial of ξ_j over $\mathbb{Z}$ is the polynomial

$$P_j(X) = (g_j X - c_j)^m - 2v_j^m,$$

which is primitive since, by condition (I_j), its leading coefficient is coprime with its constant term. Thus, recalling that $c_j \leq 2g_j$, we get that $g_j^{m_j} \leq \mathrm{H}(\xi_j) \leq (2g_j)^{m_j}$. Furthermore, for any $j \geq 1$ we have

$$\xi_j + g_j^{-v_j}/2 < \xi < \xi_j + g_j^{-v_j},$$

and the definition of v_j implies that

$$\mathrm{H}(\xi_j)^{-w_{m_j}^* - 1}/2 \leq |\xi - \xi_j| \leq 2^{v_j}\,\mathrm{H}(\xi_j)^{-w_{m_j}^* - 1}. \tag{7.25}$$

Moreover, if α_m is a real algebraic number of degree $m \leq n$ which is not equal to one of the ξ_j, then, by (IV_j) we have

$$|\xi_j - \alpha_m| \geq \lambda_m \mathrm{H}(\alpha_m)^{-\chi_m},$$

hence, as j tends to infinity,

$$|\xi - \alpha_m| \geq \lambda_m H(\alpha_m)^{-\chi_m} \geq \lambda_m H(\alpha_m)^{-w_m^* - 1}, \qquad (7.26)$$

since $\chi_m \leq w_m^* + 1$. As $m_j = n$ for infinitely many integers j, it follows from (7.25), (7.26) and from the fact that the sequence $(w_m^*)_{m \geq 1}$ is increasing that

$$w_n^*(\xi) = w_n^*.$$

It remains for us to prove that $w_n(\xi) = w_n$. This is clear for $n = 1$, thus we assume $n \geq 2$. Denote by $\xi_j = \theta_{j,1}, \theta_{j,2}, \ldots, \theta_{j,m_j}$ the roots of the polynomial $P_j(X)$. Until the end of this proof, we write $A \ll B$ when there is a positive constant $c(m_j)$, depending only on m_j, such that $|A| \leq c(m_j)|B|$, and we write $A \asymp B$ if both $A \ll B$ and $B \ll A$ hold. We have $|\xi_j - \theta_{j,k}| \asymp g_j^{\mu_j - 1}$ for $k = 2, \ldots, m_j$, and, since $|\xi_j - \theta_{j,k}| \geq 2|\xi_j - \xi|$, we get

$$|\xi - \theta_{j,k}| \asymp g_j^{\mu_j - 1} \quad \text{for } k = 2, \ldots, m_j.$$

Consequently, since $H(P_j) \asymp g_j^{m_j}$, we have

$$\begin{aligned}
|P_j(\xi)| = g_j^{m_j} \prod_{k=1}^{m_j} |\xi - \theta_{j,k}| \\
\asymp g_j^{m_j} g_j^{-m_j(w_{m_j}^* + 1)} g_j^{(m_j - 1)(\mu_j - 1)} \\
\asymp H(P_j)^{-w_{m_j}^* + ((m_j - 1)/m_j)(\mu_j - 1)} = H(P_j)^{-w_{m_j}}.
\end{aligned} \qquad (7.27)$$

Since $m_j = n$ for infinitely many j, we infer from (7.27) that $w_n(\xi) \geq w_n$. In order to show that we have equality, let $P(X)$ be an integer polynomial of degree at most n, which we write under the form

$$P(X) = aR_1(X) \ldots R_p(X) \cdot Q_1(X) \ldots Q_s(X),$$

where a is an integer and the polynomials $R_i(X)$ and $Q_j(X)$ are primitive and irreducible. We moreover assume that the $R_i(X)$ do not have a root equal to one of the ξ_ℓs, but that each $Q_j(X)$ has a root equal to some ξ_ℓ. If k denotes the degree of the polynomial $R_i(X)$, then, by Lemma A.8, it has a root θ satisfying

$$\begin{aligned}
|R_i(\xi)| \gg H(R_i)^{2-k} |\xi - \theta| \\
\gg \lambda_n H(R_i)^{-\chi_k - k + 2} = \lambda_n H(R_i)^{-w_k} \gg \lambda_n H(R_i)^{-w_n}. \quad (7.28)
\end{aligned}$$

If ℓ denotes the degree of $Q_j(X)$, then (7.27) shows that

$$|Q_j(\xi)| \asymp H(Q_j)^{-w_\ell} \geq H(Q_j)^{-w_n}.$$

Together with (7.28) and Lemma A.3, this gives

$$|P(\xi)| \gg \big(\mathrm{H}(R_1)\dots\mathrm{H}(R_p)\mathrm{H}(Q_1)\dots\mathrm{H}(Q_s)\big)^{-w_n} \gg \mathrm{H}(P)^{-w_n},$$

and we get $w_n(\xi) = w_n$, as claimed.

7.4 On the gap between w_n^* and w_n

Many ideas from the construction of T-numbers have been used by Bugeaud [128] to prove that, for any integer $n \geq 3$, the set of values taken by the function $w_n - w_n^*$ includes the interval $[0, n/4]$.

THEOREM 7.2. *Let $n \geq 3$ be an integer. Let w_n and w_n^* be real numbers such that*

$$w_n^* \leq w_n \leq w_n^* + n/4, \quad w_n > 2n^3 + 2n^2 + 3n - 1.$$

Then there exists real numbers ξ such that

$$w_n^*(\xi) = w_n^* \quad and \quad w_n(\xi) = w_n.$$

Exercise 7.2 is devoted to the proof of Theorem 7.2. The main idea is how-ever easy to explain. As pointed out after the statement of Proposition 7.1, the algebraic numbers $2^{1/m_j}$ occurring there have no particular importance for ob-taining a T-number, and they could be replaced by other algebraic numbers of the same degree. Since we are only interested in approximation by algebraic numbers of degree at most n, we use a suitable version of Proposition 7.1 where $m_j = n$ and $\xi_j = (c_j + \gamma_j)/g_j$ for every $j \geq 1$. Here, the c_js and the g_js are positive integers and the γ_js are real algebraic numbers of degree n. The differences $|\xi - \xi_j|$ are precisely controlled and $w_n^*(\xi)$ satisfies

$$w_n^*(\xi) = \lim_{j\to+\infty} \frac{-\log|\xi - \xi_j|}{\log \mathrm{H}(\xi_j)} - 1.$$

The idea consists in choosing the γ_js such that the algebraic numbers ξ_j have a complex conjugate ξ_j^σ very close to ξ_j in terms of $\mathrm{H}(\xi_j)$. It then follows that $|\xi - \xi_j^\sigma|$ is very small, and that $|P_j(\xi)|$ is much smaller than $\mathrm{H}(\xi_j)^{-w_n^*(\xi)}$, where $P_j(X)$ denotes the minimal polynomial of ξ_j over $\mathbb{Z}$. Consequently, $w_n(\xi)$ is larger than $w_n^*(\xi)$. In [128], Bugeaud took γ_j to be roots of the integer polynomials $Q_{a,n}(X) := X^n - 2(aX - 1)^2$, which were first mentioned by Mignotte [421]. It is readily verified that for any integers $n \geq 3$ and $a \geq 10$, the polynomial $Q_{a,n}(X)$ is irreducible and has two *conjugate* roots distant by at most $2^n \mathrm{H}(Q_{a,n})^{-n/4-1/2}$: this estimate yields the upper bound $n/4$ for the range of values of $w_n - w_n^*$ in Theorem 7.2.

Furthermore, by considering the family of integer polynomials $P_{a,d}(X) := (X^d - aX + 1)^2 - 2X^{2d-2}(aX - 1)^2$ mentioned below Theorem A.3, it is possible to improve Theorem 7.2. For any integer $d \geq 3$, it follows either from the results of Müller [433] or of Laurent and Poulakis [355] that the polynomial $P_{a,d}(X)$ is irreducible for any integer a sufficiently large, thus it has two *conjugate* roots distant by at most $4a^{-2d}$. Taking for γ_j one of these roots, it follows from the proof of [128] that, for any even integer $n \geq 6$, the set of values taken by the function $w_n - w_n^*$ contains the interval $[0, n/2[$, see [133].

Unlike Theorem 7.1, Theorem 7.2 is effective. Indeed, it does not rest on Theorem 2.7, but on Liouville's inequality. As explained in Section 8 of [128], the construction of Theorem 7.2 does not ensure the existence of T-numbers ξ with $w_n(\xi) - w_n^*(\xi) > (n-1)/n$ for some positive integer n.

7.5 Hausdorff dimension and Hausdorff measure

It has been proved (Corollary 5.2) that the Hausdorff dimension of the set of T-numbers is zero. However, thanks to the multiple possible choices for the integers c_j occurring in the proof of Proposition 7.1, R. C. Baker [47] showed that this set is not 'too small'. Namely, its $\mathcal{H}^f$-measure is infinite for any dimension function f increasing in a neighbourhood of the origin faster than any power function, that is, for any dimension function f such that

$$f \prec (x \mapsto x^\delta) \quad \text{for any } \delta > 0. \tag{7.29}$$

It is possible to rework R. C. Baker's argument to prove the existence of S-numbers ξ for which $w_n(\xi)$ and $w_n^*(\xi)$ are different for finitely many integers n.

THEOREM 7.3. *Let* $n_1 < \ldots < n_k$ *be positive integers. Let* $w_1^* \leq \ldots \leq w_k^*$ *and* $w_1 \leq \ldots \leq w_k$ *be real numbers satisfying*

$$w_i^* \leq w_i \leq w_i^* + n_i/4, \quad w_i > 2n_i^3 + 2n_i^2 + 3n_i - 1 \quad (1 \leq i \leq k).$$

Then there exists a real S-number ξ *such that*

$$w_{n_i}^*(\xi) = w_i^* \quad \text{and} \quad w_{n_i}(\xi) = w_i \quad (1 \leq i \leq k). \tag{7.30}$$

As explained in [128], the proof of Theorem 7.2 can easily be adapted to assert the existence of real numbers ξ satisfying (7.30). To prove that some of them are S-numbers, it is sufficient, by Corollary 5.2, to show that they form a set of positive Hausdorff dimension. We use a version of Proposition 7.1 in which the sequence $(m_j)_{j \geq 1}$ takes only the values $n_1, \ldots, n_k$, and each of these appears infinitely often. Further, proceeding as in Section 7.4, we take $v_j = n_i(w_i^* + 1)$ if m_j is equal to n_i. Denote by v the maximum of the v_j and let ε be a real number with $0 < \varepsilon < 1/2$. We can choose the sequence $(g_j)_{j \geq 1}$ in such a way that, besides the requirements of Proposition 7.1, we have

$$g_j^\varepsilon > 32n_k \, g_{j-1}^v,$$

for any integer $j \geq 2$. Then, (7.14) implies that, at each step, there are at least $g_j^{1-\varepsilon}$ suitable choices for the integer c_j. We thus have a Cantor-type construction (see Section 5 from [47]), at the j-th stage of which there are at least $g_j^{1-\varepsilon}$ intervals of length $g_j^{-v}/4$ and distant by at least $1/g_j$. It follows from Proposition 5.2 that the Hausdorff dimension of the residual set obtained is positive and even at least equal to $(1 - \varepsilon)/v$. Consequently, this set contains S-numbers, and even S-numbers of type less than $v + 2v\varepsilon$, by Theorem 5.8.

7.6 On U-numbers

As mentioned in Chapter 3 and in Section 7.1, there are several ways to define subclasses of the set of U-numbers. We consider in the present book the most classical one, introduced by LeVeque [361] and recalled below.

DEFINITION 7.1. *Let m be a positive integer. The U_m-numbers are precisely the U-numbers of type m, that is, the real numbers ξ such that $w_m(\xi) = +\infty$ and $w_n(\xi) < +\infty$ for any integer n with $1 \leq n < m$.*

Since for any positive integer n both functions w_n and w_n^* take simultaneously finite values or are simultaneously infinite, we may replace the functions w_m and w_n in Definition 7.1 by w_m^* and w_n^*.

The first examples of U_2-numbers are due to Maillet [403] (see also Perron [454], pp. 143–148). In the first paper entirely devoted to U-numbers, LeVeque [361] proved, among other deeper results, that none of the sets of U_m-numbers is empty.

THEOREM 7.4. *For any positive integer m, there exist uncountably many real U_m-numbers.*

PROOF. For any positive integer j, let a_j be an element of $\{2, 4\}$. We prove that, for any integer $m \geq 1$, the positive real m-th root ξ of $(3 + \sum_{j \geq 1} a_j \, 10^{-j!})/4$ is a U_m-number. For $k \geq 1$, set

$$p_k = 10^{k!}\left(3 + \sum_{j=1}^{k} a_j \, 10^{-j!}\right), \quad q_k = 4 \cdot 10^{k!}, \quad \text{and} \quad \alpha_k = \left(\frac{p_k}{q_k}\right)^{1/m}.$$

Then, we have $H(\alpha_k) = 4 \cdot 10^{k!}$ and

$$|\xi - \alpha_k| \le |\xi^m - \alpha_k^m| \le 2 \cdot 10^{-(k+1)!} \le 2^{2k+3} H(\alpha_k)^{-k-1}. \quad (7.31)$$

Consequently, ξ is a U-number, and its type does not exceed m.

Let β be a non-zero real algebraic number of degree n strictly less than m and of height greater than $H(\alpha_1)$. Then, there exists a positive integer k such that

$$H(\alpha_k) \le H(\beta)^{2m} \le H(\alpha_{k+1}) \le H(\alpha_k)^{k+1}. \quad (7.32)$$

It follows from Corollary A.2 that

$$|\beta - \alpha_k| \ge 2 (m+1)^{-5m/2} H(\beta)^{-m} H(\alpha_k)^{-n}$$
$$\ge 2 (m+1)^{-5m/2} H(\alpha_k)^{-m-(k+1)/2}. \quad (7.33)$$

By taking $H(\beta)$ large enough, the index k satisfies

$$H(\alpha_k)^{-m+(k+1)/2} > 2^{2k+3} (m+1)^{5m/2},$$

and it follows from (7.31) and (7.33) that $|\beta - \alpha_k| > 2|\xi - \alpha_k|$. Thus, except for finitely many algebraic numbers β of degree strictly less than m, we have

$$|\xi - \beta| \ge |\beta - \alpha_k| - |\xi - \alpha_k| > |\beta - \alpha_k|/2$$
$$\ge (m+1)^{-5m/2} H(\beta)^{-m} H(\alpha_k)^{-n} \ge (m+1)^{-5m/2} H(\beta)^{-m-2mn},$$

by (7.32). We conclude that

$$w_n^*(\xi) \le 2mn + m - 1 \quad (7.34)$$

for $n = 1, \ldots, m-1$, hence, ξ is not a U-number of type strictly less than m.

Using the idea explained in Section 7.5, R. C. Baker [47] proved that, for any positive integer m, the set of U_m-numbers has infinite $\mathcal{H}^f$-measure for any dimension function f satisfying (7.29). In particular, the $\mathcal{H}^f$-measure of the set of Liouville numbers is infinite if f satisfies (7.29), and is zero if there exists a positive real number δ such that $(x \mapsto x^\delta) \prec f$ (see also Exercise 6.5).

In Theorem 7.4, we constructed explicitly U_m-numbers ξ and gave effective upper bounds for $w_n^*(\xi)$ for any positive integer n smaller than m. These are somehow unsatisfactory, since they are quite large (a slight improvement of (7.34) has been obtained by Güting ([269], Satz 10) who, essentially, replaced $2mn$ by mn). Alniaçik, Avci, and Bugeaud [18] improved upon Theorem 7.4 in the sense that they established the existence of U_m-numbers ξ with sharper upper bounds for $w_n^*(\xi)$, where $n = 1, \ldots, m-1$. Like Theorem 7.4, Theorem 7.5 below is of an effective nature, since it ultimately rests on

Theorem 2.11. The underlying idea is to construct ξ as the limit of a rapidly converging sequence of real algebraic numbers of degree m, using the fact that these numbers can be quite well approximated by algebraic integers of same degree.

THEOREM 7.5. *Let $m \geq 2$ be an integer. There are uncountably many U_m-numbers ξ with*

$$w_n^*(\xi) \leq m + n - 1, \qquad for \; n = 1, \ldots, m - 1.$$

The proof of Theorem 7.5 relies on the following consequence of Theorem 2.11.

LEMMA 7.1. *Let $n \geq 2$ be an integer and ξ be a real algebraic number of degree n. There exist an effectively computable constant $\kappa(\xi)$ and infinitely many algebraic integers α of degree n such that*

$$0 < |\xi - \alpha| < \kappa(\xi) \, \mathrm{H}(\alpha)^{-n}.$$

Moreover, we use repeatedly an easy consequence of the triangle inequality.

LEMMA 7.2. *Let $(\alpha_1, \alpha_2, \beta)$ be a triple of real numbers. Let H, N, and c be positive real numbers such that*

$$0 < |\alpha_1 - \alpha_2| < N^{-1} \quad and \quad |\alpha_1 - \beta| \geq c^{-1} H^{-1}.$$

Then, for any positive real number δ, we have

$$|\alpha_2 - \beta| > (c + \delta)^{-1} H^{-1} \quad as \; soon \; as \quad H \leq c^{-1}\delta(c + \delta)^{-1}N.$$

PROOF. Assuming that $H \leq c^{-1}\delta(c + \delta)^{-1}N$, we have

$$\begin{aligned}|\alpha_2 - \beta| &\geq |\alpha_1 - \beta| - |\alpha_1 - \alpha_2| \\ &> c^{-1} H^{-1} - N^{-1} \\ &\geq H^{-1}\left(c^{-1} - c^{-1}\delta(c + \delta)^{-1}\right) = H^{-1}(c + \delta)^{-1},\end{aligned}$$

as claimed.

PROOF OF THEOREM 7.5. Let $m \geq 2$ be an integer. Thanks to repeated use of Liouville's inequality, we construct by induction a converging sequence of real algebraic integers of degree m, whose limit is a U_m-number with the given property. Throughout the proof, for $n = 1, \ldots, m - 1$, we denote by β_n a real algebraic number of degree n.

Let ξ_1 be a real algebraic number of degree m. By Corollary A.2, we have

$$|\xi_1 - \beta_n| > c_1^{-1}\mathrm{H}(\xi_1)^{-n} \, \mathrm{H}(\beta_n)^{-m} \geq c_1^{-1}\mathrm{H}(\xi_1)^{-m} \, \mathrm{H}(\beta_n)^{-m}, \qquad (7.35)$$

with $c_1 = (m + 1)^{2m}$. Furthermore, Lemma 7.1 implies that there is an effectively computable constant $\kappa(\xi_1) > 1$ and infinitely many real algebraic integers ξ_2 of degree m with

$$0 < |\xi_1 - \xi_2| < \kappa(\xi_1) \, H(\xi_2)^{-m}. \tag{7.36}$$

We infer from (7.35), (7.36), and Lemma 7.2 applied to the triple (ξ_1, ξ_2, β_n) with $H = H(\beta_n)^m$, $N = \kappa(\xi_1)^{-1} H(\xi_2)^m$, $c = c_1 H(\xi_1)^m$, and $\delta = 1/2$ that

$$|\xi_2 - \beta_n| > (c_2 + 1/2)^{-1} H(\beta_n)^{-m}, \tag{7.37}$$

with $c_2 = c_1 H(\xi_1)^m$, provided that

$$H(\beta_n) \le (c_2(2c_2 + 1))^{-1/m} \kappa(\xi_1)^{-1/m} H(\xi_2) := A.$$

Furthermore, we deduce from Corollary A.2 that

$$|\xi_2 - \beta_n| > c_1^{-1} H(\xi_2)^{-n} H(\beta_n)^{-m}, \tag{7.38}$$

and, if $H(\beta_n) > A$, we infer from (7.38) that

$$\begin{aligned}|\xi_2 - \beta_n| &> c_1^{-1} \left(c_2(2c_2 + 1)\kappa(\xi_1)\right)^{-n/m} H(\beta_n)^{-m-n} \\ &> H(\beta_n)^{-m-n} \left(\log 3H(\beta_n)\right)^{-1}\end{aligned} \tag{7.39}$$

holds if $H(\beta_n)$ exceeds some effectively computable constant c_3 depending only on ξ_1. We choose ξ_2 such that we have $H(\xi_2) > \kappa(\xi_1)H(\xi_1)$ and $A > c_3$. Then, by (7.37) and (7.39), we see that

$$|\xi_2 - \beta_n| > (c_2 + 1/2)^{-1} H(\beta_n)^{-m-n} \left(\log 3H(\beta_n)\right)^{-1}$$

holds for any algebraic number β_n.

Let $k \ge 2$ be an integer. Assume that there are real algebraic integers $\xi_2, \ldots, \xi_k$ of degree m and real numbers $\kappa(\xi_1), \ldots, \kappa(\xi_{k-1})$ such that

$$|\xi_{i+1} - \beta_n| > \left(c_2 + \frac{1}{2} + \ldots + \frac{1}{2^i}\right)^{-1} H(\beta_n)^{-m-n} \left(\log 3H(\beta_n)\right)^{-1},$$

for any β_n,

$$0 < |\xi_i - \xi_{i+1}| < \kappa(\xi_i) \, H(\xi_{i+1})^{-m} \quad \text{and} \quad H(\xi_{i+1}) > \kappa(\xi_i) \, H(\xi_i)^i$$

for $i = 1, \ldots, k - 1$.

By Lemma 7.1, there exists a real number $\kappa(\xi_k) > 1$ and a real algebraic number ξ_{k+1} of degree m satisfying

$$H(\xi_{k+1}) > \kappa(\xi_k) \, H(\xi_k)^k \quad \text{and} \quad |\xi_k - \xi_{k+1}| < \kappa(\xi_k) \, H(\xi_{k+1})^{-m}. \tag{7.40}$$

By assumption, we have

$$|\xi_k - \beta_n| \geq c_4^{-1} H(\beta_n)^{-m-n} \left(\log 3H(\beta_n)\right)^{-1},$$

with $c_4 = c_2 + 2^{-1} + \ldots + 2^{-k+1}$, and we infer from Lemma 7.2 applied to the triple $(\xi_k, \xi_{k+1}, \beta_n)$ with $\delta = 2^{-k}$ that

$$|\xi_{k+1} - \beta_n| \geq (c_4 + 2^{-k})^{-1} H(\beta_n)^{-m-n} \left(\log 3H(\beta_n)\right)^{-1} \qquad (7.41)$$

holds provided that

$$H(\beta_n)^{m+n} \left(\log 3H(\beta_n)\right) \leq c_4^{-1} (2^k c_4 + 1)^{-1} \kappa(\xi_k)^{-1} H(\xi_{k+1})^m. \qquad (7.42)$$

By Corollary A.2, we have

$$|\xi_k - \beta_n| \geq c_1^{-1} H(\xi_k)^{-n} H(\beta_n)^{-m},$$

and we apply Lemma 7.2 to the triple $(\xi_k, \xi_{k+1}, \beta_n)$ with the parameters $H = H(\beta_n)^m$, $N = \kappa(\xi_k)^{-1} H(\xi_{k+1})^m$, $c = c_1 H(\xi_k)^n$ and $\delta = H(\xi_k)^n$ to get

$$|\xi_{k+1} - \beta_n| > (c_1 + 1)^{-1} H(\xi_k)^{-n} H(\beta_n)^{-m}$$

provided that

$$H(\beta_n) \leq \left(c_1^{-1} (c_1 + 1)^{-1} \kappa(\xi_k)^{-1} H(\xi_k)^{-n}\right)^{1/m} H(\xi_{k+1}) := B. \qquad (7.43)$$

Hence, we have

$$|\xi_{k+1} - \beta_n| > H(\beta_n)^{-m-1} \qquad (7.44)$$

as soon as β_n satisfies

$$(c_1 + 1)H(\xi_k)^n \leq H(\beta_n) \leq B. \qquad (7.45)$$

By choosing ξ_{k+1} with $H(\xi_{k+1})$ large enough in terms of ξ_1 and ξ_k, (7.42) is satisfied for all algebraic numbers β_n with $H(\beta_n) < (c_1 + 1)H(\xi_k)^n$. Thus, by (7.41), (7.44), and (7.45), we have

$$|\xi_{k+1} - \beta_n| > (c_4 + 2^{-k})^{-1} H(\beta_n)^{-m-n} \left(\log 3H(\beta_n)\right)^{-1} \qquad (7.46)$$

when $H(\beta_n) \leq B$.

Finally, assuming that $H(\beta_n) > B$, we use

$$|\xi_{k+1} - \beta_n| > c_1^{-1} H(\xi_{k+1})^{-n} H(\beta_n)^{-m},$$

given by Corollary A.2, to get from (7.43) that, for $H(\xi_{k+1})$ large enough in terms of m, ξ_1, and ξ_k, we have

$$|\xi_{k+1} - \beta_n| > c_1^{-1} \left(c_1(c_1 + 1)\kappa(\xi_k)H(\xi_k)^n\right)^{-n/m} H(\beta_n)^{-m-n}$$
$$> H(\beta_n)^{-m-n} (\log B)^{-1} > H(\beta_n)^{-m-n} \left(\log 3H(\beta_n)\right)^{-1}.$$
$$(7.47)$$

The desired conclusion follows from (7.46) and (7.47): for any algebraic number β_n, we have

$$|\xi_{k+1} - \beta_n| > \left(c_2 + \frac{1}{2} + \ldots + \frac{1}{2^k}\right)^{-1} H(\beta_n)^{-m-n} \left(\log 3H(\beta_n)\right)^{-1}.$$
$$(7.48)$$

We deduce from (7.40) that, for any integers i, j with $2 \leq i < j$, we have

$$|\xi_j - \xi_i| \leq 2 H(\xi_i)^{-i}. \qquad (7.49)$$

Thus, the sequence $(\xi_j)_{j \geq 1}$ is a Cauchy sequence, and we denote its limit by ξ. Letting j tend to infinity in (7.49), we get $|\xi - \xi_i| \leq 2 H(\xi_i)^{-i}$, whence $w_m^*(\xi) = +\infty$. Moreover, by (7.48), we get that $w_n^*(\xi) \leq m + n - 1$, for any $n = 1, \ldots, m-1$, as claimed. Finally, we observe that, at each step, we have in fact infinitely many choices for the algebraic integer ξ_k, thus we can construct uncountably many real numbers satisfying the required properties.

Unfortunately, Theorem 7.5 does not assert the existence of U_m-numbers ξ with a prescribed value of $w_n(\xi)$ or $w_n^*(\xi)$ for some integer n with $1 \leq n < m$. However, using Lagrange's Theorem 1.16 (asserting that the continued fraction expansion of a real number is ultimately periodic if, and only if, this number is a quadratic irrationality), we can construct real U_2-numbers ξ with any prescribed value for $w_1(\xi)$ (see Exercise 7.3). Moreover, again with the use of continued fractions, Alniaçik [9] gave effective constructions of U-numbers which are not well approximable by rational numbers.

THEOREM 7.6. *For any integer* $m \geq 2$, *there exist* U_m-*numbers* ξ *with* $w_1(\xi) = 1$.

Exercise 7.4 is devoted to a proof of Theorem 7.6. Actually, the U_m-numbers ξ constructed by Alniaçik in [9] satisfy $w_1(\xi) = 1$ and $w_n^*(\xi) \leq c_5 m^6$ for any integer n with $2 \leq n < m$ and some absolute constant c_5.

7.7 A method of Güting

The main idea behind LeVeque work [361] mentioned in Section 7.6 can be phrased as follows: *Any real number with sufficiently many very good algebraic approximations of given degree cannot have infinitely many very, very good algebraic approximants of smaller or comparable degree.*

This can be seen by applying Liouville's inequality under the form given by Corollary A.2. Indeed, assume that the algebraic numbers α_1 and α_2 are very close to the transcendental real number ξ. Suppose that the algebraic number β satisfies

$$H(\alpha_1) < H(\beta) < H(\alpha_2).$$

For $i = 1, 2$ we have

$$|\xi - \beta| \geq |\alpha_i - \beta| - |\alpha_i - \xi|,$$

and Corollary A.2 gives a lower bound for $|\alpha_i - \beta|$. Thus, if $|\alpha_i - \xi|$ is sufficiently small, say smaller than $|\alpha_i - \beta|/2$, then $|\xi - \beta|$ must be greater than $|\alpha_i - \beta|/2$, thus greater than $|\alpha_i - \xi|$; consequently, β cannot be a very good approximation of ξ. Of course, the above argument is rather rough, since we have omitted any reference to the heights and to the degrees of the algebraic numbers involved. However, it lies at heart of Güting's work [269], where the existence of real numbers ξ with prescribed values for $w_n(\xi)$ for some integers n is established in a constructive way.

THEOREM 7.7. *Let $n \geq 1$ and $k \geq 0$ be integers. Let d be a real number with $(d + 1 - n)(d + 1 - n - k) > n(n + k)(d + 1)$. Then, there exist real numbers ξ with*

$$w_n(\xi) = w_n^*(\xi) = \ldots = w_{n+k}(\xi) = w_{n+k}^*(\xi) = d.$$

In particular, the functions w_n and w_n^* take any value $d > (2n + 1 + \sqrt{4n^2 + 1})/2$.

PROOF. Let d, n, and k be as in the statement of the theorem. Throughout the proof, the numerical constants implied by $\ll$ and $\gg$ depend, at most, on n, k, and d. Let $(n_i)_{i \geq 1}$ be a strictly increasing sequence of positive integers such that n_{i+1}/n_i tends to $d + 1$ and $\gcd(n_i, n) = 1$ for any $i \geq 1$. Define the positive real number ξ by

$$\xi^n = \sum_{j \geq 1} 2^{-n_j}.$$

Let ε be a real number with $0 < \varepsilon < 1$. Let $i_0 \geq 1$ be such that $d + 1 - \varepsilon <$

$n_{i+1}/n_i < d + 1 + \varepsilon$ holds for any integer $i \geq i_0$. For $i \geq i_0$, the polynomial

$$P_i(X) := 2^{n_i} X^n - 2^{n_i} \sum_{j=1}^{i} 2^{-n_j}$$

satisfies $H(P_i) = 2^{n_i}$ and, since n_i and n are coprime, a result by Dumas [214] (see also [585], page 76) asserts that $P_i(X)$ is irreducible. It follows from

$$P_i(\xi) = \sum_{j>i} 2^{n_i - n_j} = \sum_{j>i} H(P_i)^{1 - n_j/n_i}, \quad P_i'(\xi) = n\xi^{n-1} H(P_i),$$

and Lemma A.5 that $P_i(X)$ has a root α_i such that $H(\alpha_i) = H(P_i)$ and

$$H(\alpha_i)^{-d-1-\varepsilon} \ll |\xi - \alpha_i| \ll H(\alpha_i)^{-d-1+\varepsilon}. \tag{7.50}$$

It is immediate that, for $i \geq i_0$, we have

$$H(\alpha_i) \leq H(\alpha_{i+1}) \leq H(\alpha_i)^{d+1+\varepsilon}. \tag{7.51}$$

Let α be a real algebraic number of degree $n + k$. There exists an integer i with

$$H(\alpha_i) \leq H(\alpha) < H(\alpha_{i+1}). \tag{7.52}$$

We may assume that $H(\alpha)$ is sufficiently large in order to ensure that $i \geq i_0$. Thus, (7.50) and (7.51) are satisfied. We distinguish two cases and introduce a real number $u > n$, which will be specified later on.

First, we assume that $\alpha \neq \alpha_i$ and

$$H(\alpha)^n \leq H(\alpha_i)^u. \tag{7.53}$$

Corollary A.2 asserts that $|\alpha_i - \alpha| \gg H(\alpha)^{-n} H(\alpha_i)^{-n-k}$, whence, by (7.53), we get

$$|\alpha_i - \alpha| \gg H(\alpha_i)^{-n-k-u}. \tag{7.54}$$

By (7.50), (7.52), (7.54), and the triangle inequality $|\xi - \alpha| \geq |\alpha_i - \alpha| - |\xi - \alpha_i|$, we have $|\xi - \alpha| \geq H(\alpha_i)^{-d-1+\varepsilon}$, and thus $|\xi - \alpha| \geq H(\alpha)^{-d-1+\varepsilon}$, as soon as

$$u < d + 1 - n - k - \varepsilon \tag{7.55}$$

and $H(\alpha)$ is large enough.

We assume now that $\alpha \neq \alpha_{i+1}$ and $H(\alpha)^n > H(\alpha_i)^u$. We then get $|\alpha_{i+1} - \alpha| \gg H(\alpha)^{-n} H(\alpha_{i+1})^{-n-k}$ by Corollary A.2, and we infer from (7.50), $H(\alpha) < H(\alpha_{i+1})$, and the triangle inequality $|\xi - \alpha| \geq |\alpha_{i+1} - \alpha| - |\xi - \alpha_{i+1}|$ that $|\xi - \alpha| \geq H(\alpha)^{-d-1+\varepsilon}$ holds if $H(\alpha)^{-n} H(\alpha_{i+1})^{-n-k} \gg H(\alpha)^{-d-1+\varepsilon}$.

This condition is fulfilled as soon as

$$H(\alpha_{i+1})^{n+k} \ll H(\alpha_i)^{(d+1-n-\varepsilon)u/n},$$

thus, by (7.51), as soon as

$$u > \frac{n(n+k)(d+1+\varepsilon)}{d+1-n-\varepsilon}. \tag{7.56}$$

We can select a real number u such that (7.55) and (7.56) hold simultaneously if ε satisfies

$$\frac{n(n+k)(d+1+\varepsilon)}{d+1-n-\varepsilon} < d+1-n-k-\varepsilon.$$

Together with (7.50), by letting ε tend to 0, we obtain that $w_{n+k}^*(\xi) = d$ as soon as

$$(d+1-n-k)(d+1-n) > n(n+k)(d+1), \tag{7.57}$$

as claimed. The last statement of the theorem follows by applying the first assertion to the pair $(n, k) = (1, n-1)$. Indeed, condition (7.57) then becomes $d(d-n+1) > n(d+1)$, that is, precisely $d > (2n-1+\sqrt{4n^2+1})/2$.

A similar argument yields the result for the function w_n, as proved by Güting [269]. Instead of taking an arbitrary algebraic number α with (7.52), we take a polynomial $P(X)$ and we consider the index i for which $H(P_i) \leq H(P) < H(P_{i+1})$. We then apply Theorem A.1 and Rolle's Theorem to get that $w_{n+k}(\xi) = d$.

Güting ([269], Satz 7 and Satz 8) gave analogous, but slightly weaker, statements for roots of real numbers defined by their continued fraction expansions. In that case, there is a new difficulty, since the polynomials playing the roles of $P_i(X)$ may not be irreducible.

It follows from a general result of Nishioka [438, 439] that, for any positive integer d, the number $\sum_{j \geq 1} 2^{-(d+1)^j}$ is an S-number.

7.8 Brief summary of the results towards the Main Problem

In Chapters 5, 6 and in Sections 7.1 to 7.7 of the present Chapter, we have developed different methods yielding to results towards the Main Problem introduced in Chapter 3.

MAIN PROBLEM. *Let $(w_n)_{n \geq 1}$ and $(w_n^*)_{n \geq 1}$ be two non-decreasing sequences in $[1, +\infty]$ such that*

$$n \leq w_n^* \leq w_n \leq w_n^* + n - 1, \quad \text{for any } n \geq 1.$$

Then there exists a real transcendental number ξ such that

$$w_n(\xi) = w_n \quad and \quad w_n^*(\xi) = w_n^* \quad for\ any\ n \geq 1.$$

We first summarize the results concerning a single function w_n or w_n^*. As emphasized in Chapter 1, the function w_1 (which is equal to w_1^*) is well understood thanks to the theory of continued fractions. Further, Theorem 7.7 asserts, in an explicit way, that w_n and w_n^* take any value w with $w > (2n + 1 + \sqrt{4n^2 + 1})/2$. In the range $n \leq w \leq (2n + 1 + \sqrt{4n^2 + 1})/2$, using Hausdorff dimension theory, it is proved in Chapter 5 that there exist real numbers ξ and ξ^* with $w_n(\xi) = w$ and $w_n^*(\xi^*) = w$. Moreover, by means of a Cantor-type construction based on the effective, explicit result of Beresnevich [60] (see below Proposition 6.1), it is possible to construct real numbers ξ with prescribed values for $w_2^*(\xi)$.

The Main Problem is partially solved when the sequence $(w_n^*)_{n\geq1}$ increases sufficiently rapidly, that is, faster than n^3+2n^2+4n+3 (see Theorem 7.1). This result follows ultimately from Schmidt's Subspace Theorem, hence, it does not yield explicit examples of transcendental numbers ξ with the required property. Moreover, this approach is up to now the only one allowing us to confirm the existence of real numbers ξ with $w_n(\xi) \neq w_n^*(\xi)$ for some integer $n \geq 2$. The real numbers obtained in this way are limits of sequences of algebraic numbers which converge very rapidly. A similar idea is also used in Section 7.6, but the construction is simpler.

For given integers $n < n'$ and real numbers $n \leq w_n \leq w_{n'}$ with $(w_n + 1)(n' + 1) = (w_{n'} + 1)(n + 1)$, we can use the Hausdorff dimension theory to ensure the existence of real numbers ξ with $w_n^*(\xi) = w_n$ and $w_{n'}^*(\xi) = w_{n'}$. However, this method is ineffective.

Alternatively, fully explicit examples of real numbers ξ with (very) specific values of $w_{n_1}(\xi), \ldots, w_{n_k}(\xi)$ for some integers $n_1 < \ldots < n_k$ have been obtained as n-th roots of suitable convergent series, see Section 7.7.

7.9 Exercises

EXERCISE 7.1. Modify (very) slightly the proof of Theorem 7.1 to show that if $0 < w_1 \leq w_2 \leq \ldots$ is a sequence in $[1, +\infty]$ such that $w_n > n^3+2n^2+3n+1$ for any $n \geq 1$, then there exists a real number ξ such that $w_n^*(\xi) = w_n$ and $w_n(\xi) = w_n$ for any $n \geq 1$.

Hint. Observe that the exponent of g_j in (7.8) is replaced by $-(n^2 + n + 2n\varepsilon_n)$ since we have $g_j = v_j$. Consequently, (7.7), (7.21), and (7.23) have to be modified accordingly. Furthermore, (7.24) can be replaced by $\chi_n = w_n + 1$.

EXERCISE 7.2. Prove Theorems 7.2 and 7.3.

Hint. For δ in $[(n-1)/n, n/4]$, use for γ_j a root close to $[g_j^\mu]^{-1}$ of the polynomial $X^n - 2([g_j^\mu]X - 1)^2$, where $\mu = 2(n\delta - n + 1)/(n-2)$. To get Theorem 7.2, argue as in the proof of Theorem 7.1, replacing, however, the use of Theorem 2.7 by that of Corollary A.2 and (7.24) by $\chi_n = w_n - n + 2$. With a slight modification, prove Theorem 7.3.

EXERCISE 7.3. Let $w \geq 1$ be a real number. Apply Theorem 1.16 to prove that there exist uncountably many U_2-numbers ξ with $w_1(\xi) = w$.

EXERCISE 7.4. Proof of Theorem 7.6.

Let $m \geq 2$ be an integer and let $\alpha := [0; a_1, a_2, a_3, \ldots]$ be a real algebraic number of degree m. For any $j \geq 1$, denote by q_j the denominator of the j-th convergent of α. Let $(r_k)_{k \geq 1}$ be an increasing sequence of positive integers such that the ratio $(\log q_{r_{k+1}})/(\log q_{r_k})$ tends to infinity with k. For any $j \geq 1$, set $b_j = a_j$ if j does not belong to the sequence $(r_k)_{k \geq 1}$ and $b_j = a_j + 1$ otherwise. Our aim is to prove that the number $\xi := [0; b_1, b_2, b_3, \ldots]$ is a U_m-number with $w_1(\xi) = 1$.

First, use Roth's Theorem 2.1 to show that $w_1(\xi) = 1$. Next, for any integer $k \geq 1$, set $\alpha_k := [0; b_1, b_2, \ldots, b_{r_k}, a_{r_k+1}, a_{r_k+2}, \ldots]$, which is an algebraic number of degree m. Show that $-(\log |\xi - \alpha_k|)/(\log H(\alpha_k))$ tends to infinity with k. It remains for us to prove that ξ is not a U_n-number for some integer $n < m$. Let β be a real algebraic number of degree at most $m-1$ and with large height. Let k be the integer such that $q_{r_k} \leq H(\beta) < q_{r_{k+1}}$. Bound $|\xi - \beta|$ from below in terms of $H(\beta)$ by using the inequality $|\xi - \beta| \geq |\beta - \alpha_\ell| - |\xi - \alpha_\ell|$ with $\ell = k$ or $k+1$ according as $H(\beta) \leq q_{r_{k+1}}^{1/m^3}$ or not. Conclude.

7.10 Notes

• Güting [269] gave a sufficient condition for a real number to be a T-number.

• Alniaçik [11] used continued fractions to construct real numbers with specific properties. He claimed that, for any positive real number ε, there exist T-numbers ξ with $w_1(\xi) \leq 1 + \varepsilon$. However, it seems that his work contains serious gaps (not only because he did not use the correct definition of T^*-numbers). His idea was to construct ξ as a limit of algebraic numbers $\xi_\ell = [a_0; a_1, \ldots, a_{n_\ell}, \alpha_\ell]$, where the a_is are integers and the α_ℓs are algebraic numbers. Compared with Schmidt who considered a limit of numbers of the shape $\alpha_\ell + p_\ell/q_\ell$, the advantage of Alniaçik's approach is that the continued fraction expansion allows us to control $w_1(\xi)$ in a satisfactory way. The remaining problem is to ensure that ξ has good, but not too good, algebraic approximants. To this end, one needs to establish a precise lower bound for the height of ξ_ℓ, essentially in terms of the denominator of the n_ℓ-th convergent of ξ_ℓ. This does not seem to be easy, and Alniaçik's assertion still remains unproved.

• LeVeque [361] introduced the notion of strong (and weak) Liouville numbers. This has been further studied and refined by Alniaçik [8, 10, 15] who defined semi-strong U_m-numbers and irregular semi-strong U_m-numbers. For any positive integers m and k, Alniaçik [15] proved that every algebraic number of degree m can be represented by the sum and the product of two U_{mk}-numbers. Further, he established [12] that every real number, except possibly Liouville numbers, can be represented as the sum of two U_2-numbers. His approach, somehow intricate, depends on the theory of continued fractions and is explicit. Shortly thereafter, Pollington [459] adapted the method used by Schmidt to confirm the existence of T-numbers to show that, for any positive integer m, every real number can be expressed as a sum of two U_m-numbers. He also briefly outlined the proof that every real number can be written as a sum of two T-numbers. Petruska [455] showed that the sum or the product of an arbitrary number of strong Liouville numbers is either rational or a Liouville number.

• Results on the distribution of the sequence $(n\alpha)$ for a U_2-number α have been established by Baxa [54].

• Burger and Struppeck [150] have investigated the statistical behaviour of the partial quotients of real U_2-numbers. They proved the existence of U_2-numbers with the property that if translated by any non-negative integer and then squared, the result is a Liouville number (see Burger [148] for refined statements). They further showed that there exists a real positive Liouville number whose square root has, in its continued fraction expansion, only the partial quotients 1 and 2, the partial quotient 2 occurring with probability 0, for some given probability measure on the set of positive integers.

• Other constructions of U_m-numbers have been obtained by Oryan [442, 444, 446, 447] and by Yilmaz [606]. Results of Zeren [610, 611] (see also Mahler [392] and Braune [117]) on values of gap power series at algebraic points have been extended by Gürses [264, 265].

• We have been concerned with the Lebesgue measure and with Hausdorff measures. However, a rather different measure for the size of sets of real numbers arises in harmonic analysis. A set E of real numbers is called an M_0 set if it carries a probability measure μ, the Fourier–Stieltjes transform of which

$$\hat{\mu}(x) = \int_E \exp(-2i\pi xy)\, \mathrm{d}\mu(y)$$

vanishes at infinity. Aside from the Riemann–Lebesgue Lemma, no purely metric property of E can ensure that E is an M_0 set. Further, there exist sets of full Lebesgue measure which do not share this property (see, for example, [304]). This is also the case for the triadic Cantor set (see, for example, [305] and [415], p. 168). Kaufman [314, 315] proved that the set $\mathcal{B}$ of badly approximable real numbers carries a measure in M_0, and so does any set $\mathcal{K}_1^*(\tau)$ for $\tau > 1$ (see also the notes at the end of Chapter 5). The former result has been used by Pollington and Velani [460] towards Littlewood's Conjecture (see Chapter 10). Moran, Pearce, and Pollington [428], following Schmidt's approach, proved that the set of T-numbers is an M_0 set and claimed that, for any positive integer m, the set of U_m-numbers is an M_0 set.

• Haseo Ki [324] studied the sets of A-, S-, T- and U-numbers from the point of view of Descriptive Set Theory. He established their possible locations in the Borel hierarchy (see, for example, the book of Kechris [316] for an introduction to that topic). He followed the main steps of Schmidt's construction and proved that the set of T-numbers is Π_3^0-hard, while the set of U-numbers is Σ_3^0-complete.

• Let w be a positive real number and $m \geq 2$ be an integer. Feldman [240] constructed real numbers ξ with $w < w_n(\xi) \leq m(m+2)w$ for any integer $n = 1, \ldots, m-1$.

• Amou [25] determined the values of $w_n(\xi)$ and $w_n^*(\xi)$ for the Champernowne number and any integer $n = 1, \ldots, 4$. He used the fact, observed by Mahler [380], that ξ has a sequence of very good rational approximants whose heights do not increase too rapidly. Lemma 1 of [25] rests on an idea of Güting [269].

8

Other classifications of real and complex numbers

In 1932, Mahler [376] introduced the first relevant classification of complex numbers into several classes. To this end, for given positive integers n and H and for any complex number ξ, he considered the minimum of the real numbers $|P(\xi)|$, where $P(X)$ runs through the (finite) set of integer polynomials of degree at most n and height at most H, which do not vanish at ξ. Then, he let first H tend to infinity, and then n. This order is arbitrary, and we may as well do the converse, or let tend to infinity some given function of the height and the degree. The former suggestion has been proposed by Sprindžuk [532] in 1962, and the latter one by Mahler [393] in 1971. Both yield new classifications of complex numbers, to which Sections 8.1 and 8.2 are devoted, respectively. In Section 8.3, we present further results on the approximation by algebraic numbers, which, to some extend, refine Wirsing's Theorem 3.4.

Unlike in the previous chapters, we approximate *complex* numbers, and not only real numbers. The main reason for doing this is that the results obtained here are not sharp enough to ensure that, when we start with a real number, the approximants we construct are also real numbers.

We warn the reader that in all the metric statements below the expression 'almost all' refers to the Lebesgue measure on the line (*resp.* on the plane) if the set under consideration is a real set (*resp.* a complex set). A similar remark applies for the Hausdorff dimension.

8.1 Sprindžuk's classification

According to Sprindžuk [532], for a complex number ξ and for positive integers n and H, we set

$$w_n(\xi, H) := \min\{|P(\xi)| : P(X) \in \mathbb{Z}[X], \mathrm{H}(P) \le H, \deg(P) \le n,$$

$$P(\xi) \ne 0\}, \qquad \tilde{w}(\xi, H) = \limsup_{n \to \infty} \frac{\log \log(1/w_n(\xi, H))}{\log n},$$

and

$$\tilde{w}(\xi) = \sup_{H \ge 1} \tilde{w}(\xi, H).$$

In case $\tilde{w}(\xi) = +\infty$ and if there exists a positive integer H with $\tilde{w}(\xi, H) = +\infty$, we denote by $H_0(\xi)$ the smallest *integer* with this property and, otherwise, we put $H_0(\xi) = +\infty$. Further, we set

$$\tilde{\mu}(\xi, H) = \limsup_{n \to \infty} \frac{-\log w_n(\xi, H)}{n^{\tilde{w}(\xi)}}$$

and

$$\tilde{\mu}(\xi) = \limsup_{H \to \infty} \frac{\tilde{\mu}(\xi, H)}{\log H}.$$

The quantities $\tilde{w}(\xi)$ and $\tilde{\mu}(\xi)$ are called the $\tilde{\ }$*-order* and the $\tilde{\ }$*-type* of ξ, respectively.

DEFINITION 8.1. *Let ξ be a complex number. We say that ξ is an*

$\tilde{A}$*-number, if* $0 \le \tilde{w}(\xi) < 1$ *or if* $\tilde{w}(\xi) = 1$ *and* $\tilde{\mu}(\xi) = 0$;

$\tilde{S}$*-number, if* $1 < \tilde{w}(\xi) < +\infty$ *or if* $\tilde{w}(\xi) = 1$ *and* $\tilde{\mu}(\xi) > 0$;

$\tilde{T}$*-number, if* $\tilde{w}(\xi) = +\infty$ *and* $H_0(\xi) = +\infty$;

$\tilde{U}$*-number, if* $\tilde{w}(\xi) = +\infty$ *and* $H_0(\xi) < +\infty$.

Two algebraically dependent complex numbers belong to the same class (see Exercise 8.1), thus this classification satisfies the second requirement stated in the Introduction to Chapter 3.

Sprindžuk [532] established that the $\tilde{A}$-numbers are exactly the algebraic numbers (see Exercise 8.1) and that the $\tilde{\ }$-type of any real (*resp.* complex) transcendental number of $\tilde{\ }$-order 1 is at least 1 (*resp.* at least 1/2).

Let $(\tilde{w}_H)_{H \ge 1}$ be a sequence of numbers in $[0, +\infty]$ such that there is a positive integer H_0 with

$$\tilde{w}_H = 0 \text{ for } H < H_0 \text{ and } 1 \le \tilde{w}_H \le \tilde{w}_{H+1} \le +\infty \text{ for } H \ge H_0. \tag{8.1}$$

If ξ is a complex transcendental number, then the sequence $(\tilde{w}(\xi, H))_{H \ge 1}$ satisfies (8.1), and, conversely, we may ask whether, for any sequence $(\tilde{w}_H)_{H \ge 1}$

with (8.1), there exists a complex number ξ such that $\tilde{w}(\xi, H) = \tilde{w}_H$ for any $H \geq 1$. This is the analogue of the Main Problem stated in Chapter 3. Amou [28] gave in 1996 an affirmative answer to that question.

THEOREM 8.1. *Let* $(\tilde{w}_H)_{H \geq 1}$ *be a sequence of numbers in* $[0, +\infty]$ *satisfying (8.1). Then, there exist real numbers* ξ *such that* $\tilde{w}(\xi, H) = \tilde{w}_H$ *for any* $H \geq 1$.

The proof is involved and will not be reproduced here. We merely give some of the main ideas. While we were previously interested in the distribution of algebraic numbers of bounded degree, the proof of Theorem 8.1 requires information on the distribution of algebraic numbers of fixed height. Using an effective version of a theorem of Kornblum on the polynomial analogue of Dirichlet's theorem on primes in arithmetic progressions, Amou obtained the following auxiliary result. Let H be a positive integer and let ξ be a real number with $|\xi| \neq 1$ and $(H + 1)^{-1} < |\xi| < H + 1$. Then there exists a positive constant $c(\xi, H)$, depending only on ξ and H, and an infinite set $\mathcal{N}(\xi, H)$ of positive integers such that, for each n in $\mathcal{N}(\xi, H)$, there are at least $(2H + 1)^{n/9}$ real algebraic numbers α of degree n, height H, and with $|\xi - \alpha| < \exp\{-c(\xi, H)n\}$. The lower estimate for the number of real algebraic numbers α with these properties is sharp and appears to be crucial in the remaining part of the proof, which essentially rests on the method developed by Schmidt for constructing T^*-numbers (see Chapter 7).

COROLLARY 8.1. *For any real number* $\tilde{w} \geq 1$, *there exist* $\tilde{S}$-*numbers* ξ *with* $\tilde{w}(\xi) = \tilde{w}$. *There exist* $\tilde{T}$-*numbers. For any positive integer* H, *there exist* $\tilde{U}$-*numbers* ξ *with* $H_0(\xi) = H$.

Corollary 8.1 shows that Sprindžuk's classification is non-trivial. As for metric results, Sprindžuk [532] established that almost all complex numbers are $\tilde{S}$-numbers of $\tilde{\ }$-order at most 2 and conjectured that 2 may be replaced by 1. This was claimed by Chudnovsky [164], who only sketched the proof. Amou [28] supplied a complete proof and gave some improvements in a subsequent work [29].

THEOREM 8.2. *Almost all complex (resp. real) numbers are* $\tilde{S}$-*numbers of* $\tilde{\ }$-*order 1.*

Theorem 8.2 is a straightforward consequence of Theorem 8.3, due to Amou and Bugeaud [31].

THEOREM 8.3. *Let* ε *be a positive real number. Then, for almost all complex (resp. real) numbers* ξ, *there exists a positive constant* $c_1(\xi, \varepsilon)$, *depending only on* ξ *and on* ε, *such that*

$$|P(\xi)| > \exp\{-(3 + \varepsilon)n \log(nH) - n \log n\} \qquad (8.2)$$

for all integer polynomials $P(X)$ of degree n and height H satisfying $\max\{n, H\} \geq c_1(\xi, \varepsilon)$.

PROOF. We only give a proof for the real case, since the complex case follows exactly the same lines. Let I be a real interval of length 1. Let $n_0 \geq 4$ be an integer which will be chosen later. In view of Sprindžuk's Theorem 4.2, for almost all real numbers ξ in I, there are only finitely many integer polynomials $P(X)$ of degree n less than n_0 and height H with

$$|P(\xi)| \leq \exp\{-(1 + \varepsilon)n \log H\}.$$

Consequently, in order to prove the theorem, it is sufficient to show that the set $\mathcal{E}$ of real numbers ξ in I for which there exist infinitely many integer polynomials $P(X)$ of degree n at least n_0 and height H such that

$$|P(\xi)| \leq \exp\{-(3 + \varepsilon)n \log(nH) - n \log n\}$$

is a null set. To this end, for any positive integers n, s, and H with $1 \leq s \leq n$ and $n \geq n_0$, we consider the set $\mathcal{A}(H, n, s)$ of complex algebraic numbers α satisfying $P(\alpha) = 0$ with multiplicity s for some integer polynomial $P(X)$ of degree at most n and height H. Let $\mathcal{E}(H, n, s)$ denote the set of all real numbers ξ for which there exists an algebraic number α in $\mathcal{A}(H, n, s)$ such that

$$|\xi - \alpha| \leq \exp\left\{\left(\frac{2}{s^2} - \frac{3 + \varepsilon}{s}\right) n \log H + \left(\frac{3}{2s^2} - \frac{3 + \varepsilon}{s}\right) n \log n\right\}, \quad (8.3)$$

and set $\mathcal{E}(H, n) := \mathcal{E}(H, n, 1) \cup \ldots \cup \mathcal{E}(H, n, n)$. By Lemma A.7, each ξ in $\mathcal{E}$ belongs to infinitely many sets $\mathcal{E}(H, n)$, thus

$$\mathcal{E} \subset \bigcap_{N \geq n_0} \bigcup_{n=N}^{+\infty} \bigcup_{H=1}^{+\infty} \mathcal{E}(H, n) \cup \bigcap_{H_0 \geq 1} \bigcup_{n=n_0}^{+\infty} \bigcup_{H=H_0}^{+\infty} \mathcal{E}(H, n). \quad (8.4)$$

Observe that the cardinality of $\mathcal{A}(H, n, s)$ is bounded by n times the number of integer polynomials of degree at most n and of height H, hence, we get

$$\text{Card}\,\mathcal{A}(H, n, s) \leq n\big((2H + 1)^{n+1} - (2(H - 1) + 1)^{n+1}\big)$$
$$\leq 2n(n + 1)(2H + 1)^n \leq 2^{3n} H^n. \quad (8.5)$$

Furthermore, for any element α in $\mathcal{A}(H, n, s)$, the minimal polynomial of α over $\mathbb{Z}$, denoted by $P_\alpha(X)$, divides some integer polynomial of degree n and height H. By Lemma A.3, we have $\mathrm{H}(P_\alpha) \leq 2^{n/s} H^{1/s}$, hence,

$$\text{Card}\,\mathcal{A}(H, n, s) \leq \frac{n}{s}(2^{1+n/s} H^{1/s} + 1)^{1+n/s} \leq 2^{7n^2/s^2} H^{2n/s^2}. \quad (8.6)$$

Let $\mathcal{D}(H, n, s)$ be the domain in the complex plane consisting of numbers whose distance from the interval I is less than the number in the right-hand side of (8.3). It follows from Theorems A.1 and A.2 that

$$|\alpha - \beta| \geq 2^{-n/s} n^{-5n/(2s)} H^{-2n/s}$$

holds for any α, β in $\mathcal{A}(H, n, s)$ with $\alpha \neq \beta$. For $s \geq 3$, this yields the upper estimate

$$\text{Card}\,(\mathcal{A}(H, n, s) \cap \mathcal{D}(H, n, s)) \leq 2^{3n/s} n^{5n/(2s)} H^{2n/s}, \qquad (8.7)$$

if n is large enough.

To bound the Lebesgue measure of $\mathcal{E}(H, n, s)$, we combine (8.3) with (8.5) (*resp.* with (8.6), with (8.7)) if $s = 1, 2$ (*resp.* if $n\sqrt{7/\log n} \leq s \leq n$, if $3 \leq s < n\sqrt{7/\log n}$), and, choosing n_0 sufficiently large, we get for any $s = 1, \ldots, n$ the upper bound $\lambda(\mathcal{E}(H, n, s)) \leq (nH)^{-(2+\varepsilon)}$ as soon as $n \geq n_0$. Thus, we obtain

$$\lambda(\mathcal{E}(H, n)) \leq (nH)^{-(1+\varepsilon)}$$

and the double sum

$$\sum_{n \geq n_0} \sum_{H \geq 1} \lambda(\mathcal{E}(H, n))$$

converges. The Borel–Cantelli Lemma 1.2 and (8.4) then show that the set $\mathcal{E}$ has zero Lebesgue measure.

Maybe, it is possible to replace (8.2) by

$$|P(\xi)| > \exp\{-(1 + \varepsilon)n \log H - f(n)\},$$

for some suitable function $n \mapsto f(n)$. Such a result seems however difficult to prove.

A refinement of the proof of Theorem 8.3 yields a strengthening of Theorem 8.2, whose proof is left as Exercise 8.2.

THEOREM 8.4. *There exists a complex (resp. real) set $\mathcal{E}_c$ (resp. $\mathcal{E}_r$) of Hausdorff dimension zero such that, for all complex (resp. real) numbers ξ not in $\mathcal{E}_c$ (resp. $\mathcal{E}_r$), there exists a positive constant κ, depending only on ξ, such that*

$$|P(\xi)| > \exp\{-\kappa\, n \log(nH)\}$$

for all integer polynomials $P(X)$ of degree n and height H. In particular, the complex (resp. real) $\tilde{S}$-numbers of $\tilde{}$-order strictly larger than 1, the complex (resp. real) $\tilde{T}$-numbers and the complex (resp. real) $\tilde{U}$-numbers form sets of Hausdorff dimension zero.

We may as well define an analogous classification of the complex numbers in terms of their approximation properties by algebraic numbers. For a complex number ξ and positive integers H and n, we set

$$\tilde{w}_n^*(\xi, H) := \min\{|\xi - \alpha| : \alpha \text{ algebraic}, \deg(\alpha) \leq n, H(\alpha) = H, \alpha \neq \xi\}.$$

Observe that $\tilde{w}_n^*(\xi, H)$ does not coincide with $w_n^*(\xi, H)$ defined in Chapter 3, which is equal to the minimum of the $\tilde{w}_n^*(\xi, h)$ for h varying between 1 and H. We then set

$$\tilde{w}^*(\xi, H) = \limsup_{n \to \infty} \frac{\log \log(1/\tilde{w}_n^*(\xi, H))}{\log n}.$$

The function $H \mapsto \tilde{w}^*(\xi, H)$ describes how well ξ can be approximated by algebraic numbers of height H.

The method used to prove Theorem 8.1 allowed Amou [28] to obtain a similar statement for the functions $H \mapsto \tilde{w}^*(\xi, H)$.

THEOREM 8.5. *Let $(\tilde{w}_H^*)_{H \geq 1}$ be a sequence of numbers in $[0, +\infty]$ satisfying*

$$\tilde{w}_H^* = 0 \text{ for } H < H_0, \ 1 \leq \tilde{w}_H^* \leq +\infty \text{ for } H \geq H_0,$$

for some positive integer H_0. There exist real numbers ξ such that $\tilde{w}^*(\xi, H) = \tilde{w}_H^*$ for any $H \geq 1$.

We refer to [28] for a proof of Theorem 8.5. We point out that in the above theorem the sequence $(\tilde{w}_H^*)_{H \geq 1}$ is not assumed to be non-decreasing, unlike the sequence $(\tilde{w}_H^*)_{H \geq 1}$ in Theorem 8.1.

8.2 Another classification proposed by Mahler

In 1971, Mahler [393] introduced a new classification of complex numbers in terms of their approximation properties by algebraic numbers. Unlike his first classification and Sprindžuk's one, he allowed degree and height to vary simultaneously. In this Section, we study this classification, and we report on various results obtained by Mahler [393], Durand [216, 219], Nesterenko [436], and Amoroso [21].

For an integer polynomial $P(X) = a_n X^n + \ldots + a_1 X + a_0$ of degree n, we set $L(P) := |a_0| + \ldots + |a_n|$ its *length* and

$$\Lambda(P) := 2^n L(P) = 2^n (|a_0| + \ldots + |a_n|) \tag{8.8}$$

its *size*. We may replace the number 2 in (8.8) by any real number strictly greater than 1 and $L(P)$ by the naive height $H(P)$ of $P(X)$ without any notable change in the results below. Although the length is at present rarely used, we choose to keep the original definition of Mahler. For a non-zero algebraic number α, we define $L(\alpha)$ and $\Lambda(\alpha)$ as $L(P)$ and $\Lambda(P)$, respectively, where $P(X)$ denotes the minimal polynomial of α over $\mathbb{Z}$. Setting $\Lambda(0) = 2$, we have

$\Lambda(\alpha) \geq 2$ for any algebraic number α, and we point out that, for any positive real number M, there exist only finitely many complex algebraic numbers α with $\Lambda(\alpha) \leq M$.

The idea of Mahler was to associate to a given complex number ξ a non-negative valued non-decreasing function $u \mapsto O(u \mid \xi)$ of the integer variable u, called the *order function* of ξ.

DEFINITION 8.2. *Let ξ be a complex number. The order function $u \mapsto O(u \mid \xi)$ of ξ is defined for any integer $u \geq 2$ by*

$$O(u \mid \xi) = \sup\{-\log|P(\xi)| : P(X) \in \mathbb{Z}[X], \Lambda(P) \leq u, P(\xi) \neq 0\}.$$

Analogously, we define the order function $u \mapsto O^*(u \mid \xi)$ by

$$O^*(u \mid \xi) = \sup\{-\log|\xi - \alpha| : \alpha \text{ algebraic}, \quad \Lambda(\alpha) \leq u, \alpha \neq \xi\}.$$

Both functions $u \mapsto O(u \mid \xi)$ and $u \mapsto O^*(u \mid \xi)$ are non-decreasing. We check that $u \mapsto O(u \mid \xi)$ vanishes identically if, and only if, ξ is an algebraic integer in an imaginary quadratic field. Otherwise, $O(u \mid \xi)$ is positive as soon as u is sufficiently large.

Mahler [393] defined a partial ordering and an equivalence relation on the set of positive valued non-decreasing functions. Let a and b be two such functions of the integer variable u. If there exist two positive integers c and u_0 and a positive real number γ such that

$$a(u^c) \geq \gamma b(u) \quad \text{for } u \geq u_0,$$

then we write $a(u) \gg b(u)$ and $b(u) \ll a(u)$. The relation $\gg$ defines a partial ordering. If $a(u) \gg b(u)$ and $b(u) \gg a(u)$ hold simultaneously, then we write $a(u) \asymp b(u)$, which defines the equivalence relation $\asymp$. With respect to this relation, the order functions can be distributed into disjoint classes, and $\gg$ defines a partial ordering on these classes, which is not a total ordering. By definition, two complex numbers ξ and η belong to the same class if, and only if, their order functions $u \mapsto O(u \mid \xi)$ and $u \mapsto O(u \mid \eta)$ are in the same class.

Mahler [393] (see Exercise 8.3) proved that two algebraically dependent complex transcendental numbers belong to the same class, thus this classification satisfies the second requirement stated in the Introduction of Chapter 3 (if we adopt the convention that all algebraic numbers belong to the same class).

The next lemma is used in the proofs of Theorems 8.6, 8.9, and 8.11.

LEMMA 8.1. *Let ξ be a complex number. Let n be an integer with $n \geq 2$ and H be a real number. There exist a positive constant c_2, depending only on ξ,* *and an absolute positive constant c_3 such that, for any $H \geq c_2$, there is a non-zero integer polynomial $P(X)$ with*

$$\deg P \leq n, \quad \mathrm{H}(P) \leq H, \quad and \quad |P(\xi)| \leq H^{-c_3 n}.$$

Moreover, if $n \geq 50$, suitable values for c_2 and c_3 are given by $(4 + |\xi|)^{50}$ and 0.455, respectively.

PROOF. Assume $H \geq 30$ and set $H' = [\sqrt{H}]^2/2$ if $[\sqrt{H}]$ is even, and $H' = ([\sqrt{H}] - 1)^2/2$ otherwise. Put $c_4 = 1 + |\xi| + \ldots + |\xi|^n$. The $(2H' + 1)^{n+1}$ points $a_n \xi^n + \ldots + a_1 \xi + a_0$, where $a_0, \ldots, a_n$ are integers with $-H' \leq a_0, \ldots, a_n \leq H'$, lie in the square centered at the origin and of sidelength $2H' c_4$. We divide this square into $(2H')^{n+1}$ small squares of sidelength $2\,H'\,(2H')^{-(n+1)/2}\,c_4$. By Dirichlet's *Schubfachprinzip*, two among these $(2H' + 1)^{n+1}$ points lie in the same small square, thus we get a non-zero integer polynomial $P(X)$ of degree at most n and height at most $2H'$ such that

$$|P(\xi)| \leq 2^{1-n/2}\, H'^{-(n-1)/2}\, c_4,$$

and the desired estimate follows from $H - 4\sqrt{H} \leq 2H' \leq H$ and $n \geq 2$. The last assertion is an easy computation and reproduces the numerical values obtained in [356]. We observe that we may alternatively use Theorem B.2 instead of Dirichlet's *Schubfachprinzip* (see the proof of Proposition 3.1).

Theorem 8.6, the proof of which is left as Exercise 8.4, derives from Lemma 8.1.

THEOREM 8.6. *Let ξ be a complex number which is not an algebraic integer in an imaginary quadratic field. Then $O(u \mid \xi) \asymp \log u$ if ξ is algebraic, otherwise we have $O(u \mid \xi) \gg (\log u)^2$.*

We infer from Theorem 8.6 that all algebraic numbers which are not integers in an imaginary quadratic field belong to the same class.

At the end of [393], Mahler addressed some (at that time) open problems, including the following ones. Do there exist uncountably many distinct classes of real numbers? Do there exist real transcendental numbers ξ and η such that the functions $O(u \mid \xi)$ and $O(u \mid \eta)$ are not comparable? Given a positive valued non-decreasing function a, establish necessary and sufficient conditions for the existence of a real number ξ with $O(u \mid \xi) \asymp a(u)$. Does there exist such a function a with $O(u \mid \xi) \asymp a(u)$ for almost all real numbers ξ?

In the following text, we answer (at least partially) these questions and we investigate the relationship between $O(u \mid \xi)$ and $O^*(u \mid \xi)$.

PROPOSITION 8.1. *For any transcendental complex number ξ and any real number $\tau \geq 2$, we have $O^*(u \mid \xi) \ll O(u \mid \xi)$ and*

$$\limsup_{u \to \infty} \frac{O^*(u \mid \xi)}{(\log u)^\tau} = +\infty \quad \textit{iff} \quad \limsup_{u \to \infty} \frac{O(u \mid \xi)}{(\log u)^\tau} = +\infty.$$

The proof of Proposition 8.1 is left as Exercise 8.5. As noticed by Mahler [393], the works of Koksma [333] and Wirsing [598] suggest that the results for $O^*(u \mid \xi)$ should be 'completely analogous to those for $O(u \mid \xi)$'. However, this is surprisingly not the case: Corollary 8.4 below asserts that $O^*(u \mid \xi) \asymp O(u \mid \xi)$ does not hold for all transcendental real numbers ξ.

DEFINITION 8.3. *Let $A_{-\infty}$ denote the set composed of the rational integers and the non-real quadratic integers. For any complex number ξ not in $A_{-\infty}$, the transcendence type of ξ is*

$$\tau(\xi) = \sup\{\tau \geq 0 : O(u \mid \xi) \gg (\log u)^\tau\}.$$

Furthermore, for τ in $[0, +\infty]$, we set $A_\tau = \{\xi \in \mathbb{C} : \tau(\xi) = \tau\}$.

By Theorem 8.6, any complex transcendental number ξ satisfies $\tau(\xi) \geq 2$ and A_1 consists of all algebraic numbers which are not in $A_{-\infty}$. Further, we observe that the transcendence type of ξ is the infimum of the real numbers τ such that $\log |P(\xi)| > -(\log \Lambda(P))^\tau$ holds for any integer polynomial $P(X)$ of sufficiently large size (see Amoroso [21], Chudnovsky [164], and Waldschmidt [588]; there are some subtle differences between the definitions of 'transcendence type').

Theorems 8.7 and 8.8 are concerned with the sets A_τ. The first assertion of Theorem 8.7 has been proved independently by Durand [216] and Nesterenko [436], while the second assertion is due to Amoroso [21, 24].

THEOREM 8.7. *The set A_2 has full Lebesgue measure, that is, almost all complex numbers (resp. real numbers) have transcendence type 2. Furthermore, there exists a complex set $\mathcal{E}_c$ (resp. a real set $\mathcal{E}_r$) of Hausdorff dimension zero such that, for all complex (resp. real) numbers not in $\mathcal{E}_c$ (resp. not in $\mathcal{E}_r$), we have $O(u \mid \xi) \asymp (\log u)^2$.*

PROOF. Let ξ be a complex transcendental number. If $O(u \mid \xi) \ll (\log u)^2$, we have $O(u \mid \xi) \asymp (\log u)^2$ by Theorem 8.6. Otherwise, for any positive real number κ, there exist infinitely many integer polynomials $P(X)$ such that

$$\log |P(\xi)| < -\kappa \left(\log(\Lambda(P))\right)^2 < -\frac{\kappa}{2} (\deg P) \left(\log(\mathrm{H}(P) \deg(P))\right).$$

Thus, by Theorem 8.4, the set of complex (*resp.* real) transcendental numbers ξ for which $O(u \mid \xi) \not\ll (\log u)^2$ has Hausdorff dimension zero, and the proof of Theorem 8.7 is complete.

THEOREM 8.8. *Let $q \geq 2$ and $\tau \geq (3 + \sqrt{5})/2$ be real numbers. For any positive integer k, set*

$$b_k = \left[q^{\tau^k}\right] \quad \text{and} \quad \sigma_k = \sum_{j=1}^{k} 2^{-b_j}.$$

The transcendence type of the Liouville number

$$\xi_\tau^{(q)} := \sum_{k \geq 1} 2^{-b_k} = \lim_{k \to +\infty} \sigma_k$$

is exactly equal to τ. Furthermore, for any real number $\tau > 2$, there are uncountably many real numbers in A_τ.

PROOF. Write ξ instead of $\xi_\tau^{(q)}$. Let k be a positive integer. The minimal polynomial over $\mathbb{Z}$ of σ_k is $2^{b_k} X - 2^{b_k} \sigma_k$, thus we have

$$b_k/2 < \log \Lambda(\sigma_k) < b_k. \tag{8.9}$$

Further, since the b_js are pairwise distinct, we get

$$2^{-b_{k+1}} \leq \xi - \sigma_k \leq 2^{1-b_{k+1}}. \tag{8.10}$$

Combining (8.9), (8.10), and the inequalities

$$b_k^\tau/2 < b_{k+1} < (2b_k)^\tau, \tag{8.11}$$

we deduce that there exist absolute, positive constants c_5 and c_6 such that

$$\exp\left\{-c_5(\log \Lambda(\sigma_k))^\tau\right\} \leq \xi - \sigma_k \leq \exp\left\{-c_6(\log \Lambda(\sigma_k))^\tau\right\}. \tag{8.12}$$

Thus, we have $\tau(\xi) \geq \tau$ by Proposition 8.1.

We now deal with the reverse inequality. Let α be an algebraic number with $\alpha \neq \sigma_j$ for any integer $j \geq 1$. By Corollary A.2, we have $|\alpha - \sigma_k| \geq \exp\{-2\log \Lambda(\alpha) \log \Lambda(\sigma_k)\}$, whence, by (8.9), we get

$$|\alpha - \sigma_k| \geq \Lambda(\alpha)^{-2b_k}. \tag{8.13}$$

By (8.10), (8.11), and the triangle inequality $|\xi - \alpha| \geq |\alpha - \sigma_k| - |\xi - \sigma_k|$, we infer from (8.13) that, under the assumption

$$b_k^{\tau-1} \geq 9 \log \Lambda(\alpha), \tag{8.14}$$

we have

$$|\xi - \alpha| \geq \frac{1}{2} \Lambda(\alpha)^{-2b_k}. \tag{8.15}$$

Let ℓ be the smallest positive integer k for which (8.14) is satisfied. If $\ell > 1$, we have

$$b_{\ell-1}^{\tau-1} < 9 \log \Lambda(\alpha) \le b_\ell^{\tau-1}.$$

Together with (8.11), this yields

$$b_\ell \le 2^\tau 9^{\tau/(\tau-1)} \big(\log \Lambda(\alpha)\big)^{\tau/(\tau-1)},$$

and it follows from (8.15) that there exists a positive constant c_7, depending only on τ and q, such that

$$|\xi - \alpha| \ge \exp\big\{-c_7 (\log \Lambda(\alpha))^{1+\tau/(\tau-1)}\big\}. \tag{8.16}$$

Taking c_7 large enough, (8.16) also holds if $\ell = 1$. Since $1 + \tau/(\tau - 1) \le \tau$, it follows from (8.12) and (8.16) that there exists a positive constant c_8, depending only on τ and q, such that

$$|\xi - \alpha| \ge \exp\big\{-c_8 (\log \Lambda(\alpha))^\tau\big\}$$

holds for any algebraic number α. Consequently, we have

$$\limsup_{u\to\infty} \frac{O^*(u \mid \xi)}{(\log u)^\tau} < +\infty,$$

hence, $\tau(\xi) \le \tau$ by Proposition 8.1, and $\tau(\xi) = \tau$, by (8.12).

For the last assertion of the theorem, the reader is directed to Amoroso [21] (notice that Amoroso used the size function $\max\{\log H(P), \deg(P)\}$ rather than $\Lambda(P)$). For any given real number $\tau > 2$, he constructed inductively (and in an effective way) uncountably many real transcendental numbers ξ with prescribed type τ as the limits of sequences of algebraic numbers with increasing degrees.

The first assertion of Theorem 8.8, due to Durand [216], allowed him to answer Problem 1 of Mahler [393], since any two functions $u \mapsto (\log u)^{t_1}$ and $u \mapsto (\log u)^{t_2}$, where t_1 and t_2 are positive, distinct real numbers, belong to different classes. Furthermore, as observed by Durand [219], the set $A_{+\infty}$ of complex numbers of infinite type is uncountable, for it contains the Liouville numbers given by the series $\sum_{j\ge0} 2^{-b_j}$, where $(b_j)_{j\ge1}$ is any increasing sequence of distinct integers with

$$\limsup_{j\to\infty} \frac{\log b_{j+1}}{\log b_j} = +\infty.$$

This can be shown by a suitable modification of the beginning of the proof of Theorem 8.8.

COROLLARY 8.2. *There are uncountably many distinct classes of real numbers.*

The construction given in Theorem 8.8 also implies the existence of real numbers whose order functions are not comparable.

COROLLARY 8.3. *Let $t \geq 3$ be an integer and, for any $k \geq 1$, set*

$$b_k = 2^{t^k} \quad and \quad d_k = 4^{t^{2k}}.$$

Then the Liouville numbers

$$\xi = \sum_{k\geq 1} 2^{-b_k} \quad and \quad \eta = \sum_{k\geq 1} 2^{-d_k}$$

do not satisfy either one of the following order relations: $O(u \mid \xi) \ll O(u \mid \eta)$, $O(u \mid \eta) \ll O(u \mid \xi)$, $O^*(u \mid \xi) \ll O^*(u \mid \eta)$, $O^*(u \mid \eta) \ll O^*(u \mid \xi)$.

PROOF. By Theorem 8.8, we have $\tau(\xi) = t$ and $\tau(\eta) = t^2$, thus the relations $O(u \mid \eta) \ll O(u \mid \xi)$ and $O^*(u \mid \eta) \ll O^*(u \mid \xi)$ cannot hold. The remaining part of the proof is left as Exercise 8.6.

Theorems 8.9 and 8.10 are needed to show that the functions $u \mapsto O(u \mid \xi)$ and $u \mapsto O^*(u \mid \xi)$ behave differently for some complex numbers ξ. They have been announced by Durand [219] (up to the values of the numerical constants in Theorem 8.9).

THEOREM 8.9. *For any transcendental complex number ξ and for any real number Λ with $\Lambda \geq (4+|\xi|)^{1000}$, there exists an algebraic number α satisfying $\Lambda(\alpha) \leq \Lambda$ and*

$$\log |\xi - \alpha| \leq -\frac{(\log \Lambda)(\log \Lambda(\alpha))}{5000}.$$

In particular, we have

$$\limsup_{u \to \infty} \frac{O^*(u \mid \xi)}{(\log u)^2} > 0.$$

We do not give a specific proof of Theorem 8.9 here, since it is an immediate consequence of Corollary 8.5, established in Section 8.3.

THEOREM 8.10. *Let ξ be a transcendental complex number. Let v and τ be positive real numbers with*

$$\liminf_{u \to \infty} \frac{O^*(u \mid \xi)}{(\log u)^v} > 0 \qquad (8.17)$$

and

$$\limsup_{u \to \infty} \frac{O(u \mid \xi)}{(\log u)^{\tau}} = +\infty. \tag{8.18}$$

We then have $v(\tau - 1) < \tau$.

PROOF. Since ξ is transcendental, we may assume that $\tau \geq 2$. By Proposition 8.1, (8.18) is then equivalent to

$$\limsup_{u \to \infty} \frac{O^*(u \mid \xi)}{(\log u)^{\tau}} = +\infty.$$

Thus, there exists a sequence $(\alpha_k)_{k \geq 1}$ of algebraic numbers with

$$10 \leq 2\Lambda(\alpha_k) < \Lambda(\alpha_{k+1}) \quad \text{and} \quad \log |\xi - \alpha_k| \leq -3k \big(\log \Lambda(\alpha_k)\big)^{\tau},$$

for any $k \geq 1$. Let k be a positive integer and α be a non-zero algebraic number such that $\alpha \neq \alpha_k$ and

$$\frac{k}{2} \big(\log \Lambda(\alpha_k)\big)^{\tau-1} \leq \log \Lambda(\alpha) \leq k \big(\log \Lambda(\alpha_k)\big)^{\tau-1} \tag{8.19}$$

holds. By Corollary A.2, we have $\log |\alpha - \alpha_k| \geq -2 \log \Lambda(\alpha) \log \Lambda(\alpha_k)$. Since $|\xi - \alpha| \geq |\alpha - \alpha_k| - |\xi - \alpha_k|$, we infer from (8.19) that

$$\log |\xi - \alpha| \geq -5 k^{-1/(\tau-1)} \big(\log \Lambda(\alpha)\big)^{\tau/(\tau-1)}. \tag{8.20}$$

However, by (8.17), there exists an absolute positive constant c_9 such that

$$\log |\xi - \alpha| \leq -c_9 \big(\log \Lambda(\alpha)\big)^{v},$$

which, combined with (8.20), implies that

$$c_9 \big(\log \Lambda(\alpha)\big)^{v} \leq 5 k^{-1/(\tau-1)} \big(\log \Lambda(\alpha)\big)^{\tau/(\tau-1)}.$$

Since k can be taken arbitrarily large, we get $v(\tau - 1) < \tau$, as claimed.

Theorem 8.6 immediately implies that

$$\limsup_{u \to \infty} \frac{O(u \mid \xi)}{(\log u)^2} \geq \liminf_{u \to \infty} \frac{O(u \mid \xi)}{(\log u)^2} > 0$$

holds for any transcendental complex number ξ. The analogous statement, with the function $O(u \mid \cdot)$ replaced by $O^*(u \mid \cdot)$, is, however, not true.

COROLLARY 8.4. *For any complex transcendental number* ξ, *we have*

$$\liminf_{u \to \infty} \frac{O^*(u \mid \xi)}{(\log u)^2} > 0 \quad \textit{iff} \quad \limsup_{u \to \infty} \frac{O(u \mid \xi)}{(\log u)^2} < +\infty$$

$$\textit{iff} \quad \limsup_{u \to \infty} \frac{O^*(u \mid \xi)}{(\log u)^2} < +\infty.$$

Thus, any complex transcendental number ξ of type strictly greater than 2 satisfies $O^*(u \mid \xi) \not\asymp O(u \mid \xi)$. Furthermore, apart from a set of Hausdorff dimension zero, any complex number ξ satisfies

$$O^*(u \mid \xi) \asymp O(u \mid \xi) \asymp (\log u)^2.$$

PROOF. Let ξ be a complex transcendental number. By Theorem 8.6, the second and the third inequalities of the corollary are equivalent. Furthermore, it follows directly from Theorem 8.10 that (8.18) does not hold with $\tau = 2$ if (8.17) holds for $\nu = 2$. Hence, it only remains for us to prove that the second inequality of the corollary implies the first one. Assume that there exists a positive constant c_{10} such that $\log |\xi - \alpha| \geq -c_{10} (\log \Lambda(\alpha))^2$ for any algebraic number α. By Theorem 8.9, for any sufficiently large integer u, there exists an algebraic number α_u satisfying

$$\Lambda(\alpha_u) \leq u \quad \text{and} \quad \log |\xi - \alpha_u| \leq -10^{-4} (\log u) \left(\log \Lambda(\alpha_u)\right).$$

Consequently, we get $\log u \leq 10^4 c_{10} \log \Lambda(\alpha_u)$ and $-\log |\xi - \alpha_u| \geq (10^8 c_{10})^{-1} (\log u)^2$. This yields $O^*(u \mid \xi) \geq (10^8 c_{10})^{-1} (\log u)^2$ for u large enough, and completes the proof of the first assertion of the corollary.

As an immediate consequence, if the type of ξ strictly exceeds 2, we then have

$$\limsup_{u \to \infty} \frac{O(u \mid \xi)}{(\log u)^2} = +\infty \quad \text{and} \quad \liminf_{u \to \infty} \frac{O^*(u \mid \xi)}{(\log u)^2} = 0.$$

This implies that $O^*(u \mid \xi) \not\gg (\log u)^2$, and we deduce from Theorem 8.6 that $O^*(u \mid \xi) \not\asymp O(u \mid \xi)$ holds. Finally, the last assertion of the corollary follows from Theorem 8.7.

Many interesting questions on the order functions $u \mapsto O(u \mid \xi)$ and $u \mapsto O^*(u \mid \xi)$ have not been investigated up to now. For instance, we may introduce, for any complex transcendental number ξ, the quantity

$$i(\xi) := \inf\left\{\nu \geq 0 : \liminf_{u \to \infty} \frac{O^*(u \mid \xi)}{(\log u)^\nu} = 0\right\}.$$

We infer from Corollary 8.4 that $i(\xi) = 2$ for almost all complex numbers ξ and from Theorem 8.10 that $i(\xi) \leq \tau/(\tau - 1)$ if $\tau(\xi) \geq 2$, hence, $i(\xi) = 1$ if ξ has infinite type. However, it seems to be difficult to construct explicit examples of complex numbers ξ with prescribed values for $\tau(\xi)$ and $i(\xi)$.

8.3 Transcendence measures and measures of algebraic approximation

Let ξ be a given complex transcendental number. If, for any integer polynomial $P(X)$, we establish explicit (even up to the numerical constants) lower bounds for $|P(\xi)|$ in terms of the degree of $P(X)$ and its height, we then get immediately information on the classes to which ξ belongs in the different classifications of numbers we have considered. This motivates the introduction of the notion of *transcendence measure*.

DEFINITION 8.4. *Let ξ be a complex transcendental number. A function Φ : $\mathbb{R}_{\geq 1} \times \mathbb{R}_{\geq 1} \to \mathbb{R}_{>0}$ is a transcendence measure for ξ if, for any sufficiently large integer n and any sufficiently large real number H, we have $|P(\xi)| \geq \exp\{-\Phi(n, \log H)\}$, for all non-zero integer polynomials $P(X)$ of degree at most n and height at most H.*

For instance, a transcendence measure for π is (see Waldschmidt [589])

$$2^{40} n (\log H + n \log n) (1 + \log n),$$

which implies that π is either an S-number or a T-number, and that its transcendence type is 2.

It follows from Definitions 8.2 and 8.4 that the transcendence type $\tau(\xi)$ of a given complex number ξ is the infimum of the real numbers τ for which there exists a positive constant $c(\xi, \tau)$ such that $c(\xi, \tau) (n + \log H)^{\tau}$ is a transcendence measure for ξ.

We may as well adopt Koksma's point of view and introduce an analogue of Definition 8.4. It becomes in that context more convenient to use Mahler's measure (defined in Appendix A) instead of the naive height, although this is not followed by all authors.

DEFINITION 8.5. *Let ξ be a complex transcendental number. A function $\Psi : \mathbb{R}_{\geq 1} \times \mathbb{R}_{\geq 1} \to \mathbb{R}_{>0}$ is a measure of algebraic approximation for ξ if there exists a positive constant κ with the following property: for any positive integer n and any real number M with $n \geq \kappa$ and $\log M \geq \kappa n$, we have $|\xi - \alpha| \geq \exp\{-\Psi(n, \log M)\}$ for all algebraic numbers α of degree at most n and Mahler's measure at most M.*

For instance, a measure of algebraic approximation for π is (see [589])

$$3 \cdot 2^{38} n (\log M + n \log n) (1 + \log n).$$

From the knowledge of a measure of algebraic approximation for a complex number ξ we immediately get information on its location in Koksma's classification.

Connection between transcendence measures and measures of algebraic approximation are discussed in [589] and in Chapter 15 of [591]. A useful tool is Lemma A.7 [189]. Examples of transcendence measures and measures of algebraic approximation can be found in [164], pp. 41–47, and in [244], pp. 163–178.

Up to the present Section, we have used a function of the naïve height and the degree in order to measure the size of an algebraic number. However, a refined notion of height, called the *absolute height*, is also frequently used in transcendental number theory. It has its own advantages and disadvantages and yields, for various kinds of problems, very precise results. The *absolute height* of an algebraic number α is denoted by $h(\alpha)$ and is defined as

$$h(\alpha) = \frac{1}{\deg(\alpha)} \log M(\alpha),$$

where $M(\alpha)$ is the Mahler measure of α.

The remaining part of this Section is devoted to some results by Diaz [188], Laurent and Roy [357], and Roy and Waldschmidt [483], in which both the degrees and the heights of the approximants are allowed to vary simultaneously.

The most interesting feature in Theorem 8.11, due to Diaz [188] (see also Philippon [457]) and inspired by ideas of Laurent and Roy [356], is the assumption $M \geq n+1$, which advantageously replaces the condition $\log M \gg n$ occurring in [356]. This is a direct consequence of the good dependence on the degree n (essentially n^n instead of 2^{n^2}) in Lemma A.8.

THEOREM 8.11. *Let ξ be a complex number. Let n be an integer and M a real number with $n \geq 50$, $M \geq n + 1$, and $M \geq (4 + |\xi|)^{100}$. Then, there exists an algebraic number α with*

$$\deg(\alpha) \leq n, \quad M(\alpha) \leq M, \quad and$$

$$|\xi - \alpha| \leq \exp\left\{ -\frac{6}{1000} \left(n \log M(\alpha) + \deg(\alpha) \log M \right) \right\}.$$

PROOF. We follow step by step the proof of Diaz [188]. By Lemma 8.1 applied with ξ, n, and $H := M(n + 1)^{-1/2}$, there exists a non-zero integer polynomial $P(X)$ of degree at most n and height at most H satisfying $|P(\xi)| \leq \exp\{-0.455\, n \log H\}$. By Lemma A.3, we have $M(P) \leq (n+1)^{1/2} H(P) \leq M$ and

$$|P(\xi)| \leq \exp\{-0.227\, n \log M\}, \tag{8.21}$$

since $M \geq n + 1$. Write $P(X)$ as a product $P(X) = a P_1(X) \ldots P_k(X)$ of irreducible integer polynomials, where a is a non-zero integer and $P_j(X)$ has degree $n_j \geq 1$ for $j = 1, \ldots, k$. By the additivity of the degrees and the multiplicativity of the Mahler measure, we have

$$n \log M = (n_1 + \ldots + n_k) \log M \quad \text{and}$$
$$n \log M \geq n(\log M(P_1) + \ldots + \log M(P_k)),$$

and (8.21) yields that

$$\prod_{1 \leq j \leq k} |P_j(\xi)| \leq |P(\xi)| \leq \exp\left\{-0.113 \sum_{1 \leq j \leq k} (n_j \log M \right.$$
$$\left. + n \log M(P_j)) \right\}.$$

Consequently, there exists an index j with $1 \leq j \leq k$ such that

$$|P_j(\xi)| \leq \exp\left\{-0.113 \left(n_j \log M + n \log M(P_j)\right)\right\}. \tag{8.22}$$

Let α be a root of $P_j(X)$ such that $|\xi - \alpha|$ is minimal. For any positive integer ℓ, Lemma A.8 and (8.22) yield that

$$|\xi - \alpha|^{\ell(\ell+1)/2} \leq \exp\left\{-0.113 \, \ell \, (n_j \log M + n \log M(P_j)) \right.$$
$$\left. + \ell n_j \log 2 + n_j \log\big(n_j^{1/2} M(P_j)\big) \right\}.$$

Since $n_j \leq n \leq M$ and $M(P_j) \leq M$, we get

$$|\xi - \alpha|^{\ell(\ell+1)/2} \leq \exp\left\{-(0.113\ell - 0.5)n_j \log M \right.$$
$$\left. + (-0.113\ell + 1)n \log M(P_j) + \ell n_j \log 2 \right\}.$$

We take $\ell = 18$ and bound $\log 2$ by $(\log M)/200$ in order to obtain

$$|\xi - \alpha|^{171} \leq \exp\left\{-1.034\big(\deg(\alpha) \log M + n \log M(\alpha)\big)\right\},$$

since $\deg(\alpha) = n_j$ and $M(\alpha) = M(P_j)$. This proves the theorem.

COROLLARY 8.5. *Let ξ be a complex number. Let n be an integer and Λ be a real number with $n \geq 50$ and $\Lambda \geq 2^{15n}(4 + |\xi|)^{100}$. Then there exists an algebraic number α satisfying*

$$\deg(\alpha) \leq n, \quad \Lambda(\alpha) \leq \Lambda, \quad \text{and}$$

$$|\xi - \alpha| \leq \exp\left\{-\frac{3}{1000} \left(n \log \Lambda(\alpha) + \deg(\alpha) \log \Lambda\right)\right\}.$$

PROOF. Set $M := 2^{-3n} \Lambda$. We infer from Theorem 8.10 that there exists an algebraic number α with $\deg(\alpha) \leq n$, $M(\alpha) \leq M$, and

$$|\xi - \alpha| \leq \exp\left\{-\frac{6}{1000} \left(n \log M(\alpha) + \deg(\alpha) \log M\right)\right\}. \tag{8.23}$$

Lemma A.2 implies that $\Lambda(\alpha) \leq 2^{3 \deg(\alpha)} M(\alpha)$, hence, we get

$$n \log M(\alpha) + \deg(\alpha) \log M \geq n \log \Lambda(\alpha) + \deg(\alpha) \log \Lambda - 5n \deg(\alpha)$$
$$\geq \left(n \log \Lambda(\alpha) + \deg(\alpha) \log \Lambda\right)/2, \qquad (8.24)$$

by the assumption $\log \Lambda \geq 10n$. The corollary follows from (8.23) and (8.24), since $\Lambda(\alpha) \leq 2^{3n} M \leq \Lambda$.

Theorem 8.9 follows from Corollary 8.5 applied with n and Λ satisfying $\Lambda \geq (4 + |\xi|)^{1000}$ and $n = [(\log \Lambda)/14]$. Actually, Durand [221], pp. 95–96, asserted a stronger result than Theorem 8.9: namely that, for any transcendental complex number ξ and any positive real number ε, there exist infinitely many algebraic numbers α such that

$$|\xi - \alpha| \leq \exp\left\{-\frac{(\log \Lambda(\alpha))^2}{48 \log 2 + \varepsilon}\right\}.$$

Conversely, he claimed that, for almost all transcendental complex numbers ξ and any positive real number ε, there exist only finitely many algebraic numbers α such that

$$|\xi - \alpha| \leq \exp\left\{-\frac{(\log \Lambda(\alpha))^2}{8 \log 2 - \varepsilon}\right\}.$$

Theorem 8.12 below, due (up to the numerical constants) to Laurent and Roy [357], provides a more general statement than Corollary 8.5.

THEOREM 8.12. *Let ξ be a transcendental complex number. Let $(n_k)_{k \geq 0}$ be a non-decreasing sequence of positive integers and let $(t_k)_{k \geq 0}$ be an unbounded non-decreasing sequence of positive real numbers. Assume that, for any $k \geq 0$, we have*

$$n_k \geq 50, \quad t_k \geq 2n_k, \quad n_{k+1} \leq 2n_k, \quad t_{k+1} \leq 2t_k.$$

Then, there exist infinitely many integers $k \geq 0$ with the following property: for each of these k, there exists an algebraic number α satisfying

$$\frac{n_k}{4000} \leq \deg(\alpha) \leq n_k, \quad \log M(\alpha) \leq t_k, \quad \text{and} \quad \log |\xi - \alpha| \leq -\frac{n_k t_k}{22000}.$$

The proof of Theorem 8.12 requires many steps, which are displayed in Exercise 8.7. An interesting feature of Theorem 8.12 is the lower bound for the degree of the approximants.

Corollary 8.6 should be compared with Theorem 3.4, which however provides no lower bound for the degrees of the algebraic approximants. Although superseded by Theorem 3.12 (which depends on a statement not proved in the present book), we quote it.

COROLLARY 8.6. *Let ξ be a transcendental complex number. For any integer $n \geq 50$, there exist infinitely many algebraic numbers α with*

$$\frac{n}{4000} \leq \deg(\alpha) \leq n \quad and \quad |\xi - \alpha| \leq M(\alpha)^{-n/22000}.$$

PROOF. Set $n_k = n$ and $t_k = k$ for each integer $k \geq 2n$. Since ξ is transcendental, the set of algebraic numbers given by Theorem 8.12 applied with these two sequences is necessarily infinite.

Up to the numerical constants, Corollary 8.7 is due to Roy and Waldschmidt [483] (see also Corollaire 2 from [357]).

COROLLARY 8.7. *Let ξ be a transcendental complex number and $\kappa \geq 8000$ be a real number. There exist infinitely many algebraic numbers α with*

$$\mathrm{h}(\alpha) \leq \kappa \quad and \quad \log |\xi - \alpha| \leq -10^{-8} \kappa \deg(\alpha)^2.$$

PROOF. Set $n_k = k$ and $t_k = \kappa k / 4000$ for any $k \geq 1$. Since $\kappa \geq 8000$, we have $t_k \geq 2n_k$ for any $k \geq 1$. Applying Theorem 8.12, we get for infinitely many positive integers k an algebraic number α_k with

$$\mathrm{h}(\alpha_k) = \frac{\log M(\alpha_k)}{\deg(\alpha_k)} \leq \frac{t_k}{n_k/4000} = \kappa$$

and

$$\log |\xi - \alpha_k| \leq -\frac{n_k t_k}{22000} = -\frac{\kappa k^2}{88 \times 10^6} \leq -10^{-8} \kappa \deg(\alpha_k)^2.$$

Infinitely many of the α_ks are distinct since ξ is trancendental.

Theorem 8.12 asserts the existence of infinitely many integers k with a given property, but it gives no information regarding the distribution of these integers. In fact, U-numbers show that we cannot hope for a much more precise result than Theorem 8.12 (Brownawell [118], Bugeaud [123], and Laurent [353]), although a refined statement holds for almost all complex numbers [123].

8.4 Exercises

EXERCISE 8.1. Use Theorems A.1 and B.2 to prove that the $\tilde{A}$-numbers are exactly the algebraic numbers. Use estimate (3.3) of Chapter 3 to show that two algebraically dependent complex numbers belong to the same class in Sprindžuk's classification.

EXERCISE 8.2. Let $\kappa > 5$ be a real number and denote by $\mathcal{E}(\kappa)$ the set of real numbers ξ for which the equation $|P(\xi)| < \exp\{-\kappa n \log(nH)\}$ has infinitely many solutions in integer polynomials $P(X)$ of degree at most n and height at most H. Prove that the Hausdorff dimension of $\mathcal{E}(\kappa)$ is less than or equal to $5/(\kappa - 3)$ and establish Theorem 8.4.

EXERCISE 8.3. Prove that the order functions of two algebraically dependent transcendental complex numbers are equivalent.

EXERCISE 8.4. Use Theorem A.1, Dirichlet's *Schubfachprinzip*, and Lemma 8.1 to prove Theorem 8.6.

EXERCISE 8.5. Use Lemmas A.6 and A.8 to prove Proposition 8.1.

EXERCISE 8.6. Completion of the proof of Corollary 8.3.

We keep the notation of Corollary 8.3. For any integer $k \geq 1$, set $\xi_k = \sum_{j=1}^{k} 2^{-b_j}$ and $u_k = \Lambda(\xi_{2k})$. Show that there exists a positive constant c_{11}, depending only on t, such that $O^*(u_k \mid \xi) \geq c_{11} b_{2k+1}$. Let r be a positive integer and let α be an algebraic number with $\Lambda(\alpha) \leq u_k^r$. Argue as in the proof of Theorem 8.8 to show that $O^*(u_k^r \mid \eta) \leq c_{12} b_{2k+1}^{(1+t^2/(t^2-1))/t}$ holds for a constant c_{12}, depending only on t and r, and for k large enough. Prove that $O^*(u \mid \xi) \ll O^*(u \mid \eta)$ does not hold and make the suitable adaptations to establish that $O(u \mid \xi) \ll O(u \mid \eta)$ does not hold either.

EXERCISE 8.7. Proof of Theorem 8.12. Let ξ be a transcendental complex number. Proceed as in the beginning of the proof of Theorem 8.11 to show that for any sufficiently large integer k there exists a non-zero integer polynomial $P_k(X)$ such that

$$\deg(P_k) \leq n_k, \quad \log M(P_k) \leq t_k/2, \quad \text{and} \quad \log |P_k(\xi)| \leq -n_k t_k/10.$$

Then, prove that (at least) one of the following two statements holds.

(i) *There exists a factor $Q_k(X)$ of $P_k(X)$ in $\mathbb{Z}[X]$ which is a power of an irreducible, integer polynomial and satisfies*

$$\deg(Q_k) \leq n_k, \quad \log M(Q_k) \leq t_k/2, \quad \text{and} \quad \log |Q_k(\xi)| \leq -n_k t_k/15;$$

(ii) *There exist two factors $F_k(X)$ and $G_k(X)$ of $P_k(X)$ in $\mathbb{Z}[X]$ which are coprime and satisfy*

$$\deg(F_k) + \deg(G_k) \leq n_k, \quad \log M(F_k) + \log M(G_k) \leq t_k/2$$

and

$$\log \max\{|F_k(\xi)|, |G_k(\xi)|\} \leq -n_k t_k/30.$$

We further need a first auxiliary result (Lemme 3.4 and Proposition 3.5 from [483]).

Let ξ be a complex number and r be a positive real number. Let $F(X) = a_m \prod_{i=1}^{m} (X - \alpha_i)$ and $G(X) = b_n \prod_{j=1}^{n} (X - \beta_j)$ be non-constant complex polynomials of degree m and n, respectively. Let f (resp. g) denote the number of roots of $F(X)$ (resp. of $G(X)$) in the closed disc of radius r centered at ξ and set

$$\rho := \min\{|\xi - \gamma| : \gamma \text{ root of } FG(X)\}.$$

We then have

$$\rho^{fg}|\mathrm{Res}(F, G)| \leq 2^{mn} M(F)^{n-g} M(G)^{m-f} |F(\xi)|^{g} |G(\xi)|^{f}. \qquad (8.25)$$

If furthermore $|a_m| \geq 1$ and $|b_n| \geq 1$ hold, then, for any non-negative integer s, we have

$$\min\{1, \rho\}^{s^2/4}|\mathrm{Res}(F, G)| \leq 2^{mn} M(F)^{n} M(G)^{m} \max\{|F(\xi)|, |G(\xi)|\}^{s}. \qquad (8.26)$$

Hint. For inequality (8.25), set $p_i := |\xi - \alpha_i|$ and $q_j := |\xi - \beta_j|$, in such a way that we have $p_1 \leq \ldots \leq p_m$ and $q_1 \leq \ldots \leq q_n$. Bound $\prod_{1 \leq i \leq f, g < j \leq n} |\alpha_i - \beta_j|$ from above in terms of the q_js and $\prod_{f < i \leq m, 1 \leq j \leq g} |\alpha_i - \beta_j|$ from above in terms of the p_is. Further, show that we have

$$(2\rho)^{fg} \prod_{\substack{1 \leq i \leq f \\ 1 \leq j \leq g}} |\alpha_i - \beta_j| \leq \left(\prod_{1 \leq i \leq f} 2p_i \right)^{g} \left(\prod_{1 \leq j \leq g} 2q_j \right)^{f},$$

and

$$\prod_{\substack{f < i \leq m \\ g < j \leq n}} |\alpha_i - \beta_j| \leq 2^{(m-f)(n-g)}$$

$$\times \left(\prod_{f < i \leq m} \max\{1, |\alpha_i|\} \right)^{n-g} \left(\prod_{g < j \leq n} \max\{1, |\beta_j|\} \right)^{m-f}.$$

Prove (8.25) by combining the above inequalities.

If $s \leq m + n$ and all the $|\xi - \gamma|$ are distinct, where γ runs over the roots of $FG(X)$, then observe that a suitable choice of r yields (8.26). Use a continuity argument to treat the case when all the $|\xi - \gamma|$ are not distinct. Finally, deal with the case $s > m + n$.

The second auxiliary result, asserted below, follows from (8.26) with a suitable choice of s.

Let n be an integer and t be a real number with $t \geq n \geq 1$. Let ξ be a real number. Assume that there exist coprime integer polynomials $F(X)$ and $G(X)$ such that

$$\deg(F) \leq n, \quad \log M(F) \leq t \quad and \quad \log |F(\xi)| \leq -nt/62,$$
$$\deg(G) \leq n, \quad \log M(G) \leq t \quad and \quad \log |G(\xi)| \leq -nt/62.$$

Then, there exists an algebraic number α such that

$$(FG)(\alpha) = 0 \quad and \quad \log |\xi - \alpha| \leq -nt/11000.$$

The third auxiliary result, asserted below, is a consequence of the second one.

Let n be an integer and t be a real number with $t \geq n \geq 7$ and $tn \geq 650$. Let ξ be a real number. Assume that there exist coprime integer polynomials $F(X)$ and $G(X)$ such that

$$\deg(F) \leq n, \quad \log M(F) \leq t, \quad and \quad \log |F(\xi)| \leq -nt/30$$
$$\deg(G) \leq n, \quad \log M(G) \leq t, \quad and \quad \log |G(\xi)| \leq -nt/30.$$

Then, there exists a root α of the polynomial $FG(F + G)(X)$ which satisfies

$$n/4000 \leq \deg(\alpha) \leq n, \quad \log M(\alpha) \leq 2t,$$
$$and \quad \log |\xi - \alpha| \leq -nt/11000.$$

Hint. The second auxiliary result applied to the pair of polynomials (F, G) yields an algebraic approximant α_1 of ξ. Without any loss of generality, we may assume that $F(\xi) = 0$. Use Lemma A.3 to check that $\log M(F+G) \leq 2t$ and apply the second auxiliary result with $2t$ in place of t to the pair of polynomials $(G, F+G)$. We then get another algebraic approximant α_2 of ξ. Observe that α_1 and α_2 are distinct. Show that any non-zero algebraic number α satisfies $\log |\alpha| \geq -\deg(\alpha)h(\alpha)$. Set $\delta := \max\{\deg(\alpha_1), \deg(\alpha_2)\}$ and derive that $\log |\alpha_1 - \alpha_2| \geq -3.7\delta t$. Give an upper bound for $\log |\alpha_1 - \alpha_2|$ using the triangle inequality and conclude.

Finish the proof of Theorem 8.12.

Hint. You may proceed as follows. If condition (ii) is satified for infinitely many integers k, conclude by applying the third auxiliary result with $n = n_k$ and $t = t_k/2$ to the polynomials $F_k(X)$ and $G_k(X)$. If condition (i) is satisfied for any k sufficiently large, then observe that there are infinitely many integers k such that the polynomials $Q_{k-1}(X)$ and $Q_k(X)$ are coprime.

Conclude by applying the third auxiliary result with $n = n_k$ and $t = t_k/2$ to these polynomials.

8.5 Notes

• Clearly, we may define other classifications of transcendental complex numbers as well, by simultaneously varying the upper bounds for the degree and height of the approximants, denoted by n and H, respectively. For many reasons, it turns out to be convenient to regard n and $\log H$ as equivalent. Another example of size has been proposed by Mahler [395], where, for an integer polynomial $P(X) = a_n X^n + \ldots + a_1 X + a_0$, the quantity $\Lambda(P)$ is replaced by the product $(2 + |a_0|) \ldots (2 + |a_n|)$.

• Philippon [456] investigated classifications of complex numbers by using non-standard analysis. He defined the equivalence relation $\sim$ on the field of complex numbers by saying that $\alpha \sim \beta$ if, and only if, α is algebraic over the field $\mathbb{Q}(\beta)$ and β is algebraic over the field $\mathbb{Q}(\alpha)$. The quotient set $\mathbb{C}/\sim$ is the set of algebraically closed subfields of $\mathbb{C}$ of transcendence degree ≤ 1; the special point in this set is the class of algebraic numbers. Using Diophantine approximation properties of complex numbers, Philippon defined a non-standard local distance called 'dam' (acronym of 'distance d'approximation mutuelle', that is, mutual approximation distance translated into French) on $\mathbb{C}/\sim$ which endows this space with a non-discrete Hausdorff topology. He recovered the classification defined by Mahler [393] in 1971 by considering the dam to the special point.

• Chudnovsky ([164], Theorem 2.8, page 45) proved that e^π is a $\tilde{S}$-number of $\tilde{\ }$-order 1. Other examples of $\tilde{S}$-numbers of $\tilde{\ }$-order 1 have been given by Amou [26].

• Nesterenko [436] and Durand [217] introduced a multidimensional generalization of the classification defined by Mahler in [393]. They defined a function $\Theta(u \mid \xi_1, \ldots, \xi_n)$ extending to $\mathbb{C}^n$ the order function $O(u \mid \xi)$. Durand showed some properties of Θ which appropriately generalize those of $O(u \mid \xi)$. A conjecture of Chudnovsky ([163], Problem 1.3, page 178) has been confirmed by Amoroso [24]. As for the real case, the metric statement of Nesterenko [436], Theorem 2, gives a weaker result than expected and the conjecture proposed on page 237 of [436] is still open.

• Gelfond's criterion ([257], Chapter III, Section 4, Lemma VII) asserts that the existence of a, in some sense, gap-free infinite sequence of non-zero integer polynomials taking small values at a given complex number ξ implies that ξ is algebraic. It has been subsequently refined by Brownawell [118] and Waldschmidt [587] (see also Chapitre 5 of [588]). A Gelfond–type criterion with multiplicities has been established by Laurent and Roy [356]. Theorem 2b of Davenport and Schmidt [182] (see also [32]) and Theorem 4.2 of Roy and Waldschmidt [484] are further variants of Gelfond's criterion.

• Proposition 3.1 and Theorem 3.1 yield a necessary and sufficient condition for a real number ξ to be transcendental. Mahler [394, 396] obtained a simplified criterion, depending only on the approximation behaviour of a single sequence of integer polynomials of arbitrary degrees. Mahler's result can be compared to transcendence criteria of Durand [215, 218].

• Cijsouw [166], Waldschmidt [589], Chudnovsky [164], Diaz [187], Nesterenko and Waldschmidt [437], and others obtained transcendence measures for some families of numbers, including $\log \alpha$, e^{β} and α^{β}, for non-zero algebraic numbers α, β with $\log \alpha \neq 0$ and β irrational. Galočkin [254], Miller [424], Becker-Landeck [58], Wass [595], Becker [56], Nishioka [438, 439], Nishioka and Töpfer [440], Töpfer [569, 570], and others established transcendence measures for values of Mahler functions at algebraic points. Lang [348], Brownawell [119], and others (for further references and results, see, for example, Shidlovskii [522] and Section 5.2 of Chapter 5 of [244]) obtained transcendence measures for values of E-functions at algebraic points.

• Let n and t be two real numbers with $t \geq n \geq 1$ and set

$$\overline{\mathbb{Q}}(n, t) := \{\alpha \in \overline{\mathbb{Q}} : [\mathbb{Q}(\alpha) : \mathbb{Q}] \leq n, \ \log M(\alpha) \leq t\}.$$

The distribution of the sets $\overline{\mathbb{Q}}(n, t)$ in the complex plane has been studied by Waldschmidt [590], who defined specific sets of pairs (n, t). His results are in some respect not entirely satisfactory and Bugeaud [123] was led to consider subsets of $\mathbb{R}$ (or $\mathbb{C}$) rather than sets of pairs (n, t). For any $\kappa > 0$, he introduced

$$\mathcal{F}_{\kappa} := \bigcap_{\kappa' < \kappa} \bigcup_{n_0 \geq 1} \bigcup_{h_0 \geq 1} \bigcap_{n \geq n_0} \bigcap_{t \geq h_0 n}$$
$$\bigcup_{\alpha \in \overline{\mathbb{Q}}(n,t) \cap \mathbb{R}}]\alpha - e^{-\kappa' nt}, \alpha + e^{-\kappa' nt}[$$

and

$$\mathcal{F}'_{\kappa} := \bigcap_{\kappa' > \kappa} \bigcup_{n_0 \geq 1} \bigcup_{h_0 \geq 1} \bigcap_{n \geq n_0} \bigcap_{t \geq h_0 n}$$
$$\bigcap_{\alpha \in \overline{\mathbb{Q}}(n,t) \cap \mathbb{R}}]\alpha - e^{-\kappa' nt}, \alpha + e^{-\kappa' nt}[^c.$$

Further, for a real number ξ, he defined

$$\kappa_0(\xi) := \max\{\kappa > 0 : \xi \in \mathcal{F}_{\kappa}\} \text{ and } \kappa'_0(\xi) := \min\{\kappa > 0 : \xi \in \mathcal{F}'_{\kappa}\},$$

with the convention that $\kappa_0(\xi) = 0$ (*resp.* $\kappa_0'(\xi) = +\infty$) if ξ belongs to none of the sets $\mathcal{F}_\kappa$ (*resp.* $\mathcal{F}_\kappa'$). The relationship between Waldschmidt's and Bugeaud's definitions are explained in [123]. It is proved in [123] that there exist two real numbers κ and κ' with $1/850 \le \kappa \le \kappa' \le 1$, such that almost all real numbers ξ satisfy $\kappa_0(\xi) = \kappa$ and $\kappa_0'(\xi) = \kappa'$. This statement expresses that the quality of the algebraic approximation is the same for almost all real numbers. Presumably, we have $\kappa = \kappa' = 1$.

• Laurent [353] studied the location of the algebraic approximants of the Liouville number $\sum_{j \ge 1} 2^{-j!}$. Simultaneous algebraic approximation of Liouville numbers has been considered by Roy [478].

• Results in a similar spirit as Theorem 8.12, but without a lower bound for the degree of the approximants have been obtained by Laurent and Roy [356] (see also [358] for deep generalizations to, for example, approximation by hypersurfaces) and Laurent [352] (see his Theorem 4 for a weaker condition between t_k and n_k).

9

Approximation in other fields

In Chapters 1 to 7, we have exclusively considered approximation of real numbers. However, Mahler [376] and Koksma [333] defined their classifications for complex numbers as well, and Mahler [378] also introduced an analogous classification for the transcendental numbers in the field $\mathbb{Q}_p$, the completion of $\mathbb{Q}$ with respect to the prime number p. Furthermore, approximation in the field of formal power series has also been investigated, for example, by Sprindžuk [534, 539]. In the present Chapter, we consider each of these settings, and we briefly describe the state of the art for the problems corresponding to those studied in Chapters 1 to 7. Roughly speaking, it is believed (and it often turns out to be true) that Diophantine approximation results in the real case have got their complex and p-adic analogues, the proofs of which are a (more or less) straightforward adaptation of those in the real case. This however does not hold true anymore for Diophantine approximation in fields of power series. For instance, the analogue of Roth's Theorem 2.1 does not exist when the ground field has positive characteristic, see, for example, the surveys by Lasjaunias [351] and by Schmidt [515] for additional information.

9.1 Approximation in the field of complex numbers

Let ξ be a complex non-real number and let n be a positive integer. Following Mahler [376] and Koksma [333], we define the quantities $w_n(\xi)$ and $w_n^*(\xi)$, and the classes A, S, T, U, A^*, S^*, T^*, and U^*, exactly as in Chapter 3. Notice (see, for example, (16) of [598]) that we have $w_1(\xi) = w_1^*(\xi) = 0$, $w_2(\xi) = w_2^*(\xi)$, and $w_3(\xi) = w_3^*(\xi)$. Both classifications turn out to be equivalent, see, for example, Schneider [517]. Using Dirichlet's *Schubfachprinzip* it is easy to show that $w_n(\xi) \geq (n-1)/2$ holds for any complex non-real number ξ which is not algebraic of degree at most n. Further, by a suitable modification of the proof of Theorem A.1 (see the Notes at the end of Appendix A), we have $w_n(\xi) \le (d-2)/2$ if ξ is algebraic of degree d, regardless of the positive integer n. This implies that $w_n(\xi) = (n-1)/2$ holds true for any complex non-real algebraic number ξ of degree $n+1$. Moreover, proceeding exactly as in the proof of Proposition 3.2, we get $w_n(\xi) \ge w_n^*(\xi)$ for any complex number ξ and any positive integer n. Wirsing [598] proved the complex analogue of Theorem 3.4.

THEOREM 9.1. *Let $n \ge 2$ be an integer and ξ be a complex non-real number which is not algebraic of degree at most n. Then, we have*

$$w_n^*(\xi) \ge w_n(\xi) - \frac{n-1}{2},$$

$$w_n^*(\xi) \ge \frac{w_n(\xi)}{2}, \tag{9.1}$$

$$w_n^*(\xi) \ge \frac{w_n(\xi)}{2w_n(\xi) - n + 2}, \tag{9.2}$$

and

$$w_n^*(\xi) \ge \frac{n}{4}. \tag{9.3}$$

Inequality (9.3) is a consequence of (9.1), (9.2), and $w_n(\xi) \ge (n-1)/2$. It has been slightly improved by Tishchenko [562].

It follows from (9.2) that $w_n^*(\xi) = (n-1)/2$ holds as soon as we have $w_n(\xi) = (n-1)/2$. Combined with the observation above Theorem 9.1, this implies that $w_n^*(\xi) = (n-1)/2$ holds for any complex non-real algebraic number ξ of degree $n+1$. A partial result concerning the approximation of complex non-real algebraic numbers follows from Schmidt's Subspace Theorem. Namely, Evertse [233] showed that, for a complex non-real algebraic number ξ of degree d and a positive integer n, we have $w_n(\xi) = w_n^*(\xi) = \min\{(n-1)/2, (d-2)/2\}$ either if $[\mathbb{Q}(\xi) : \mathbb{Q}(\xi) \cap \mathbb{R}] \ge \lceil (n+3)/2 \rceil$ or if $[\mathbb{Q}(\xi) \cap \mathbb{R} : \mathbb{Q}] \le \lfloor (n+1)/2 \rfloor$. It is likely that this assumption could be removed.

At the end of [377], Mahler conjectured that almost all complex numbers ξ satisfy $w_n(\xi) \le n/2$ for all positive integers n. The fact that the equality $w_n(\xi) = (n-1)/2$ should hold almost everywhere has been conjectured by Kasch [310] and established for $n = 2$ in [310]. Shortly thereafter, Volkmann [581] confirmed Kasch's conjecture for $n = 3$. Using his powerful method of essential and inessential domains, Sprindžuk [538, 539] established the complex analogues of Theorems 4.1 and 4.2.

THEOREM 9.2. *Almost all complex numbers ξ satisfy $w_n(\xi) = w_n^*(\xi) = (n-1)/2$ for every positive integer n.*

The complex analogue of Theorem 4.3 is due to A. Baker [41]. Bernik and Vasiliev [96] proved the complex analogue of Theorem 4.4 and deduced from their result that complex algebraic numbers of bounded degree form a regular system with parameters of regularity slightly better than in [94]. Kleinbock [328] established the complex analogue of Theorem 4.6.

Bernik and Sakovich [94] used regular systems of complex algebraic numbers to establish the complex analogue of Theorem 5.3. They obtained partial results towards the complex analogue of Theorem 5.7 (see also R. C. Baker [48] and Sakovich [492]).

R. C. Baker [47] proved the existence of complex non-real numbers ξ for which $w_n(\xi)$ and $w_n^*(\xi)$ differ from pre-assigned values for every even integer $n \geq 4$, by constructing T-numbers with this property. Bugeaud [133] showed that, for any given integer $n \geq 4$, there exist complex non-real numbers ξ for which $w_n(\xi)$ differs from $w_n^*(\xi)$.

9.2 Approximation in the field of Gaussian integers

The most natural complex extension of approximation to real numbers by rational integers is approximation to complex numbers by ratios of Gaussian integers. This has been considered by Hermite and Hurwitz in the nineteenth century (see [332], Chapter IV, Section 4). Unlike in the real case, a continued fraction approach did not give the analogue of Hurwitz' Theorem 1.18. This was established by Ford [247], who used additional geometrical ideas based on the Picard group **SL** $(2, \mathbb{Z})$. Theorem 4.5 of Dodson and Kristensen [205] provides an analogue of Theorem 1.1, but is presumably not best possible. The problem of generalizing the theory of continued fractions to the complex field has been studied by A. L. Schmidt [499].

LeVeque [360] established the analogue in $\mathbb{Z}[i]$ of Khintchine's Theorem 1.10. Another proof (yielding a slightly sharper result) has been given by Sullivan [547]. For extensions to other imaginary quadratic fields, see [547] and Nakada [434].

Melián and Pestana [416] proved that balls in $\mathbb{C}$ of radius $1/|w|^2$ and centered at z/w, where z, w are Gaussian integers with $w \neq 0$ form a 'well-distributed system', a notion close to that of regular system. They obtained Theorem 5.2 and its analogue in imaginary quadratic fields. Diophantine approximation in hyperbolic space has been investigated by many authors, see [205] and Section 7.7 of Bernik and Dodson [86] for references. A full proof of the analogue in $\mathbb{Z}[i]$ of Theorem 5.2 (as well as other results, including the existence of badly approximable complex numbers), independent of the hyperbolic space framework, has been given by Dodson and Kristensen [205].

9.3 Approximation in the p-adic fields

In this Section, p denotes a given prime number. Every non-zero rational number a can be expressed uniquely under the form $a = p^m a'$, where m is an integer and a' is a rational number whose numerator and denominator are prime to p. By definition, the p-adic valuation $v_p(a)$ is equal to m and we set $v_p(0) = +\infty$. It defines a p-adic metric $|\cdot|_p$ on $\mathbb{Q}$, which we normalize by setting $|p|_p = p^{-1}$. Since $v_p(a+b) \geq \min\{v_p(a), v_p(b)\}$ for any rational numbers a and b, the metric $|\cdot|_p$ is ultrametric. The completion of $\mathbb{Q}$ with respect to $|\cdot|_p$ is the p-adic field $\mathbb{Q}_p$ of p-adic numbers. Each element ξ in $\mathbb{Q}_p$ has a unique representation

$$\xi = \sum_{r=m}^{\infty} c_r \, p^r,$$

where m and the coefficients c_r are integers with c_m non-zero and $0 \leq c_r \leq p - 1$ for any $r \geq m$. We then have $|\xi|_p = p^{-m}$. In order to simplify the notation, we write in this Section $|\cdot|$ instead of $|\cdot|_p$ for the p-adic absolute value. Furthermore, we denote by $\overline{\mathbb{Q}}_p$ a fixed algebraic closure of $\mathbb{Q}_p$ and we recall that $|\cdot|$ extends uniquely to $\overline{\mathbb{Q}}_p$. For more information and results on p-adic fields, see, for example, the books of Amice [20], Mahler [398], Robert [473], and Schickhof [494].

In analogy with his classification of complex numbers, Mahler [378] proposed a classification of p-adic numbers. Let ξ be an element of $\mathbb{Q}_p$. For given $n \geq 1$ and $H \geq 1$, define the quantity

$$w_n(\xi, H) := \min\{|P(\xi)| : P(X) \in \mathbb{Z}[X], \mathrm{H}(P) \leq H, \deg(P) \leq n, P(\xi) \neq 0\}$$

and set

$$w_n(\xi) = \limsup_{H \to \infty} \frac{-\log(H w_n(\xi, H))}{\log H} \quad \text{and} \quad w(\xi) = \limsup_{n \to \infty} \frac{w_n(\xi)}{n}. \quad (9.4)$$

In analogy with Koksma's classification of complex numbers, we define the quantity

$$w_n^*(\xi, H) := \min\{|\xi - \alpha| : \alpha \text{ algebraic in } \mathbb{Q}_p, \deg(\alpha) \leq n,$$
$$\mathrm{H}(\alpha) \leq H, \alpha \neq \xi\}, \quad (9.5)$$

where $\mathrm{H}(\alpha)$ (*resp.* $\deg(\alpha)$) denotes the height (*resp.* the degree) of the p-adic number α, that is, by definition, the height (*resp.* the degree) of its minimal polynomial over $\mathbb{Z}$. Then, we set

$$w_n^*(\xi) = \limsup_{H \to \infty} \frac{-\log(Hw_n^*(\xi, H))}{\log H} \quad \text{and} \quad w^*(\xi) = \limsup_{n \to \infty} \frac{w_n^*(\xi)}{n}. \quad (9.6)$$

In other words, $w_n(\xi)$ (*resp.* $w_n^*(\xi)$) is the upper limit of the real numbers w for which there exist infinitely many integer polynomials $P(X)$ (*resp.* algebraic numbers α in $\mathbb{Q}_p$) of degree at most n satisfying

$$0 < |P(\xi)| \le \mathrm{H}(P)^{-w-1} \quad (resp. \ 0 < |\xi - \alpha| \le \mathrm{H}(\alpha)^{-w-1}).$$

Exactly as in Chapter 3, we call ξ an A-, S-, T- or U-number and an A^*-, S^*-, T^*- or U^*-number according to the behaviour of the sequences $(w_n(\xi))_{n \ge 1}$ and $(w_n^*(\xi))_{n \ge 1}$, respectively. As in the real and complex cases, both classifications turn out to be equivalent.

Mahler [378] proved that two algebraically dependent elements ξ and η in $\mathbb{Q}_p$ belong to the same class. Furthermore, if there is an integer polynomial $F(X, Y)$ of degree M in X and degree N in Y such that $F(\xi, \eta) = 0$, then, for any integer $n \ge 1$, we have

$$w_n(\xi) + 1 \le M(w_{nN}(\eta) + 1) \quad \text{and} \quad w_n(\eta) + 1 \le N(w_{nM}(\xi) + 1).$$

Actually, the definitions of the quantities $w_n(\xi)$ and $w_n^*(\xi)$ given here differ from those used by Mahler [378], Sprindžuk [539], and Schlickewei [496]. Indeed, for these authors, the numerator of the first fraction in (9.4) is $-\log w_n(\xi, H)$ instead of $-\log(Hw_n(\xi, H))$ (the same applies to (9.6)). This means that there is a shift by 1 in the value of the critical exponent, which however does not imply any change regarding the class of a given p-adic number. We have adopted this choice in order to have, as will be stated below, $w_n(\xi) = w_n^*(\xi) = n$ for almost all p-adic numbers ξ, with respect to the Haar measure on $\mathbb{Q}_p$ (see, for example, [539] or [86] for definition).

The shift by 1 is however not the only difference between our definition of $w_n^*(\xi)$ and the previous definitions. Indeed, in the previous literature, $w_n^*(\xi)$ is defined as in (9.6), with however $w_n^*(\xi, H)$ replaced by the minimum of $|\xi - \alpha|$ over *all* numbers $\alpha \ne \xi$ which are zero of an integer polynomial of degree at most n and height at most H. The point is that not every such α is in $\mathbb{Q}_p$. However, as can be seen by using the p-adic analogue of Rolle's Theorem, both definitions coincide (compare with the discussion before Lemma 3.1). To prove this claim, let $n \ge 1$ be an integer, $H > 1$ be a real number and ξ be a p-adic number not algebraic of degree at most n. Let α_1 be an algebraic number in $\overline{\mathbb{Q}}_p$ of height at most H and degree n_1 at most n, such that

$$|\xi - \alpha_1| = \min\{|\xi - \alpha| : \alpha \text{ algebraic in } \overline{\mathbb{Q}}_p, \deg(\alpha) \le n, \mathrm{H}(\alpha) \le H, \alpha \ne \xi\}.$$

We may assume that α_1 is not in $\mathbb{Q}_p$, otherwise there is nothing to prove. Denote by $\alpha_1^{(1)} := \alpha_1, \alpha_1^{(2)}, \dots, \alpha_1^{(n_1)}$ the conjugates of α_1 numbered in such a way that

$$|\xi - \alpha_1| \le |\xi - \alpha_1^{(2)}| \le \dots \le |\xi - \alpha_1^{(n_1)}|.$$

If $|\xi - \alpha_1| < |\xi - \alpha_1^{(2)}|$, then Krasner's Lemma (see, for example, [473], page 130) implies that α_1 lies in $\mathbb{Q}_p$, that we have excluded. Consequently, the minimal polynomial $P_1(X)$ of α_1 over $\mathbb{Z}$ has two roots α_1 and $\alpha_1^{(2)}$ with $|\xi - \alpha_1| = |\xi - \alpha_1^{(2)}|$. Let ℓ be the largest integer such that

$$|p^{-\ell}\xi - p^{-\ell}\alpha_1| = |p^{-\ell}\xi - p^{-\ell}\alpha_1^{(2)}| < p^{-1/(p-1)}.$$

The polynomial $P_1(p^\ell X + \xi)$ has then two roots in the open disc of radius $p^{-1/(p-1)}$. By the p-adic version of Rolle's Theorem (see, for example, [473], page 316), we deduce that the polynomial $P_1'(p^\ell X + \xi)$ has a root in the open unit disc. Thus, the integer polynomial $P_1'(X)$ has a root α_2 with $\mathrm{H}(\alpha_2) \le 2^{n_1}\mathrm{H}(P_1') \le 2^n n H$ and

$$|\xi - \alpha_2| < p^{-\ell} \le p^2 |\xi - \alpha_1|,$$

by our choice of ℓ. We do not know whether α_2 is a p-adic number, but, if this is not the case, we iterate this process as soon as we end up with an approximant lying in $\mathbb{Q}_p$. This always happen since the degrees of the algebraic numbers we construct from a strictly decreasing sequence. Consequently, there exists a p-adic number α with

$$\mathrm{H}(\alpha) \le 2^{n^2} n^n H \quad \text{and} \quad |\xi - \alpha| \le p^{2n} |\xi - \alpha_1|.$$

By the same argument used to conclude the proof of Lemma 3.1, this shows that the value of $w_n^*(\xi)$ does not increase if the minimum in (9.5) is taken over all algebraic numbers in $\overline{\mathbb{Q}}_p$ of degree at most n and height at most H. This proves our claim.

Proposition 9.1, due to Mahler [378], provides the p-adic analogue of Proposition 3.1.

PROPOSITION 9.1. *Let $n \ge 1$ be an integer and let ξ be a p-adic number which is not algebraic of degree at most n. Then, we have $w_n(\xi) \ge n$ and, if ξ is transcendental, $w(\xi) \ge 1$.*

Proceeding exactly as in the proof of Proposition 3.2, we also get $w_n(\xi) \geq w_n^*(\xi)$ for any p-adic number ξ and any positive integer n. Conversely, we have the following analogue of Theorem 3.4.

THEOREM 9.3. *Let $n \geq 1$ be an integer and ξ be a p-adic number which is not algebraic of degree at most n. Then, we have*

$$w_n^*(\xi) \geq w_n(\xi) - n + 1, \tag{9.7}$$

$$w_n^*(\xi) \geq \frac{w_n(\xi) + 1}{2}, \tag{9.8}$$

$$w_n^*(\xi) \geq \frac{n}{w_n(\xi) - n + 1} \tag{9.9}$$

and

$$w_n^*(\xi) \geq \frac{n}{4} + \frac{\sqrt{n^2 + 8n}}{4}. \tag{9.10}$$

Theorem 9.3 slightly improves Theorem 1 of Morrison [432], who got (9.7) and $w_n^*(\xi) \geq \min\{n - 1, (w_n(\xi) + 1)/2\}$ instead of (9.8). Throughout this paragraph, the numerical constants implied by $\ll$ depend only on ξ and on n. Morrison's proof follows exactly the same lines as that of (3.12) of Theorem 3.4, and splits into several cases. In one case, using the notation of the bottom of page 342 of [432], he obtained that there exist infinitely many primitive, integer polynomials $P(X) = a_t X^t + \ldots + a_1 X + a_0 = a_t(X - \alpha_1) \ldots (X - \alpha_t)$ of degree t at most n such that $|P(\xi)| \ll \mathrm{H}(P)^{-n-1}$, where the roots of $P(X)$ are numbered in such a way that $|\xi - \alpha_1| \leq |\xi - \alpha_2| \leq \ldots \leq |\xi - \alpha_t|$ and $|\xi - \alpha_1| < 1$, $|\xi - \alpha_2| > 1$. According to [432], this implies that $w_n(\xi) \geq n - 1$, but this yields actually the sharper estimate $w_n(\xi) \geq n$, as we show now. Without loss of generality, we may assume that $|\xi| = 1$. Hence, we get $|\alpha_1| = 1$ and $|\xi - \alpha_j| = |\alpha_j| > 1$ for $j = 2, \ldots, t$. Consequently, $|\xi^{-1} - \alpha^{-1}| = |\xi - \alpha|$ and $|\xi^{-1} - \alpha_j^{-1}| = 1$ holds for $j = 2, \ldots, t$. Setting $Q(X) := X^t P(1/X)$, we have $|Q(\xi^{-1})| = |a_0(\xi^{-1} - \alpha_1^{-1}) \ldots (\xi^{-1} - \alpha_t^{-1})| \ll \mathrm{H}(P)^{-n-1}$, thus $|a_0(\xi - \alpha)| \ll \mathrm{H}(P)^{-n-1}$. Furthermore, relations between coefficients and roots of $Q(X)$ yield that $v_p(a_j) \geq v_p(a_0)$ for $j = 0, \ldots, t - 1$. Since $P(X)$ is primitive, we get $|a_0| = 1$, hence $|\xi - \alpha| \ll \mathrm{H}(P)^{-n-1}$, as expected. Consequently, we obtain $w_n^*(\xi) \geq \min\{n, (w_n(\xi) + 1)/2\}$, which, combined with (9.7), gives (9.8).

Estimate (9.9) is due to Bugeaud and Teulié [548], page 62, and rests essentially on [550], where Teulié carried the approach of Davenport and Schmidt [182] to the field of p-adic numbers. Combined with (9.8) and Proposition 9.1, this gives (9.10). Tishchenko [566] obtained a slight refinement of (9.10).

The definitive result on the approximation of p-adic algebraic numbers by p-adic algebraic numbers of bounded degree is a consequence of (9.9) and the p-adic version of Schmidt's Subspace Theorem, due to Schlickewei [495].

THEOREM 9.4. *Let ξ be an algebraic p-adic number of degree d and let $n \geq 1$ be an integer. Then we have*

$$w_n(\xi) = w_n^*(\xi) = \min\{n, d - 1\}.$$

Teulié [551] improved an earlier result of Morrison [432] by showing that $w_2^*(\xi) \geq 2$ holds for every p-adic number ξ not algebraic of degree at most 2, thus establishing the p-adic analogue of Theorem 3.7 (this has been previously claimed by Guntermann, page 347 of [263]).

Metric results over the p-adic fields are discussed in [539] and in Chapter 6 of [86]. The Haar measure replaces the Lebesgue measure, while the Hausdorff dimension is defined exactly as the Hausdorff dimension on $\mathbb{R}$.

THEOREM 9.5. *Almost all p-adic numbers ξ satisfy $w_n(\xi) = w_n^*(\xi) = n$ for every positive integer n.*

Theorem 9.5 follows from (9.9) and the p-adic analogue of Theorem 4.2 due to Sprindžuk [539]. Previously, Turkstra [573] proved in his dissertation that almost all p-adic numbers are S-numbers. Further, partial results towards the resolution of the p-adic version of Mahler's conjecture were due to Lock [371], Kasch and Volkmann [312, 313] (however, there is a gap in [313]), and Sprindžuk [535].

The p-adic analogue of Theorem 5.2 has been established by Melničuk [417] and that of Theorem 5.7 by Morotskaya [431].

THEOREM 9.6. *For any integer $n \geq 1$ and any real number $\tau \geq 1$, the Hausdorff dimension of any of the sets*

$$\{\xi \in \mathbb{Q}_p : w_n(\xi) \geq \tau(n + 1) - 1\}, \quad \{\xi \in \mathbb{Q}_p : w_n^*(\xi) \geq \tau(n + 1) - 1\},$$

$$\{\xi \in \mathbb{Q}_p : w_n(\xi) = \tau(n + 1) - 1\}, \quad \{\xi \in \mathbb{Q}_p : w_n^*(\xi) = \tau(n + 1) - 1\}$$

is equal to $1/\tau$.

Theorem 9.6 is a restatement of Theorem 6.19 of [86]. Partial results were obtained by Bernik and Morotskaya [91].

Beresnevich, Bernik and Kovalevskaya [69] proved that algebraic numbers of bounded degree in $\mathbb{Q}_p$ form an optimal regular system and they established a complete p-adic analogue of Theorem 4.5 (see also Kovalevskaya [339, 340] for the convergence case and the p-adic analogue of Theorem 4.4, and Beresnevich and Kovalevskaya [72] for the p-adic analogue of [62]). The p-adic analogue of Theorem 4.6 is due to Kleinbock and Tomanov [331].

Schlickewei [496] (see also [279]) proved that p-adic T-numbers exist by adapting to the p-adic case the proof of Schmidt [507]. Alniaçik [14] tried to carry the proof of Schmidt [508] to the p-adic case, but the definition of T-numbers he used is not the correct one. Hernandez [279] adapted [128] to the p-adic case and proved that, for any integer $n \geq 2$, there exist p-adic numbers ξ for which $w_n(\xi)$ differs from $w_n^*(\xi)$.

The existence of p-adic U_m-numbers has been proved by Alniaçik [8, 13], who carried some of LeVeque's results [361] to the p-adic case. Zeren [610], Oryan [445], Xin [602], and Yilmaz [606] have constructed p-adic U_m-numbers.

9.4 Approximation in fields of formal power series

Let $\mathbf{k}$ be a (finite or infinite) field of arbitrary characteristic and denote by $\mathbf{k}[x]$ the ring of polynomials with coefficients in $\mathbf{k}$ and by $\mathbf{k}(x)$ the quotient field of $\mathbf{k}[x]$. We define a non-Archimedean absolute value $|\cdot|$ on $\mathbf{k}(x)$ by setting $|0| = 0$ and

$$|f/g| := \exp\{\deg_x(f) - \deg_x(g)\}, \quad \text{for any non-zero}$$
$$\text{polynomials } f, g \text{ in } \mathbf{k}[x],$$

where $\deg_x$ denotes the degree of a polynomial. The completion of $\mathbf{k}(x)$ for this absolute value is the field of formal power series $\mathbb{K}$ with coefficients in $\mathbf{k}$. We denote by $\hat{\mathbb{K}}$ an algebraic closure of $\mathbf{k}(x)$ contained in the algebraic closure of $\mathbb{K}$.

Let n be a positive integer and ξ be in $\mathbf{k}$. In analogy with Chapter 3, we define $w_n(\xi)$ as the supremum of the real numbers w for which there exist infinitely many polynomials $P(X)$ in $\mathbf{k}[x][X]$ of degree at most n satisfying

$$0 < |P(\xi)| \leq \mathrm{H}(P)^{-w},$$

and $w_n^*(\xi)$ as the supremum of the real numbers w for which there exist infinitely many algebraic numbers α in $\hat{\mathbb{K}}$ of degree at most n and height arbitrarily large satisfying

$$0 < |\xi - \alpha| \leq \mathrm{H}(\alpha)^{-w-1}.$$

Here, the height of $P(X)$ is the maximum of the absolute values of its coefficients (these are elements of $\mathbf{k}[x]$) and the height of α is the height of its minimal polynomial over $\mathbf{k}[x]$. We require the height of the approximants to be arbitrarily large to avoid trivialities: when the field $\mathbf{k}$ is infinite, there are infinitely many algebraic elements of bounded degree and bounded height in $\hat{\mathbb{K}}$, unlike in the complex and p-adic cases.

With the above definitions of $w_n(\xi)$ and $w_n^*(\xi)$, we define the classes A, S, T, U, A^*, S^*, T^*, and U^* exactly as in Chapter 3. For $\mathbf{k}$ a finite field, this classification has been introduced by Bundschuh [141].

Mahler [382] worked out an analogue in fields of formal power series to Minkowski's theory of geometry of numbers. It follows from the analogue of Theorem B.2 that $w_n(\xi) \geq n$ holds for any ξ in $\mathbb{K}$ which is not algebraic of degree at most n (when $\mathbf{k}$ is finite, see Amou [30] for a best possible statement obtained via Dirichlet's *Schubfachprinzip*).

When ξ is algebraic of degree d and $\mathbf{k}$ has characteristic zero, Ratliff [469] (see also Dubois [211]) proved that $w_n(\xi) = \min\{n, d-1\}$. This does not hold any more when $\mathbf{k}$ is a finite field, as shown by de Mathan [412].

Guntermann [263] investigated the analogue of Theorem 3.4 in fields of power series and proved Theorem 9.7 below.

THEOREM 9.7. *Let n be a positive integer. For any ξ in $\mathbb{K}$ which is not algebraic of degree at most n, we have $w_n^*(\xi) \geq (n+1)/2$ and $w_2^*(\xi) \geq 2$ if $n = 2$.*

When $\mathbf{k}$ is a field of characteristic zero, Theorem 9.7 and the inequality $w_n^*(\xi) \geq w_n(\xi) - n + 1$ were previously obtained by Sprindžuk [534].

Sprindžuk [539] proved the analogue of Theorem 4.1 when $\mathbf{k}$ is a finite field. Like in the p-adic case, the Haar measure replaces the Lebesgue measure (see [539], Part II, Chapter I).

THEOREM 9.8. *When $\mathbf{k}$ is a finite field, almost all ξ in $\mathbb{K}$ satisfy $w_n(\xi) = n$ for every positive integer n.*

At the end of [263], Guntermann stated without proof that one can deduce from Theorem 9.8 that almost all ξ in $\mathbb{K}$ satisfy $w_n^*(\xi) = n$ for every integer $n \geq 1$. However, this presumably needs an estimate like $w_n^*(\xi) \geq n/(w_n(\xi) - n + 1)$, which does not seem to be in the literature; most likely, results of [382] combined with the method of proof of Theorem 2.10 (see Exercise 3.3) should yield such an estimate.

As noticed by Sprindžuk [534] (see also [539], page 150), taking for $\mathbf{k}$ the field of complex numbers, we get a classification of analytic functions in terms of their behaviour with respect to approximation by algebraic functions.

When $\mathbf{k}$ is a finite field, Kristensen [341] established the analogues of Theorems 1.10 and 5.2, as well as multidimensional extensions.

Examples of U_2-numbers for $\mathbf{k}$ arbitrary have been given by Burger and Dubois [149] and Dubois [212]. Oryan [443] gave explicit constructions of U_m-numbers when $\mathbf{k}$ is a finite field.

9.5 Notes

• As for multiplicative approximation, metric results in the complex domain involving the function Π_+ introduced in Chapter 4 have been established by Yu [608].

• Bernik and Morozova [92, 93] studied approximation of complex numbers by lacunary polynomials. They established [92] the exact value of the Hausdorff dimension of the set of complex numbers at which the lacunary polynomials of the form $a_3 X^m + a_2 X^n + a_1 X^\ell + a_0$ approximate 0 with a given error term.

• Generalizations of Dirichlet's Theorem 1.1 and Khintchine's Transference Theorem B.5 to S-integer approximation in a number field have been worked out by Burger [143].

• For simultaneous Diophantine approximation and improvements of Dirichlet's Theorem 1.1 in real quadratic fields, see Burger [146, 147].

• The analogues in $\mathbb{Z}[i]$ of Theorems 6.6 and 6.8 have been established in [71].

• Hightower [280] proved the analogue of Theorem 3.7 for the approximation of complex numbers by algebraic numbers of degree at most 2 over a given imaginary quadratic field.

• Markovich [407] proved the case $n = 2$ of Theorem 4.2 for polynomials with coefficients in a given real number field.

• A. L. Schmidt [498] studied the approximation of quaternions.

• Teulié [552] established the p-adic analogue of a result of Peck [450].

• Abercrombie [1] established the existence of badly approximable p-adic integers and thus provided a p-adic analogue of Jarník's results [289]. He used an approximation scheme of Mahler [381], which is a substitute for the continued fraction algorithm. His work can be viewed as the p-adic analogue of Schmidt's paper [503].

• Abercrombie [2] generalized the case $n = 1$ of Theorem 9.6 to systems of linear forms and established the p-adic analogue of a result of Bovey and Dodson [116]. This has been further extended by Dickinson, Dodson, and Yuan [194].

• Diophantine approximation by conjugate algebraic integers in p-adic fields has been studied by Roy and Waldschmidt [484].

• Jarník [297] proved a p-adic generalization of Khintchine's Theorem 1.10. It has been extended to systems of linear forms by Lutz [373].

• A p-adic inhomogeneous analogue of Theorem 4.2 has been established by Bernik, Dickinson, and Yuan [85]. For $n = 1$ (that is, for rational inhomogeneous approximation), the result is due to Lutz [373].

• Zheludevich [614] proved Sprindžuk's Conjecture H_3 [541] on simultaneous approximation in the real, complex, and p-adic fields. This generalizes an earlier result of Bernik [76] (see Notes to Chapter 4) and allowed him to extend [612, 613, 615] Theorem 4.2 to simultaneous approximation by algebraic numbers. See the survey of Bernik [80] for further references. Zheludevich's results have been extended by Kleinbock and Tomanov [331], in particular to multiplicative approximation. See also Kalosha [308].

• The p-adic analogues of Theorems 6.6 and 6.8 have been established by Beresnevich, Dickinson, and Velani [71].

• Jarník [296] established a p-adic transference theorem.

• Slesoraĭtene [526] worked out an analogue of Sprindžuk's theorem 4.2 for polynomials of degree two in two p-adic variables. For arbitrary degrees, a weaker result is due to Yanchenko [605].

• Menken [418] studied semi-strong p-adic U-numbers. He proved the p-adic analogue of Erdös' result [227], that is, that any p-adic number is the sum of two p-adic Liouville numbers.

• Väänänen [574] established transcendence measures for the values of the p-adic exponential function at algebraic points. Xu [603] proved that the values of p-adic E-functions at algebraic points are S-numbers and he established a trancendence measure. Molchanov [426] showed that values at some p-adic algebraic points of Mahler functions satisfying certain functional equations are p-adic S-numbers.

• Diophantine approximation over the ring of adeles has been studied by D. G. Cantor [151], who established an analogue of Khintchine's Theorem 1.10.

• Zong [617] worked out analogues of Mahler's results [401] in p-adic fields and in fields of formal power series.

• Zero–infinity laws for Diophantine approximation over p-adic fields and fields of formal power series over a finite field have been obtained by Bugeaud, Dodson, and Kristensen [135].

• Sprindžuk [539], page 160, claimed (and this has been later established by Kleinbock [328]) that Theorem 4.3 holds in the complex case, with the exponent n replaced by $(n-1)/2$. According to him, 'probably a similar result is true for locally compact fields with a non-Archimedean valuation'.

• Burger [144] investigated decompositions of elements from an arbitrary local field into Liouville numbers.

• An analogue of Khintchine's Theorem 1.10 for formal power series over a finite field has been established by de Mathan [410] (see also Fuchs [252], Kristensen [341], and Inoue and Nakada [285]).

• Bundschuh [141] established transcendence measures for classical elements of fields of formal power series over a finite field. His method has been applied by Özdemir [448, 449], who showed that some explicitly given elements of $\mathbb{K}$ are not U-numbers.

• Amou [30] established a formal power series analogue of Theorem 8.3 when $\mathbf{k}$ is a finite field.

• When $\mathbf{k}$ is finite, Becker [57] established transcendence measures for the values of generalized Mahler functions and gave the first explicit examples of S-numbers.

• Badly approximable linear forms over a field of formal series have been considered by Kristensen [342], who provided the analogue of Schmidt's generalization [505] of Jarník's result [288].

10

Conjectures and open questions

We begin this Chapter with a short survey on the celebrated Littlewood Conjecture. We then gather open problems encountered in the preceding Chapters with several new questions.

The reader interested in open questions in Diophantine approximation is also directed to survey papers by Schmidt [513], Waldschmidt [592], Beresnevich and Bernik [65], and to the Appendix of Montgomery's book [427].

10.1 The Littlewood Conjecture

A famous open problem in Diophantine approximation is Littlewood's Conjecture which claims that, for any given pair (ξ, η) of real numbers and for any positive real number ε, there exist integers q, r, and s with $q > 0$ such that

$$q \cdot |q\xi - r| \cdot |q\eta - s| \leq \varepsilon.$$

Denoting by $\| \cdot \|$ the distance to the nearest integer, this statement is equivalent to

$$\inf_{q \geq 1} q \cdot \|q\xi\| \cdot \|q\eta\| = 0. \tag{10.1}$$

Obviously, the conjecture holds true if ξ or η have unbounded partial quotients in their continued fraction expansions. This is also the case if the numbers 1, ξ, and η are linearly dependent over the rational integers, by Dirichlet's Theorem 1.1.

Apart from these easy remarks, very little is known towards a proof or a disproof of Littlewood's Conjecture. The first important contribution is due to Cassels and Swinnerton-Dyer [158] who showed that (10.1) holds when 1, ξ, and η belong to the same cubic field. To this end, they applied Minkowski's Theorem B.2 to get a dual formulation of the conjecture (actually, they only used and proved the 'if' part of Lemma 10.1).

LEMMA 10.1. *Let ξ and η be real numbers such that 1, ξ, and η are linearly independent over the rational integers. Then*

$$\inf_{q \geq 1} q \cdot \|q\xi\| \cdot \|q\eta\| = 0$$

holds if, and only if, we have

$$\inf\{|x + \xi y + \eta z| \cdot \max\{|y|, 1\} \cdot \max\{|z|, 1\} : x, y, z \in \mathbb{Z},$$
$$(y, z) \neq (0, 0)\} = 0.$$

An open problem is to decide whether Littlewood's Conjecture holds for pairs $(\xi, 1/\xi)$, when the real number ξ has bounded partial quotients. It follows from Lemma 10.1 that the answer is positive if ξ is not badly approximable by quadratic numbers, that is, when we have

$$\inf\{|x + \xi y + \xi^2 z| \cdot \max\{|x|, |y|, |z|\}^2 : x, y, z \in \mathbb{Z},$$
$$(x, y, z) \neq (0, 0, 0)\} = 0. \quad (10.2)$$

However, the existence of *transcendental* real numbers not satisfying (10.2) is still unproved, while it follows from Theorem A.1 that real cubic numbers do not verify (10.2). Thus, Cassels and Swinnerton-Dyer [158] provide examples of real numbers ξ such that Littlewood's Conjecture is true for the pair $(\xi, 1/\xi)$, although (10.2) does not hold. Since it remains unknown whether or not cubic real numbers have bounded partial quotients, their result does not yield examples of pairs of badly approximable real numbers for which Littlewood's Conjecture holds. However, Pollington and Velani [460] confirmed that such pairs do exist.

THEOREM 10.1. *Let ξ be a real number with bounded partial quotients. Then the Hausdorff dimension of the set of real numbers η with bounded partial quotients such that the pair (ξ, η) satisfies the Littlewood Conjecture is equal to 1.*

Chowla and DeLeon [162] observed that $(\sqrt{2}, \sqrt{3})$ satisfies the Littlewood Conjecture provided 0 is a limit point of the sequence $((\sqrt{6}/4)(\sqrt{2} + 1)^n)_{n \geq 1}$ modulo 1. The first explicit non-trivial examples of pairs of real numbers (ξ, η) such that (10.1) holds have been given by de Mathan [413].

De Mathan and Teulié [414] have proposed a very interesting 'mixed Littlewood Conjecture', claiming that, for any real number ξ and any prime number p, we have

$$\inf_{q \geq 1} q \cdot \|q\xi\| \cdot |q|_p = 0. \quad (10.3)$$

They proved that (10.3) holds (actually, their result is much more general) when ξ is a quadratic irrationality, which can be viewed as the p-adic analogue of the above quoted result from [158].

10.2 Open questions

This Section is devoted to open problems. Instead of formulating them in terms of questions, we merely prefer to propose statements whose validity is open. In most of the cases there is no evidence for, or against, the assertion claimed. We do not recall the partial results obtained towards these problems, since they can be easily found in the present book.

We begin by assertions on Mahler's and Koksma's classifications of numbers, we continue with metric number theory, and we end up with rational approximation and questions related to polynomials.

Problem 1 has been pointed out at the end of Section 3.4 of Chapter 3. No argument can reasonably be put forward against it, although there is no evidence in favour of it.

PROBLEM 1. (Main Problem) *Let* $(w_n)_{n\geq 1}$ *and* $(w_n^*)_{n\geq 1}$ *be two non-decreasing sequences in* $[1, +\infty]$ *such that*

$$n \leq w_n^* \leq w_n \leq w_n^* + n - 1, \quad for\ any\ n \geq 1.$$

Then there exists a real transcendental number ξ *such that*

$$w_n(\xi) = w_n \quad and \quad w_n^*(\xi) = w_n^* \ for\ any\ n \geq 1.$$

The first conjecture, due to Wirsing [598], deals with the approximation of real transcendental numbers by real algebraic numbers of bounded degree.

PROBLEM 2. (Wirsing's Conjecture) *For any integer* $n \geq 1$ *and for any real transcendental number* ξ, *we have* $w_n^*(\xi) \geq n$.

We propose a slightly weaker problem than Wirsing's Conjecture.

PROBLEM 3. *The* $*$-*type of any* S^*-*number is at least equal to 1, that is, we have*

$$\limsup_{n\to\infty} \frac{w_n^*(\xi) + 1}{n + 1} \geq 1$$

for any S^*-*number* ξ.

As we discussed in Chapter 3, there are several possibilities to define the type of an S-number, but two of them may coincide.

PROBLEM 4. *There exist real S-numbers ξ with*

$$\limsup_{n\to\infty} \frac{w_n(\xi)+1}{n+1} < \sup_{n\geq 1} \frac{w_n(\xi)+1}{n+1}$$

$$and/or \quad \limsup_{n\to\infty} \frac{w_n^*(\xi)+1}{n+1} < \sup_{n\geq 1} \frac{w_n^*(\xi)+1}{n+1}.$$

Corollary 3.2 asserts that $w_n^*(\xi) = n$ holds if $w_n(\xi) = n$, but the converse is an open question.

PROBLEM 5. *For any positive integer n, we have $w_n(\xi) = n$ if $w_n^*(\xi) = n$.*

Corollary 3.2 also suggests the following claim.

PROBLEM 6. *Any S-number of type 1 is an S^*-number of $*$-type 1.*

It is known (see Chapter 7) that there exist real numbers ξ with $w_n(\xi) \neq w_n^*(\xi)$ for every integer $n \geq 2$, but it is not known if there exist S-numbers with this property.

PROBLEM 7. *There exist real numbers ξ such that $w(\xi) \neq w^*(\xi)$.*

The existence of T-numbers has been proved by Schmidt (see Chapter 7). However, it is at present not known whether or not there exist T-numbers with some specific properties.

PROBLEM 8. *There exist T-numbers ξ such that*

$$\lim_{n\to+\infty} \frac{w_n(\xi)}{n} \neq +\infty.$$

PROBLEM 9. *Let τ be a real number with $1 \leq \tau < 3$. There exist real T-numbers of type τ.*

PROBLEM 10. *Give an effective proof of the existence of T-numbers.*

Let ξ be a transcendental real number and let n, k be positive integers with $k \geq 2$. At the end of Section 3.2 (*resp.* in Exercise 3.6), we established a relation between $w_n(\xi)$ and $w_n(\xi^k)$ (*resp.* between $w_n^*(\xi)$ and $w_n^*(\xi^k)$). A natural question (answered when $n = 1$ in Exercise 3.7) asks whether or not it is possible to refine these estimates.

PROBLEM 11. *For any positive integers n, k, and any non-negative real number δ, there exist real transcendental numbers ξ such that*

$$k\big(w_n(\xi^k)+1\big) = w_n(\xi)+1+\delta$$

and

$$k\big(w_n^*(\xi^k) + 1\big) = w_n^*(\xi) + 1 + \delta.$$

A similar question concerns the relations between $w_{kn}(\xi)$ and $w_n(\xi^k)$.

PROBLEM 12. *For any positive integers n, k and any non-negative real number δ, there exist real transcendental numbers ξ such that*

$$w_{kn}(\xi) = w_n(\xi^k) + \delta$$

and

$$w_{kn}^*(\xi) = w_n^*(\xi^k) + \delta.$$

Problems 11 and 12 invite us to ask whether all integer powers of an S-number have same type.

PROBLEM 13. *For any S-number ξ and any non-zero integer k we have $t(\xi) = t(\xi^k)$ and $t^*(\xi) = t^*(\xi^k)$.*

The formulation of Problem 14, closely related to Problem 4, is rather vague. There is no known result in this direction, and Theorem 7.7 suggests that Problem 14 could be very difficult.

PROBLEM 14. *Let ξ be a transcendental real number and let n, m be positive integers with $n > m$. Give a lower estimate of $w_n(\xi)$ (resp. $w_n^*(\xi)$) in terms of $n, m,$ and $w_m(\xi)$ (resp. $w_m^*(\xi)$).*

In Section 3.6, we introduced the functions w_n', $\hat{w}_n$, $\hat{w}_n'$, and $\hat{w}_n^*$. Problem 15 is motivated by Theorem 3.9 establishing a link between w_n and simultaneous rational approximation.

PROBLEM 15. *Let $n \geq 2$ be an integer and let $w_n' \leq n$ and $w_n \geq n$ be two real numbers satisfying*

$$\frac{n}{w_n - n + 1} \leq w_n' \leq \frac{(n-1)w_n + n}{w_n}.$$

There exist real numbers ξ such that $w_n(\xi) = w_n$ and $w_n'(\xi) = w_n'$.

Jarník's results [293, 295] on Khintchine's Transference Theorem B.5 allows us to think that the answer of Problem 15 could be affirmative.

PROBLEM 16. *For any integer $n \geq 2$, find relations between the six functions w_n, w_n^*, w_n', $\hat{w}_n$, $\hat{w}_n^*$, and $\hat{w}_n'$. Find the sets of values taken by each of these functions and the sets of their limit points.*

PROBLEM 17. *For any integer $n \geq 3$ and any real transcendental number ξ, we have $\hat{w}_n(\xi) = \hat{w}'_n(\xi) = n$.*

By Proposition 3.3 and the proof of Theorem 2.11, a positive answer to Problem 17 implies a positive answer to Problem 18.

PROBLEM 18. *For any integer $n \geq 4$, any positive real number ε, and any real transcendental number ξ, there exist a constant $c_1(\xi, n, \varepsilon)$ and infinitely many real algebraic integers α of degree less than or equal to n such that*

$$|\xi - \alpha| \leq c_1(\xi, n, \varepsilon) \operatorname{H}(\alpha)^{-n+\varepsilon}.$$

The assumption $n \geq 3$ in Problem 17 (*resp.* $n \geq 4$ in Problem 18) is needed because of Roy's results [480, 481] asserting the existence of real transcendental numbers ξ with $\hat{w}'_2(\xi) = (1 + \sqrt{5})/2$ and such that $|\xi - \alpha| \geq c_2 \operatorname{H}(\alpha)^{-(3+\sqrt{5})/2}$ holds for any algebraic integer α of degree at most three, for a suitable positive constant c_2 depending only on ξ. In view of this result, it may be doubtful that the answers of Problems 17 and 18 are affirmative.

Recall that an algebraic unit is an algebraic integer whose minimal polynomial over $\mathbb{Z}$ has constant coefficient equal to 1. Approximation by algebraic units has been investigated by Teulié [549].

PROBLEM 19. *For any integer $n \geq 3$, any positive real number ε and any real transcendental number ξ, there exist a constant $c_3(\xi, n, \varepsilon)$ and infinitely many real algebraic units α of degree less than or equal to n such that*

$$|\xi - \alpha| \leq c_3(\xi, n, \varepsilon) \operatorname{H}(\alpha)^{-n+1+\varepsilon}.$$

It is known that the conclusions of Problems 17, 18, and 19 hold true for almost all real numbers ξ.

Théorème 6 of [140] and Theorem 2 of [125] give links between approximation by algebraic integers of degree at most n and algebraic numbers of degree at most $n - 1$.

PROBLEM 20. *For any integer n with $n \geq 3$, compare the quality of approximation by algebraic numbers of degree at most $n - 1$ with the quality of approximation by algebraic integers of degree at most n, and with the quality of approximation by algebraic units of degree at most $n + 1$.*

We now state more precise questions on the approximation by algebraic numbers. For a given positive integer n and a given real number ξ, we observe that the quantities $w_n(\xi)$ and $w_n^*(\xi)$ have been defined as the infima of two sets, hence, they provide no information on the following problems:

(P.1) Do there exist a positive constant $c(\xi, n)$ and infinitely many integer polynomials $P(X)$ of degree at most n such that

$$|P(\xi)| \leq c(\xi, n) \, \mathrm{H}(P)^{-w_n(\xi)}?$$

(P.2) Do there exist a positive constant $c^*(\xi, n)$ and infinitely many algebraic numbers α of degree at most n such that

$$|\xi - \alpha| \leq c^*(\xi, n) \, \mathrm{H}(\alpha)^{-w_n^*(\xi)-1}?$$

After having determined the exact value of $w_n(\xi)$ and that of $w_n^*(\xi)$, it is natural to ask whether (P.1) and (P.2) have an affirmative answer. This corresponds to the notion of 'eigentlicher Index' introduced by Koksma [333]. In order to take into account these two problems, we introduce some new notation.

DEFINITION 10.1. *Let n be a positive integer and ξ be a real number. Let w be a positive real number. We write $w_n(\xi) \doteq w$ if $w_n(\xi) = w$ holds and the answer to (P.1) is positive. Likewise, we write $w_n^*(\xi) \doteq w$ if $w_n^*(\xi) = w$ holds and the answer to (P.2) is positive.*

For instance, it follows from Theorem 4.5 that almost all real numbers ξ satisfy $w_n(\xi) \doteq n$ and $w_n^*(\xi) \doteq n$ for any positive integer n. We propose a refinement of Problem 1.

PROBLEM 21. *Let $(w_n)_{n \geq 1}$ and $(w_n^*)_{n \geq 1}$ be two non-decreasing sequences in $[1, +\infty]$ such that*

$$n \leq w_n^* \leq w_n \leq w_n^* + n - 1, \quad \text{for any } n \geq 1.$$

Then there exists a real transcendental number ξ such that

$$w_n(\xi) \doteq w_n \quad \text{and} \quad w_n^*(\xi) \doteq w_n^*.$$

Some of the results of Chapter 7 show, or can be adapted to show, that the answer to Problem 21 is positive for certain pairs (n, w).

Schmidt [512], p. 258, proposed a slightly stronger conjecture than Wirsing's; however, he expressed in Summer 2003 serious doubts on the validity of his conjecture.

PROBLEM 22. (Schmidt's Conjecture) *For any positive integer n and any real transcendental number ξ, there exist a constant $c_4(\xi, n)$ and infinitely many real algebraic numbers α of degree less than or equal to n such that*

$$|\xi - \alpha| \leq c_4(\xi, n) \, \mathrm{H}(\alpha)^{-n-1}.$$

Using Definition 10.1, Schmidt's Conjecture can be rephrased as follows: For any integer $n \geq 1$ and any real transcendental number ξ, we have either $w_n^*(\xi) > n$ or $w_n^*(\xi) \doteq n$.

Davenport and Schmidt [180] gave a positive answer to Problem 22 in the case $n = 2$, but Problem 23, where we fix the exact degree of the approximants instead of an upper bound for it, remains unsolved, even for $n = 2$.

PROBLEM 23. *For any integer $n \geq 2$ and any real transcendental number ξ, there exist a constant $c_5(\xi, n)$ and infinitely many real algebraic numbers α of degree n such that*

$$|\xi - \alpha| \leq c_5(\xi, n) \, \mathrm{H}(\alpha)^{-n-1}.$$

Results from Roy [480, 481] would speak in favour of the existence of transcendental numbers which do not satisfy the conclusion of Problem 23, even for $n = 2$.

The next problem deals with transcendental numbers badly approximable by algebraic numbers.

PROBLEM 24. *Let n be a positive integer. There exist a real transcendental number ξ and positive constants $c_6(\xi, n)$ and $c_7(\xi, n)$ such that*

$$|\xi - \alpha| \geq c_6(\xi, n) \, \mathrm{H}(\alpha)^{-n-1} \quad \text{*for any real algebraic number α*} \atop \text{*of degree $\leq n$*}$$

and

$$|\xi - \alpha| \leq c_7(\xi, n) \, \mathrm{H}(\alpha)^{-n-1} \quad \text{*for infinitely many real*} \atop \text{*algebraic numbers α of degree $\leq n$.*}$$

Theorem 1.9 answers positively Problem 24 for $n = 1$, however, there is no contribution for $n \geq 2$. Furthermore, Theorem 2.9 asserts that for any integer $d \geq 2$, real algebraic numbers of degree d are badly approximable by real algebraic numbers of degree at most $d - 1$.

Problem 24 can be formulated in terms of polynomials, as well.

PROBLEM 25. *Let n be a positive integer. There exist a real transcendental number ξ and a positive constant $c_8(\xi, n)$ such that*

$$|P(\xi)| \geq c_8(\xi, n) \, \mathrm{H}(P)^{-n} \quad \text{*for any integer*} \atop \text{*polynomial $P(X)$ of degree $\leq n$.*} \qquad (10.4)$$

Khintchine [318] proved that the set of real numbers ξ for which (10.4) holds for some positive integer n and some positive constant $c_8(\xi, n)$ has Lebesgue measure zero.

Very little is known regarding transcendental numbers which are badly approximable by algebraic numbers of bounded degree.

PROBLEM 26. *Let* $n \geq 2$ *be an integer. There exist transcendental real numbers* ξ *such that, for some positive constant* $c_9(\xi, n)$, *we have*

$$|\xi - \alpha| \geq c_9(\xi, n) \, \mathrm{H}(\alpha)^{-n-1} \left(\log 3\mathrm{H}(\alpha)\right)^{-1},$$

for any non-zero real algebraic number α *of degree at most* n.

Of course, if Problem 26 could be solved affirmatively, it then would be desirable to determine the Hausdorff dimension of such exceptional sets. Problem 26 can be formulated in terms of polynomials, as well.

The next problem extends a question posed by Beresnevich, Dickinson, and Velani [70] in the case of (simultaneous) rational approximation.

PROBLEM 27. *Let* n *be a positive integer and let* $\tau > 1$ *be real. Is the set of real numbers* ξ *for which there exists a positive constant* $c_{10}(\xi)$ *such that*

$$|\xi - \alpha| \leq \mathrm{H}(\alpha)^{-\tau(n+1)} \quad \text{for infinitely many } \alpha \text{ in } \mathbb{A}_n$$

and

$$|\xi - \alpha| \geq c_{10}(\xi) \, \mathrm{H}(\alpha)^{-\tau(n+1)} \quad \text{for every } \alpha \text{ in } \mathbb{A}_n$$

non-empty? If yes, determine its Hausdorff dimension.

Problem 27 has been solved by Bugeaud [129] when $n = 1$. One may also replace the approximation functions $x \mapsto x^{-\tau(n+1)}$ by more general non-increasing functions Ψ.

Problem 28 is likely to be difficult, especially for $n \geq 3$.

PROBLEM 28. *For any integer* $n \geq 2$ *and any real number* $\tau > 1$, *determine the Hausdorff measure at the critical exponent* $1/\tau$ *of the set* $\mathcal{K}_n(\tau)$.

Problems 29 and 30 deal with the exponents of approximation introduced in Section 3.6.

PROBLEM 29. *Let* $n \geq 2$ *be an integer. Let* w'_n *be a real number with* $1 \leq w'_n < n$. *Determine the Hausdorff dimension of the set of real numbers* ξ *such that* $w'_n(\xi) = w'_n$.

PROBLEM 30. *Determine the Hausdorff dimension of the set of real numbers* ξ *such that* $\hat{w}_2(\xi) > 2$ *(resp.* $\hat{w}_2^*(\xi) > 2$*).*

There are very few results concerning the existence of real numbers with prescribed order of approximation by algebraic numbers of different degrees.

PROBLEM 31. *Let n and n′ be positive distinct integers. Let $\tau \geq 1$ and $\tau' \geq 1$ be real numbers. Then the set $\mathcal{W}_n^*(\tau) \cap \mathcal{W}_{n'}^*(\tau')$ is non-empty, and its Hausdorff dimension is equal to $\min\{1/\tau, 1/\tau'\}$.*

PROBLEM 32. *Let $n \geq 2$ be an integer and $\tau \geq 1$ be a real number. The sets $\mathcal{W}_n(\tau)$ and $\mathcal{W}_n^*(\tau)$ contain badly approximable real numbers.*

The existence of normal (*resp.* non-normal) numbers with specific approximation properties by rational numbers has been investigated by several authors (see Notes in Chapters 1 and 5).

PROBLEM 33. *Let n be a positive integer and $\tau > 1$ be a real number. The sets $\mathcal{W}_n(\tau)$ and $\mathcal{W}_n^*(\tau)$ contain normal numbers and numbers which are simply normal in no base.*

PROBLEM 34. *There exist normal T-numbers and T-numbers which are simply normal in no base.*

The next two problems concern Diophantine approximation on sets of Lebesgue measure zero. They were posed by Mahler [394, 399, 400].

PROBLEM 35. *There exist very well approximable numbers, other than Liouville numbers, in the triadic Cantor set.*

PROBLEM 36. *Every point in the triadic Cantor set is either rational or transcendental.*

It is known that the triadic Cantor set contains badly approximable real numbers (see, for example, Theorem 10.3 of [329]).

We propose three problems on multiplicative approximation. With the functions w_n^+ introduced in Section 5.7, we set

$$w^+(\xi) = \limsup_{n \to +\infty} w_n^+(\xi),$$

and we say that ξ is an

$$A^+\text{-number, if } w^+(\xi) = 0;$$
$$S^+\text{-number, if } 0 < w^+(\xi) < +\infty;$$
$$T^+\text{-number, if } w^+(\xi) = +\infty \text{ and } w_n^+(\xi) < +\infty \text{ for any } n \geq 1;$$
$$U^+\text{-number, if } w^+(\xi) = +\infty \text{ and } w_n^+(\xi) = +\infty \text{ from some } n \text{ onwards.}$$

PROBLEM 37. *What can be said about the sets of A^+-, S^+-, T^+-, and U^+-numbers?*

Problem 37 requires presumably new methods and seems to be very difficult.

PROBLEM 38. *Let* $n \geq 1$ *be an integer and* $\Psi : \mathbb{R}_{\geq 1} \rightarrow \mathbb{R}_{>0}$ *be a non-increasing continuous function such that*

$$\sum_{h \geq 1} (\log h)^{n-1} \Psi(h) = +\infty.$$

Then, for almost all real numbers ξ, *the equation*

$$|P(\xi)| < \Psi\big(\Pi_+(P)\big) \qquad (10.5)$$

has infinitely many solutions in integer polynomials $P(X)$ *of degree at most* n.

Theorem 4.7 asserts that if the above sum converges, then (10.5) has only a finite number of solutions in integer polynomials $P(X)$ of degree at most n.

PROBLEM 39. *Let* ε *be a positive real number and let* n *be a positive integer. Then the Hausdorff dimension of the set* $\mathcal{K}_n^+(\tau)$ *of real numbers* ξ *such that the inequality*

$$|P(\xi)| < \Pi_+(P)^{-2\tau+1}$$

has infinitely many solutions in integer polynomials $P(X)$ *of degree at most* n *is equal to* $1/\tau$.

This is Theorem 5.2 for $n = 1$ and has been proved for $n = 2$ (see Exercise 5.6).

   We recall a celebrated problem on rational approximation of algebraic numbers, which is perhaps due to Khintchine (see [521], p. 156).

PROBLEM 40. *Any real algebraic number* ξ *of degree at least* 3 *has unbounded partial quotients in its continued fraction expansion, that is, for any positive real number* ε, *there exists a rational number* p/q *such that* $|\xi - p/q| < \varepsilon/q^2$.

As observed in the Notes in Chapter 2, known upper bounds for the partial quotients of real algebraic numbers of degree at least 3 are presumably much larger than the truth.

PROBLEM 41. *Let* ξ *be an algebraic number of degree at least* 3. *Give sharp upper bounds for its partial quotients.*

We mentioned in Section 1.4 the Duffin–Schaeffer Conjecture.

PROBLEM 42. (*Duffin–Schaeffer's Conjecture*) *Let* $\Psi : \mathbb{R}_{\geq 1} \rightarrow \mathbb{R}_{\geq 0}$ *be some continuous function. Then the set*

$$\left\{ \xi \in \mathbb{R} : \left| \xi - \frac{p}{q} \right| < \Psi(q) \text{ for infinitely many rationals } \frac{p}{q} \right.$$
$$\left. \text{with } \gcd(p, q) = 1 \right\}$$

has full Lebesgue measure if the sum $\sum_{q=1}^{+\infty} \varphi(q)\Psi(q)$ *diverges.*

There are many open problems concerning continued fractions. One of them is related to Ridout's Theorem 2.3, and has been firstly formulated by Erdös and Mahler [229].

PROBLEM 43. *Let ξ be a real transcendental number for which there exist a positive number M and infinitely many convergents p_n/q_n such that the greatest prime factor of $p_n q_n$ is less than M. Then ξ is a Liouville number.*

The fact that such numbers ξ do exist has been asserted by Erdös and Mahler [229] and proved by Fraenkel and Borosh [250], who also showed that they form a set of Hausdorff dimension zero.

Maillet [403] proved that if ξ is a Liouville number and f a rational function with rational coefficients, then $f(\xi)$ is also a Liouville number. This motivated the following question posed by Mahler [400].

PROBLEM 44. *Which analytic functions f have the property that if ξ is any Liouville number, then so is $f(\xi)$? In particular, are there entire transcendental functions with this property?*

Bernik and Dombrovskiĭ [87] and Alniaçik [17] obtained some results related to Problem 44.

For two positive real numbers i and j with $i + j = 1$, let us denote by $\mathcal{B}(i, j)$ the set of pairs (ξ, η) of real numbers for which there exists a positive constant $c_{11}(\alpha, \beta)$ such that

$$\max\{\|q\alpha\|^{1/i}, \|q\beta\|^{1/j}\} \geq \frac{c_{11}(\alpha, \beta)}{q}$$

holds for any positive integer q. Schmidt [513] proposed the following conjecture, investigated by Pollington and Velani [461].

PROBLEM 45. (Schmidt's Conjecture) *For any pairs (i, j) and (i', j') of positive real numbers with $i + j = i' + j' = 1$, the sets $\mathcal{B}(i, j)$ and $\mathcal{B}(i', j')$ have non-empty intersection.*

As noted by Schmidt [513], a counter-example to his conjecture would imply Littlewood's Conjecture.

PROBLEM 46. (Littlewood's Conjecture) *Let ξ and η be real numbers. For any positive ε, there exist integers q, r, and s with $q > 0$ and*

$$q \cdot |q\xi - r| \cdot |q\eta - s| \leq \varepsilon.$$

We display a particular case of the Littlewood Conjecture, and a related question.

PROBLEM 47. *Let ξ be a badly approximable real number. For any positive ε, there exist integers q, r, and s with $q > 0$ and*

$$q \cdot |q\xi - r| \cdot |q\xi^{-1} - s| \le \varepsilon.$$

By Lemma 10.1, there is a connection between Problems 25 and 47.

PROBLEM 48. *Let i be a real number with $0 \le i \le 1$. There exist transcendental real numbers ξ and η and a positive constant $c_{12}(\xi, \eta)$ such that*

$$|p\xi + q\eta + r| \ge \frac{c_{12}(\xi, \eta)}{\max\{|p|, |q|, 1\}^{1+i} \min\{|p|, |q|, 1\}^{1-i}},$$

for any non-zero integer triple (p, q, r).

Davenport [175] proved that Problem 48 is true for $i = 1$. By Lemma 10.1, Problem 48 for $i = 0$ corresponds to the Littlewood Conjecture.

We recall the 'mixed Littlewood Conjecture', proposed by de Mathan & Teulié [414].

PROBLEM 49. *Let ξ be a real number and p be a prime number. For any positive ε, there exist integers q and r with $q > 0$ and*

$$q \cdot |q\xi - r| \cdot |q|_p \le \varepsilon.$$

As for restricted simultaneous Diophantine approximation, Schmidt [511] (see also Thurnheer [554, 555, 556, 557, 558]) investigated the following question.

PROBLEM 50. *Let ξ and η be real numbers. There exist a constant $c_{13}(\xi, \eta)$ and infinitely many triples (p, q, r) of integers with p and q positive and*

$$|p\xi + q\eta + r| < c_{13}(\xi, \eta) \, (\max\{p, q\})^{-2}. \tag{10.6}$$

An easy metric argument shows that Problem 50 holds true for almost all pairs (ξ, η) of real numbers. One may ask for the Hausdorff dimension of the set of pairs (ξ, η) for which (10.6) holds with an exponent of $\max\{p, q\}$ greater than 2, for infinitely many triples (p, q, r) of integers with p and q positive.

We recall a question posed in Chapter 8.

PROBLEM 51. *Let ε be a positive real number. Then, for almost all real numbers ξ, there exist a constant $c_{14}(\xi, \varepsilon)$, depending only on ξ and on ε, and a constant $c_{15}(n)$ depending only on n such that*

$$|P(\xi)| > \exp\{-(1+\varepsilon)n \log H - c_{15}(n)\}$$

for all integer polynomials $P(X)$ of degree n and height H satisfying $\max\{n, H\} \ge c_{14}(\xi, \varepsilon)$.

There are many connections between approximation by algebraic numbers and separation of roots of integer polynomials.

PROBLEM 52. *Let $n \geq 2$ and H be positive integers. Denote by $L(n, H)$ (resp. $L^*(n, H)$, $L_0(n, H)$, $L_0^*(n, H)$) the minimum of the minimal distance between two of the roots of a (resp. an irreducible, a monic, an irreducible and monic) separable integer polynomial $P(X)$ of degree n and height at most H. Give sharp upper bounds for the upper limits as H tends to infinity of the quantities*

$$\frac{-\log L(n, H)}{\log H}, \quad \frac{-\log L^*(n, H)}{\log H}, \quad \frac{-\log L_0(n, H)}{\log H},$$

$$and \quad \frac{-\log L_0^*(n, H)}{\log H}.$$

In particular, we may ask whether the dependence on $H(P)$ in Theorem A.3 is best possible.

The point of view taken in Problem 52 is that of Wirsing, since we fix an upper bound for the degree and let then the height tend to infinity. We may as well do the converse.

PROBLEM 53. *With the notation of Problem 52, for a fixed positive integer H, determine the asymptotic behaviour of the functions $n \mapsto L(n, H)$, $n \mapsto L^*(n, H)$, $n \mapsto L_0(n, H)$, and $n \mapsto L_0^*(n, H)$ as n tends to infinity.*

To conclude, we state a conjecture of Schmidt [513] on approximation in a given number field.

PROBLEM 54. (Schmidt's Conjecture) *For any number field $\mathbb{K}$ and any positive real number ε, we have*

$$|\alpha - \beta| > c_{16}(\mathbb{K}, \varepsilon)\big(\max\{H(\alpha), H(\beta)\}\big)^{-2-\varepsilon},$$

for any distinct α, β in $\mathbb{K}$, where $c_{16}(\mathbb{K}, \varepsilon)$ is some constant depending only on $\mathbb{K}$ and on ε.

Schmidt's Conjecture, if established, would provide a considerable improvement on Theorem 2.5. For results towards this problem, see the survey of Evertse [232].

Most of the questions posed in this Section can be addressed in the context of Diophantine approximation in p-adic fields or in fields of formal power series, as well.

10.3 Notes

• Davenport and Lewis [178] have proved that the analogue of the Littlewood Conjecture does not hold in the fields of formal power series over infinite fields.

Specific examples have been given by A. Baker [39] and Cusick [170, 171]. Armitage [33] showed that the analogue of Littlewood's Conjecture does not hold in the fields of formal power series over fields of characteristic at least equal to 5. De Mathan [411] established that, in any algebraic extension of the field with two elements, the Littlewood Conjecture holds true for every pair of quadratic elements. De Mathan and Teulié [414] have investigated the analogue of Problem 49 in fields of power series and they established that this conjecture does not hold if the ground field is infinite.

• Some of Cassels and Swinnerton-Dyer's results [158] have been extended by Peck [450].

• Problem 40 can be addressed also for p-adic algebraic numbers. An interesting contribution is due to Lagarias [346]. In a field or formal power series, the situation is different: there exist algebraic elements of degree at least three which have bounded partial quotients (see Baum and Sweet [52] and the surveys [351] and [515] for additional references).

• M. Einsiedler, A. Katok and E. Lindenstrauss have established that the set of exceptions to the Littlewood Conjecture has Hausdorff dimension zero.

Appendix A

Lemmas on polynomials

In the present Appendix, we gather several lemmas on polynomials, which have been used throughout the preceding Chapters. They provide us with estimates, for example, for the distance between two algebraic numbers, or between a complex number and the set of zeros of an integer polynomial. The first systematic study of such questions is due to Güting [268] and no significant progress has been made since that time, at least concerning the most useful lemmas. Apart from in Chapter 8, we have always used the most natural height, that is, the naive height (Definition 2.1), in order to measure the size of an algebraic number. In particular, we did not work with either the *logarithmic height,* or with the *Mahler measure*, since the problems investigated in Chapters 1 to 7 do not require such precise and sophisticated tools. Proofs of several classical auxiliary results can nevertheless be slightly simplified by the use of the Mahler measure, defined in Section A.1. Nevertheless, we express most of the statements in this Appendix only in terms of the naive height. Furthermore, we mention totally explicit examples, due to Bugeaud and Mignotte [139], showing that some of the classical results gathered here are near to best possible.

A.1 Definitions and useful lemmas

The *height* (or *naive height*) of a polynomial with complex coefficients

$$P(X) = a_n X^n + \ldots + a_1 X + a_0 = a_n(X - \alpha_1) \ldots (X - \alpha_n),$$

denoted by H(P), is the maximum of the moduli of its coefficients. If $P(X)$ is non-zero, its *Mahler measure* is, by definition, the quantity

$$M(P) := |a_n| \prod_{i=1}^{n} \max\{1, |\alpha_i|\}.$$

The height (*resp.* the Mahler measure) of an algebraic number α, denoted by $H(\alpha)$ (*resp.* by $M(\alpha)$), is the height (*resp.* the Mahler measure) of its minimal polynomial over $\mathbb{Z}$ (that is, the integer polynomial of lowest positive degree, with coprime coefficients and positive leading coefficient, which vanishes at α).

The Mahler measure of a polynomial appeared for the first time in a paper by Mahler [387], which contains the following lemma.

LEMMA A.1. *For any non-zero complex polynomial $P(X)$, we have*

$$M(P) = \exp\left\{ \int_0^1 \log |P(e^{2i\pi t})|\, dt \right\}. \tag{A.1}$$

PROOF. This is Jensen's formula, see, for example, Ahlfors [4], page 205, or Lemma 1.9 of Everest and Ward [231].

Unlike the naive height, the Mahler measure is a multiplicative function; for this reason, it is, in many cases, much easier to handle. It turns out, however, that these two notions of size are comparable.

LEMMA A.2. *Let $P(X)$ be a non-zero complex polynomial of degree n. We have the inequalities*

$$\binom{n}{[n/2]}^{-1} H(P) \leq M(P) \leq \sqrt{n+1}\, H(P). \tag{A.2}$$

PROOF. Write $P(X) = a_n X^n + \ldots + a_1 X + a_0 = a_n (X - \alpha_1) \ldots (X - \alpha_n)$. For any integer $j = 0, \ldots, n-1$, the relations between coefficients and roots of $P(X)$ give

$$a_j = (-1)^{n-j} a_n \sum_{1 \leq s_1 < \ldots < s_{n-j} \leq n} \alpha_{s_1} \ldots \alpha_{s_{n-j}}.$$

The above sum is composed of at most $\binom{n}{[n/2]}$ terms, each of which has modulus at most equal to $M(P)/|a_n|$. We thus get

$$|a_j| \leq \binom{n}{[n/2]} M(P).$$

Since $|a_n| \leq M(P)$, this proves the left-hand side of (A.2).

As for the right-hand side, the convexity of the exponential function together with the Cauchy–Schwarz inequality yield

$$M(P) \leq \int_0^1 |P(e^{2i\pi t})|\, dt \leq \left(\int_0^1 |P(e^{2i\pi t})|^2\, dt \right)^{1/2}$$
$$= \left(|a_n|^2 + \ldots + |a_0|^2 \right)^{1/2} \leq \sqrt{n+1}\, H(P),$$

as claimed.

Lemma A.2 allows us to recover estimate (12) of Wirsing [598].

COROLLARY A.1. *Let $P(X) = a_n(X - \alpha_1)\ldots(X - \alpha_n)$ be a non-zero complex polynomial of degree n with leading coefficient a_n in $\mathbb{R}$. Let ξ be a complex number and ρ be a positive real number. We then have*

$$2^{-n-1}(n+1)^{-1/2}\max\{1, |\xi|\}^{-n}\frac{|P(\xi)|}{H(P)} \leq \prod_{i:|\xi-\alpha_i|<\rho}|\xi - \alpha_i|$$

$$\leq 2^{n+1}\binom{n}{[n/2]}\max\{\rho, \rho^{-1}\}^n\max\{1, |\xi|\}^n\frac{|P(\xi)|}{H(P)}.$$

PROOF. The Mahler measure of the polynomial $P_\xi(X) := P(X + \xi)$ is

$$M(P_\xi) = |a_n|\prod_{i=1}^{n}\max\{1, |\xi - \alpha_i|\}.$$

Since $|P(\xi)| = M(P_\xi)\prod_{i=1}^{n}\min\{1, |\xi - \alpha_i|\}$, Lemma A.2 implies

$$\frac{1}{\sqrt{n+1}}\cdot\frac{|P(\xi)|}{H(P_\xi)} \leq \prod_{i=1}^{n}\min\{1, |\xi - \alpha_i|\} \leq \binom{n}{[n/2]}\frac{|P(\xi)|}{H(P_\xi)}.$$

The inequalities

$$2^{-n-1}\max\{1, |\xi|\}^{-n}H(P) \leq H(P_\xi) \leq 2^{n+1}\max\{1, |\xi|\}^n H(P)$$

then yield

$$2^{-n-1}(n+1)^{-1/2}\max\{1, |\xi|\}^{-n}\frac{|P(\xi)|}{H(P)} \leq \prod_{i:|\xi-\alpha_i|<1}|\xi - \alpha_i|$$

$$\leq 2^{n+1}\binom{n}{[n/2]}\max\{1, |\xi|\}^n\frac{|P(\xi)|}{H(P)},$$

and the corollary is established for $\rho = 1$. The general statement now follows straightforwardly.

The next lemma relates the height of a product of polynomials to the product of the heights of these polynomials. Koksma and Popken's original result [334], Hilfssatz 13, has been considerably refined by Gelfond [257], and Lemma A.3 is often referred to as *Gelfond's Lemma*.

LEMMA A.3. *Let $P_1(X), \ldots, P_r(X)$ be non-zero complex polynomials of degree $n_1, \ldots, n_r$, respectively, and set $n = n_1 + \ldots + n_r$. We then have*

$$2^{-n}H(P_1)\ldots H(P_r) \leq H(P_1\ldots P_r) \leq 2^n H(P_1)\ldots H(P_r). \tag{A.3}$$

PROOF. We follow an idea of Mahler [387]. Lemma A.2 and the multiplicativity property of the Mahler measure imply

$$\binom{n_1}{[n_1/2]}^{-1} \cdots \binom{n_r}{[n_r/2]}^{-1} H(P_1)\ldots H(P_r) \leq M(P_1)\ldots M(P_r)$$

$$= M(P_1 \ldots P_r) \leq \sqrt{n+1}\, H(P_1 \ldots P_r).$$

Further, we check that

$$\binom{n_1}{[n_1/2]} \cdots \binom{n_r}{[n_r/2]} \sqrt{n+1} \leq 2^n,$$

since

$$\binom{\ell}{k} \leq 2^\ell/\sqrt{\ell+1} \quad \text{for any integers } \ell \geq 0, k \geq 0$$

Thus, we get the left inequality of (A.3). Moreover, we have

$$H(P_1 \ldots P_r) \leq \prod_{j=1}^{r}(n_j+1)H(P_j) \leq \prod_{j=1}^{r} 2^{n_j}H(P_j) = 2^n \prod_{j=1}^{r} H(P_j),$$

which concludes the proof of the lemma.

LEMMA A.4. *Let α be a non-zero algebraic number of degree n. Let k, a, b, and c be integers with $c \neq 0$. We then have*

$$H(\alpha^k) \leq 2^n\,(n+1)^{|k|/2}\,H(\alpha)^{|k|}$$

and

$$H\left(\frac{a\alpha+b}{c}\right) \leq 2^{n+1}\,H(\alpha)\max\{|a|,|b|,|c|\}^n.$$

PROOF. The first assertion easily follows from Lemma A.2 and the fact that $M(\alpha^k) \leq M(\alpha)^{|k|}$. As for the second assertion, denoting by $P(X)$ the minimal polynomial of α over $\mathbb{Z}$, we see that $Q(X) := a^n P(cX/a - b/a)$ is the one of $(a\alpha+b)/c$. Since the height of $Q(X)$ is bounded from above by $2^{n+1}H(\alpha)\max\{|a|,|b|,|c|\}^n$, the proof is complete.

A.2 Liouville's inequality

The easiest version of Liouville's inequality states that the absolute value of any non-zero integer is at least equal to one. More generally, the results of this Section give non-trivial lower bounds for the distance between two distinct algebraic numbers and for the value of an integer polynomial evaluated at an algebraic number. *Liouville's inequality* is a generic name for similar estimates. Standard tools to investigate these questions are the notions of *discriminant* and *resultant*. Recall that a separable polynomial is a polynomial with distinct roots.

DEFINITION A.1. *Let* $P(X) = a_n(X - \alpha_1) \ldots (X - \alpha_n)$ *and* $Q(X) = b_m(X - \beta_1) \ldots (X - \beta_m)$ *be non-constant integer polynomials. The resultant of* $P(X)$ *and* $Q(X)$, *defined as*

$$\text{Res}(P, Q) = a_n^m b_m^n \prod_{1 \le i \le n} \prod_{1 \le j \le m} (\alpha_i - \beta_j)$$
$$= a_n^m \prod_{1 \le i \le n} Q(\alpha_i) = (-1)^{mn} b_m^n \prod_{1 \le j \le m} P(\beta_j),$$

is an integer and is non-zero if, and only if, $P(X)$ *and* $Q(X)$ *have no common root. The discriminant of* $P(X)$, *defined as*

$$\text{Disc}(P) = a_n^{2n-2} \prod_{1 \le i < j \le n} (\alpha_i - \alpha_j)^2 = (-1)^{n(n-1)/2} a_n^{-1} \text{Res}(P, P'),$$

is an integer and is non-zero if, and only if, $P(X)$ *is separable.*

Further results on resultants and discriminants can be found in most of the textbooks on algebra, for example, in [349], Chapter IV, in [586], Chapter 5, or in [156], Appendix A.

For any integer $n \ge 2$ and any non-zero integer a we have

$$\text{Res}(X^n - aX + 1, aX - 1) = 1$$

and

$$\text{Res}(X^n - aX + 1, (a + 1)X^n - X^{n-1} - aX + 1) = 1,$$

hence, it seems to be difficult to improve the trivial lower bound for the absolute value of the resultant of integer polynomials with no common root.

Theorem A.1 is of comparable strength as Theorem 6 of Güting [268]. An earlier (and less precise) version goes back to Cohn [167], who used it to construct transcendental numbers.

THEOREM A.1. *Let* $P(X)$ *and* $Q(X)$ *be non-constant integer polynomials of degree* n *and* m, *respectively. Denote by* α *a zero of* $P(X)$ *of order* s *and by* β *a zero of* $Q(X)$ *of order* t. *Assuming that* $P(\beta) \ne 0$, *we have*

$$|P(\beta)| \ge (n + 1)^{1 - m/t} (m + 1)^{-n/(2t)} \text{H}(P)^{1 - m/t} \text{H}(Q)^{-n/t}$$
$$\times \left(\max\{1, |\beta|\} \right)^n \qquad (A.4)$$

and

$$|\alpha - \beta| \geq 2^{1-n/s} (n+1)^{1/(2s)-m/(st)} (m+1)^{-n/(2st)}$$
$$\times H(P)^{-m/(st)} H(Q)^{-n/(st)} \left(\max\{1, |\alpha|\}\right) \left(\max\{1, |\beta|\}\right).$$

PROOF. Set $P(X) = a_n (X - \alpha_1)^{s_1} \ldots (X - \alpha_p)^{s_p}$ and $Q(X) = b_m (X - \beta_1)^{t_1} \ldots (X - \beta_q)^{t_q}$, where $\alpha = \alpha_1$, $\beta = \beta_1$, $s = s_1$, $t = t_1$, and the α_is (*resp.* the β_js) are pairwise distinct. Denote by $Q_1(X) = b(X - \beta_1) \ldots (X - \beta_{q_1})$ the minimal polynomial of β over $\mathbb{Z}$. Since the resultant of $P(X)$ and $Q_1(X)$ is a non-zero integer, we get

$$1 \leq |\mathrm{Res}(P, Q_1)| = |b|^n \prod_{1 \leq i \leq q_1} |P(\beta_i)|,$$

and, using that $|P(\beta_i)| \leq (n+1) H(P)(\max\{1, |\beta_i|\})^n$ for $i = 2, \ldots, q_1$, we obtain

$$1 \leq |b|^n |P(\beta)| (n+1)^{q_1-1} H(P)^{q_1-1} \left(\frac{M(Q_1)}{|b| \max\{1, |\beta|\}}\right)^n. \qquad (A.5)$$

Then, (A.4) follows from (A.5), $q_1 \leq m/t$, $M(Q_1) \leq M(Q)^{1/t}$, and $M(Q) \leq \sqrt{m+1} H(Q)$. Furthermore, by combining (A.4) with the estimates

$$|P(\beta)| = |\beta - \alpha|^s \cdot |a_n| \cdot |\beta - \alpha_2|^{s_2} \ldots |\beta - \alpha_p|^{s_p}$$
$$\leq 2^{n-s} |\beta - \alpha|^s \cdot |a_n| \left(\max\{1, |\beta|\}\right)^{n-s}$$
$$\times \left(\max\{1, |\alpha_2|\}\right)^{s_2} \ldots \left(\max\{1, |\alpha_p|\}\right)^{s_p}$$
$$\leq 2^{n-s} |\beta - \alpha|^s \left(\max\{1, |\beta|\}\right)^{n-s} \frac{M(P)}{\max\{1, |\alpha|\}^s}$$

and $M(P) \leq \sqrt{n+1} H(P)$, we get the second assertion of the theorem.

For any integer $n \geq 2$ and any non-zero integer a, set

$$P(X) = X^n - aX + 1, \quad Q(X) = aX - 1, \quad \text{and}$$
$$R(X) = (a+1)X^n - X^{n-1} - aX + 1.$$

Let α (*resp.* γ) be the root of $P(X)$ (*resp.* of $R(X)$) closest to $1/a$, and set $\beta = 1/a$. We observe that

$$|\alpha - \beta| \leq 2 |a|^{-n-1} = 2 H(P)^{-1} H(Q)^{-n}$$

and

$$|\alpha - \gamma| \leq 2 |a|^{-2n} \leq 3 H(P)^{-n} H(R)^{-n}$$

hold provided that $|a|$ is large enough in terms of n. Consequently, Theorem

A.1 is best possible regarding the dependence on the heights of the polynomials.

A lower bound for the distance between two distinct roots of the same polynomial is slightly less easy to obtain, especially when there are multiplicities. Theorem A.2, due to Amou and Bugeaud [31], strengthens Theorem 7 of Güting [268] in terms of n (a factor $2^{-n^2/st}$ is removed). A similar improvement has been obtained previously by Amou [28], but in his result the dependence on $H(P)$ is less sharp than in [268].

THEOREM A.2. *Let $P(X)$ be an integer polynomial of degree $n \geq 2$. Denote by α a zero of $P(X)$ of order s and by β a zero of $P(X)$ of order t. Assuming that $\alpha \neq \beta$, we have*

$$|\alpha - \beta| \geq 2^{-n/t} \, n^{-n/t-3n/(2st)} \, H(P)^{-2n/(st)} \, \max\{1, |\alpha|\} \, \max\{1, |\beta|\}$$

if $t > s$, while we have

$$|\alpha - \beta| \geq 2^{-n/s} \, n^{-n(2s+3)/(4s^2)} \, H(P)^{-n/s^2+1/(2s)} \, \max\{1, |\alpha|\}^{3/2}$$
$$\times \max\{1, |\beta|\}^{3/2}$$

if $s = t$.

PROOF. Let $Q(X) = a(X - \alpha_1) \ldots (X - \alpha_d)$ be a separable integer polynomial with $\alpha = \alpha_1$ and such that $Q(X)$ divides $P(X)$ in $\mathbb{Z}[X]$ and the polynomials $Q(X)$ and $P(X)/Q^s(X)$ are coprime. Since the resultant of $Q(X)$ and $P^{(s)}(X)/s!$ is a non-zero integer, we get

$$1 \leq |a|^{n-s} \prod_{1 \leq i \leq d} \frac{|P^{(s)}(\alpha_i)|}{s!}. \tag{A.6}$$

For any integer $i = 2, \ldots, d$, we obtain

$$\frac{|P^{(s)}(\alpha_i)|}{s!} \leq H(P) \sum_{k=s}^{n} \binom{k}{s} \max\{1, |\alpha_i|\}^{n-k}$$
$$\leq \binom{n+1}{s+1} H(P) \max\{1, |\alpha_i|\}^{n-s}. \tag{A.7}$$

Denoting by a_n the leading coefficient of $P(X)$, we have

$$\frac{|P^{(s)}(\alpha)|}{s!} = |a_n| \prod_{\substack{\gamma \neq \alpha \\ P(\gamma)=0}} |\alpha - \gamma|$$
$$\leq 2^{n-s-t} |a_n| \cdot |\alpha - \beta|^t \cdot \max\{1, |\alpha|\}^{n-s-t}$$
$$\cdot \prod_{\substack{\gamma \neq \alpha, \gamma \neq \beta \\ P(\gamma)=0}} \max\{1, |\gamma|\}, \tag{A.8}$$

where the roots of $P(X)$ are counted with their multiplicities in the above products. Now, the combination of (A.6), (A.7), (A.8), $d \leq n/s$ with the inequality $M(Q) \leq M(P)^{1/s}$ yields

$$|\alpha - \beta|^t \geq 2^{-n+s+t} M(P)^{-n/s} \left(\binom{n+1}{s+1} H(P) \right)^{1-n/s}$$
$$\times \max\{1, |\alpha|\}^t \max\{1, |\beta|\}^t,$$

and the first assertion of the theorem follows from Lemma A.2 since $n \geq 2$ and

$$\binom{n+1}{s+1} \leq (n+1)(n/2)^s.$$

If $s = t$, we may assume that $\alpha_2 = \beta$. We have the analogue of (A.8) for $|P^{(s)}(\beta)|/s!$, namely the upper bound

$$\frac{|P^{(s)}(\beta)|}{s!} \leq 2^{n-2s} |a_n| \cdot |\alpha - \beta|^s \cdot \max\{1, |\beta|\}^{n-2s} \cdot \prod_{\substack{\gamma \neq \alpha, \gamma \neq \beta \\ P(\gamma)=0}} \max\{1, |\gamma|\},$$

$$(A.9)$$

where, again, the roots of $P(X)$ are counted with their multiplicities in the above product. Combining (A.9) with (A.6), (A.8), and (A.7) for $i = 3, \ldots, d$, we get

$$|\alpha - \beta|^{2s} \geq 2^{-2n+4s} M(P)^{-1-n/s} \left(\binom{n+1}{s+1} H(P) \right)^{2-n/s}$$
$$\times \max\{1, |\alpha|\}^{3s} \max\{1, |\beta|\}^{3s},$$

which yields the last assertion of the theorem.

In case of separable polynomials, Theorem A.2 is superseded by the following result of Mahler [391].

THEOREM A.3. *Let $P(X)$ be a separable, integer polynomial of degree $n \geq 2$. For any two distinct zeros α and β of $P(X)$ we have*

$$|\alpha - \beta| > \sqrt{3}\,(n+1)^{-(2n+1)/2} \max\{1, |\alpha|, |\beta|\}\, H(P)^{-n+1}.$$

PROOF. Denote by a_n the leading coefficient of $P(X)$ and by $\alpha_1, \ldots, \alpha_n$ its roots, numbered in such a way that $\alpha_1 = \alpha$ and $\alpha_2 = \beta$. Without loss of generality, we may assume that $|\alpha| \geq |\beta|$. Recalling that

$$\mathrm{Disc}(P) = \pm a_n^{2n-2} (\det \mathcal{M})^2 \quad \text{with} \quad \mathcal{M} = (\alpha_i^j)_{\substack{1 \leq i \leq n \\ 0 \leq j \leq n-1}},$$

we substract the second line of $\mathcal{M}$ from the first one and we apply Hadamard's inequality to get the upper bound

$$|\mathrm{Disc}(P)| \leq |a_n|^{2n-2} \left(\sum_{j=1}^{n-1} |\alpha^j - \beta^j|^2 \right) \times$$

$$\prod_{2 \leq i \leq n} \left(1 + |\alpha_i|^2 + \ldots + |\alpha_i|^{2n-2} \right)$$

$$\leq |a_n|^{2n-2} \cdot |\alpha - \beta|^2 \cdot \left(\sum_{j=1}^{n-1} |\alpha^{j-1} + \alpha^{j-2}\beta + \ldots + \beta^{j-1}|^2 \right).$$

$$\times n^{n-1} \prod_{2 \leq i \leq n} \max\{1, |\alpha_i|\}^{2n-2}$$

$$< |\alpha - \beta|^2 \frac{n^3}{3} \max\{1, |\alpha|\}^{-2} n^{n-1} M(P)^{2n-2}. \tag{A.10}$$

Since the polynomial $P(X)$ has distinct roots, $\mathrm{Disc}(P)$ is a non-zero integer and (A.10) and Lemma A.2 yield the theorem.

As observed in [139], for any integer $d \geq 2$ and any non-zero integer a sufficiently large in terms of d, the polynomial $P_{a,d}(X) := (X^d - aX + 1)^2 - 2X^{2d-2}(aX - 1)^2$ has two real roots distant by at most $4a^{-2d}$. This example shows that, when n is even, the exponent of $H(P)$ in Theorem A.3 cannot be replaced by any number greater than $-n/2$.

We deduce from Theorems A.1 and A.3 the following useful lower estimate for the distance between two distinct algebraic numbers.

COROLLARY A.2. *Let α and β be two distinct non-zero algebraic numbers of degree n and m, respectively. Then we have*

$$|\alpha - \beta| \geq 2 (n + 1)^{-m/2} (m + 1)^{-n/2}$$
$$\times \max\{2^{-n} (n + 1)^{-(m-1)/2}, 2^{-m} (m + 1)^{-(n-1)/2}\} \times H(\alpha)^{-m} H(\beta)^{-n}.$$

It is an interesting open question to decide whether the dependence on the degrees m and n in Theorems A.1 to A.3 is optimal or not. There is apparently no contribution to this problem.

A.3 Zeros of polynomials

This Section is concerned with the following problem: if $P(X)$ is a complex polynomial which is small but does not vanish at a complex number ξ, what can be said on the distance between ξ and the set of zeros of $P(X)$?

LEMMA A.5. *Let $P(X) = a_n(X - \alpha_1) \ldots (X - \alpha_n)$ be a non-constant complex polynomial. Let ξ be a complex number with $P(\xi) \neq 0$. Assume that α_1 is such that $|\xi - \alpha_1|$ is minimal. If $P'(\xi) \neq 0$, then we have*

$$|\xi - \alpha_1| \leq n \frac{|P(\xi)|}{|P'(\xi)|}. \tag{A.11}$$

If $P'(\alpha_1) \neq 0$, then we have

$$|\xi - \alpha_1| \leq 2^{n-1} \frac{|P(\xi)|}{|P'(\alpha_1)|} \tag{A.12}$$

and, if $n \geq 2$,

$$|\xi - \alpha_1|^2 \leq 2^{n-2} \frac{|P(\xi)| \cdot |\alpha_1 - \alpha_2|}{|P'(\alpha_1)|}, \tag{A.13}$$

for any root $\alpha_2 \neq \alpha_1$ of $P(X)$. If, further, $|\xi - \alpha_1| \leq |\alpha_1 - \alpha_j|$ for $j = 2, \ldots, n$, then we have

$$|\xi - \alpha_1| \geq 2^{1-n} \frac{|P(\xi)|}{|P'(\alpha_1)|}. \tag{A.14}$$

PROOF. To prove (A.11), it is enough to note that the rational function $P'(X)/P(X)$ can be written as

$$\frac{P'(X)}{P(X)} = \sum_{1 \leq i \leq n} \frac{1}{X - \alpha_i}. \tag{A.15}$$

Taking $X = \xi$ in (A.15), we obtain

$$\frac{|P'(\xi)|}{|P(\xi)|} \leq \frac{n}{|\xi - \alpha_1|},$$

and the proof is complete. To get (A.12), we combine

$$\frac{|P(\xi)|}{|P'(\alpha_1)|} = |\xi - \alpha_1| \prod_{2 \leq j \leq n} \frac{|\xi - \alpha_j|}{|\alpha_1 - \alpha_j|} \tag{A.16}$$

with the fact that, for any integer $j = 2, \ldots, n$, we have

$$|\alpha_1 - \alpha_j| \leq |\xi - \alpha_1| + |\xi - \alpha_j| \leq 2|\xi - \alpha_j|. \tag{A.17}$$

Moreover, for any integer $j = 2, \ldots, n$, the triangle inequality ensures that $|\xi - \alpha_j| \leq 2|\alpha_1 - \alpha_j|$ holds as soon as $|\xi - \alpha_1| \leq |\alpha_1 - \alpha_j|$. Thus, (A.14) follows from (A.16).

Finally, we get (A.13) by combining

$$\frac{|P(\xi)|}{|P'(\alpha_1)|} = \frac{|\xi - \alpha_1| \cdot |\xi - \alpha_2|}{|\alpha_1 - \alpha_2|} \prod_{3 \leq j \leq n} \frac{|\xi - \alpha_j|}{|\alpha_1 - \alpha_j|}$$

$$\geq \frac{|\xi - \alpha_1|^2}{|\alpha_1 - \alpha_2|} \prod_{3 \leq j \leq n} \frac{|\xi - \alpha_j|}{|\alpha_1 - \alpha_j|}$$

and (A.17).

LEMMA A.6. *Let* $P(X) = a_n(X - \alpha_1) \ldots (X - \alpha_n)$ *be a non-constant integer polynomial of degree n. Let* ξ *be a complex number which is not a root of* $P(X)$. *Then, for any subset J of* $\{1, \ldots, n\}$, *we have*

$$|a_n| \prod_{j \in J} |\xi - \alpha_j| \leq 2^n \left(\max\{1, |\xi|\}\right)^n \sqrt{n+1}\, H(P), \qquad (A.18)$$

and, in particular,

$$|P(\xi)| \leq 2^n \left(\max\{1, |\xi|\}\right)^n \sqrt{n+1}\, H(P) \min_{1 \leq j \leq n} |\xi - \alpha_j|.$$

PROOF. We have

$$|a_n| \prod_{j \in J} |\xi - \alpha_j| \leq |a_n| \prod_{j \in J} 2 \max\{|\xi|, |\alpha_j|\}$$

$$\leq 2^n \left(\max\{1, |\xi|\}\right)^n |a_n| \prod_{j \in J} \max\{1, |\alpha_j|\}$$

$$\leq 2^n \left(\max\{1, |\xi|\}\right)^n M(P),$$

and we apply Lemma A.2 to get (A.18). The last assertion follows by taking $J = \{1, \ldots, n\} \setminus \{k\}$ where k is such that $|\xi - \alpha_k| = \min_{1 \leq j \leq n} |\xi - \alpha_j|$.

Lemma A.7, due to Diaz and Mignotte [189], slightly improves upon an earlier result of Chudnovsky [164]. To take multiplicities into account, Chudnovsky introduced the notion of *semi-discriminant*, which generalizes that of discriminant to polynomials with multiple roots. Let $P(X) = a_n(X - \alpha_1)^{s_1} \ldots (X - \alpha_p)^{s_p}$ be an integer polynomial of degree $n = s_1 + \ldots + s_p$. Assume that $\alpha_i \neq \alpha_j$ if $1 \leq i \neq j \leq p$. Then the semi-discriminant of $P(X)$, denoted by $SD(P)$, is the quantity

$$SD(P) = a_n^{n-2} \prod_{i=1}^{p} \frac{P^{(s_i)}(\alpha_i)}{s_i!}. \qquad (A.19)$$

If $P(X)$ is separable, then $SD(P)$ reduces to the discriminant of $P(X)$. Actually, Chudnovsky [164] defined the semi-discriminant with a_n^{n-1} instead of a_n^{n-2} in (A.19). He proved that $a_n SD(P)$ is a non-zero rational integer and he used this observation to get Lemma 1.12 of [164], which is slightly weaker than Lemma A.7 below. Amou [28] used the same idea to obtain a weaker version of Theorem A.2. Further, Diaz and Mignotte [189] pointed out that the use of the semi-discriminant can be replaced by the consideration of a suitable resultant. Their method, which is also applied to get Theorem A.2 above, enabled them to considerably simplify Chudnovsky's proof. Lemma A.7 follows from [189], except for the last assertion, kindly communicated by Amou.

LEMMA A.7. *Let $P(X)$ be a non-constant integer polynomial of degree n. Let ξ be a complex number and α be a root of $P(X)$ such that $|\xi - \alpha|$ is minimal. Then, denoting by s the multiplicity of α as root of $P(X)$, we have*

$$|\xi - \alpha|^s \le n^{n+3n/(2s)} \, \mathrm{H}(P)^{2(n/s-1)} \, |P(\xi)|.$$

Further, if $P(X)$ is the s-th power of a separable polynomial of degree at least 2, we get

$$|\xi - \alpha|^s \le 2^{n/2} \, n^{n/2+3n/(4s)} \, \mathrm{H}(P)^{n/s-3/2} \, |P(\xi)|.$$

PROOF. Without any restriction, we may assume that $n \ne s$, hence, $n/s \ge 2$. Let $Q(X) = a(X - \alpha_1) \dots (X - \alpha_d)$ be as in the proof of Theorem A.2, with $\alpha = \alpha_1$. For any root γ of $P(X)$ different from α, we have $|\alpha - \gamma| \le |\xi - \alpha| + |\xi - \gamma| \le 2|\xi - \gamma|$. Thus, we get

$$\frac{|P^{(s)}(\alpha)|}{s!} = |a_n| \prod_{\substack{\gamma \ne \alpha \\ P(\gamma)=0}} |\alpha - \gamma| \le 2^{n-s} \, |P(\xi)| \, |\xi - \alpha|^{-s}, \qquad \text{(A.20)}$$

where a_n denotes the leading coefficient of $P(X)$ and the roots of $P(X)$ are counted with their multiplicities in the above product. We combine (A.6), (A.7), and (A.20) to obtain

$$|\xi - \alpha|^s \le 2^{n-s} \left(\binom{n+1}{s+1} \mathrm{H}(P) \right)^{d-1} M(Q)^{n-s} \, |P(\xi)|. \qquad \text{(A.21)}$$

Since

$$\binom{n+1}{s+1} \le (n+1)(n/2)^s, \quad d \le n/s, \quad \text{and} \quad M(Q) \le M(P)^{1/s},$$

the first assertion follows from (A.21) and Lemma A.2.

For the last statement, we take $Q(X)$ such that $Q(X)^s = P(X)$. We may assume that $\beta = \alpha_2$ satisfies

$$|\xi - \alpha| \le |\xi - \beta| \le |\xi - \alpha_i|, \quad \text{for } i = 3, \dots, d,$$

hence, we get $|\beta - \gamma| \leq 2|\xi - \gamma|$ for any root γ of $P(X)$ different from α. Consequently, we have

$$
\begin{aligned}
\frac{|P^{(s)}(\beta)|}{s!} &\leq 2^{n-2s} |\alpha - \beta|^s |P(\xi)| |\xi - \alpha|^{-s} |\xi - \beta|^{-s} \\
&\leq 2^{n-s} |P(\xi)| |\xi - \alpha|^{-s},
\end{aligned}
\tag{A.22}
$$

since $|\alpha - \beta| \leq 2|\xi - \beta|$. Combining (A.6), (A.20), (A.22), and (A.7) for $i = 3, \ldots, d$, we get

$$
|\xi - \alpha|^s \leq 2^{n-s} \left(\binom{n+1}{s+1} \mathrm{H}(P) \right)^{-1+d/2} M(Q)^{(n-s)/2} |P(\xi)|,
$$

which, since $d = n/s$, completes the proof of the lemma.

For the case of a separable polynomial, Lemma A.7 has been improved upon by Feldman [243] (Chapter 7, Lemma 1.7) and Diaz [188] (for $\ell \geq 2$ in Lemma A.8, which strengthens Lemma 5 of Laurent and Roy [356]).

LEMMA A.8. *Let $P(X)$ be a non-constant, separable, integer polynomial of degree n. Let ξ be a complex number and α be a root of $P(X)$ such that $|\xi - \alpha|$ is minimal. Then, for any positive integer ℓ with $\ell < n$ we have*

$$
(2|\xi - \alpha|)^{\ell(\ell+1)/2} \leq 2^{\ell n} (n - \ell)^{(n-\ell)/2} M(P)^{n-\ell-1} |P(\xi)|^\ell.
\tag{A.23}
$$

In particular, we have

$$
|\xi - \alpha| \leq \sqrt{2} (2n)^{n-3/2} \mathrm{H}(P)^{n-2} |P(\xi)|.
\tag{A.24}
$$

If, moreover, $|P(\xi)| \leq 1$, then, for any positive integer ℓ, we have

$$
|\xi - \alpha|^{\ell(\ell+1)/2} \leq 2^{\ell n} n^{n/2} M(P)^n |P(\xi)|^\ell.
\tag{A.25}
$$

PROOF. We may assume that $\xi \neq \alpha$. We order the roots $\alpha_1, \ldots, \alpha_n$ of $P(X)$ in such a way that $\alpha = \alpha_1$ and $|\xi - \alpha_1| \leq \ldots \leq |\xi - \alpha_n|$. Set $Q_\ell(X) = a_n(X - \alpha_{\ell+1}) \ldots (X - \alpha_n)$, where a_n denotes the leading coefficient of $P(X)$. Putting

$$
\Delta_\ell := \prod_{\substack{1 \leq i < j \leq n \\ i \leq \ell}} |\alpha_i - \alpha_j|,
$$

the discriminants $\mathrm{Disc}(P)$ and $\mathrm{Disc}(Q_\ell)$ of $P(X)$ and $Q_\ell(X)$ satisfy

$$
|\mathrm{Disc}(P)|^{1/2} = |a_n|^\ell \Delta_\ell |\mathrm{Disc}(Q_\ell)|^{1/2}.
\tag{A.26}
$$

The quantity $|\mathrm{Disc}(Q_\ell)|^{1/2}$ equals $|a_n|^{n-\ell-1}$ times the absolute value of the determinant Δ of the matrix $(\alpha_i^j)_{\ell+1 \le i \le n, 0 \le j \le n-\ell-1}$. As in the proof of Theorem A.3, we infer from Hadamard's inequality that

$$|\Delta| \le (n - \ell)^{(n-\ell)/2} \prod_{\ell+1 \le i \le n} \max\{1, |\alpha_i|\}^{n-\ell-1},$$

hence, we get

$$|\mathrm{Disc}(Q_\ell)|^{1/2} \le (n - \ell)^{(n-\ell)/2} M(Q_\ell)^{n-\ell-1}, \tag{A.27}$$

and, since $\mathrm{Disc}(P)$ is a non-zero integer, it follows from (A.26) and (A.27) that

$$1 \le |a_n|^\ell \Delta_\ell (n - \ell)^{(n-\ell)/2} M(P)^{n-\ell-1}. \tag{A.28}$$

For positive integers i and j with $i < j \le n$, we have $|\alpha_i - \alpha_j| \le 2|\xi - \alpha_j|$, thus

$$|a_n|^\ell \Delta_\ell \le |a_n|^\ell 2^{n\ell - \ell(\ell+1)/2} \prod_{1 \le j \le \ell} |\xi - \alpha_j|^{j-1} \prod_{\ell < j \le n} |\xi - \alpha_j|^\ell. \tag{A.29}$$

Since

$$|a_n|^\ell \prod_{\ell < j \le n} |\xi - \alpha_j|^\ell = |P(\xi)|^\ell \prod_{1 \le j \le \ell} |\xi - \alpha_j|^{-\ell}$$

and $|\xi - \alpha_1| \le \ldots \le |\xi - \alpha_\ell|$, we infer from (A.29) that

$$|a_n|^\ell \Delta_\ell \le 2^{n\ell} |P(\xi)|^\ell (2|\xi - \alpha_1|)^{-\ell(\ell+1)/2},$$

which, together with (A.28), gives (A.23). Choosing $\ell = 1$ in (A.23), we get (A.24). Further, (A.25) is clear when $n = 1$ and it follows from (A.23) if $\ell = 1, \ldots, n - 1$. If $n \ge 2$, we infer from (A.23) with $\ell = n - 1$ that

$$|\xi - \alpha| \le 2 |P(\xi)|^{2/n}. \tag{A.30}$$

For $\ell \ge n$, we raise both sides of (A.30) to the power $\ell n/2$ and, noticing that $|\xi - \alpha| \le 1$, we get an inequality in fact stronger than (A.25).

Theorems A.2 and A.3 should be compared with Lemmas A.7 and A.8, respectively. In both cases, the result obtained for separable polynomials is considerably better than that for arbitrary polynomials: roughly speaking, we gain a factor 2 in the exponent of $\mathrm{H}(P)$. Furthermore, the polynomials $P_{a,d}(X)$ defined below the proof of Theorem A.3 show that the dependence on the height in (A.24) is close to be best possible: indeed, when n is even, the exponent of $\mathrm{H}(P)$ in (A.24) cannot be replaced by any number smaller than $-1 + n/2$. However, the dependence on the degree may possibly be improved.

A.4 Exercises

EXERCISE A.1. Let $P(X) := a_n(X - \alpha_1) \ldots (X - \alpha_n)$ be an irreducible, integer polynomial. Let $i_1, \ldots, i_t$ be positive integers with $1 \leq i_1 < \ldots < i_t \leq n$. Prove that the number $a_n \alpha_{i_1} \ldots \alpha_{i_t}$ is an algebraic integer (see, for example, [591], pp. 71–72) with absolute value at most equal to $\sqrt{n+1}\, \mathrm{H}(P)$.

EXERCISE A.2. Study how sharp Theorems A.1, A.2, and Lemma A.7 are in terms of s and t (see [31]).

EXERCISE A.3. Let $P(X)$ be a separable, integer polynomial of degree $n \geq 2$ and let $\alpha_1, \ldots, \alpha_k$ be distinct zeros of $P(X)$, with $2 \leq k \leq n$. With a suitable modification of the proof of Theorem A.3, show that there exists a positive, effective constant $c(n)$ such that

$$\prod_{1 \leq i < j \leq k} |\alpha_i - \alpha_j| \geq c(n)\, \mathrm{H}(P)^{-n+1}. \tag{A.31}$$

By considering the polynomials $P_{a,d,k}(X) := (X^d - aX + 1)^k - 2X^{dk-k}\,(aX - 1)^k$, where a and d are integers with $d \geq 2$, prove that (A.31) is near to best possible (see [139]).

EXERCISE A.4. Prove that Theorem A.3 is, up to the numerical constant, best possible for cubic, reducible, integer polynomials.

Hint. Following an idea of Deshouillers (see [419]), consider the polynomials $P_k(X) := (q_k X - p_k)(X^2 - 2)$, where p_k/q_k denotes the k-th convergent in the continued fraction expansion of $\sqrt{2}$.

A.5 Notes

• The formula (A.1) can be generalized to define the Mahler measure of a polynomial in finitely many variables, see Mahler [389] and Chapter 3 of Everest and Ward [231].

• For inequalities relating the naive height of a polynomial to other notions of size, see papers by Durand [220] and the Annex to Chapter 3 in [591].

• Theorem A.1 can be considerably improved if β is non-real, see, for example, Güting [268]. Roughly speaking, one can take the square roots of the lower bounds obtained. The interested reader may easily write down the details of the proof and improve upon several other results from this Appendix in similar particular cases.

• There are more general versions of Liouville's inequality providing non-trivial lower bounds for the non-zero quantity $|P(\alpha_1, \ldots, \alpha_n)|_v$, where $\alpha_1, \ldots, \alpha_n$ are elements of an algebraic number field $\mathbb{K}$, the polynomial $P(X_1, \ldots, X_n)$ has integer coefficients, and v is some place on $\mathbb{K}$. We refer to [591] for estimates involving the Mahler measure and the Weil's height (see especially Exercise 3.5). When α and β are real algebraic numbers with the Mahler measure of α relatively small compared with its degree, improvements upon Liouville's inequality have been obtained for $|\alpha - \beta|$ and $|\alpha - 1|$, see Mignotte [420], Amoroso [22, 23], Dubickas [210], and the last chapter of [512]. A p-adic analogue has been worked out by Bugeaud [121].

• Dubickas [209] and Mignotte [422] have slightly improved Theorem A.3. Their method also yields lower bounds for products of differences of roots of a separable polynomial (like the statement given in Exercise A.3). Polynomial root separation is an important problem in computer algebra, see, for example, Rump [485] for more references. Following Collins and Horowitz [169], for any positive integers $n \geq 2$ and H, we denote by $L(n, H)$ the minimum of the minimal distance between two of the roots of a separable, integer polynomial $P(X)$ of degree n and height at most H. Mignotte [419] and Mignotte and Payafar [423] studied $L(n, H)$ and introduced the related quantities $L^*(n, H)$, $L_0(n, H)$, and $L_0^*(n, H)$, where we restrict the consideration to irreducible, monic, and irreducible and monic polynomials, respectively. The determination of the upper limits when H tends to infinity of $-\log L(n, H)/\log H$ and of the related quantities is apparently a very difficult problem.

Supported by computational evidence, Collins [168] conjectured that the minimal distance between two real roots of an integer polynomial $P(X)$ of degree n and height H is at least equal to $n^{-n/4} H^{-n/2}$. This is compatible with the examples given below the proof of Theorem A.3.

Appendix B

Geometry of numbers

Geometry of numbers turns out to be a very useful tool in Diophantine approximation. For instance, it allows us to construct non-zero integer polynomials taking small values at prescribed points. In the course of the book, we applied several times the 'first Theorem of Minkowski' and the 'second Theorem of Minkowski', which are Theorems B.2 and B.3 below, respectively. We give a full proof of Theorem B.2, but not of Theorem B.3, which is much deeper. Throughout this Appendix, n denotes a positive integer. A set C in $\mathbb{R}^n$ having inner points and contained in the closure of its open kernel is called a *body* (or a *domain*).

THEOREM B.1. *Let C be a bounded convex body in $\mathbb{R}^n$, symmetric about the origin and of volume* $\text{vol}(C)$. *If* $\text{vol}(C) > 2^n$ *or if* $\text{vol}(C) = 2^n$ *and C is compact, then C contains a point with integer coordinates, other than the origin.*

PROOF. This proof is due to Mordell [429]. By classical arguments from elementary topology, it is enough to treat the case where $\text{vol}(C) > 2^n$. For any positive integer m, denote by C_m the set of points of C having rational coordinates with denominator m. As m tends to infinity, the cardinality of C_m becomes equivalent to $\text{vol}(C)m^n$, and is thus strictly larger than $(2m)^n$ when m is large enough. Dirichlet's *Schubfachprinzip* asserts that there exist two distinct points $A = (a_1/m, \ldots, a_n/m)$ and $B = (b_1/m, \ldots, b_n/m)$ in C_m with a_i, b_i integers and $a_i \equiv b_i \pmod{2m}$ for $i = 1, \ldots, n$. By symmetry and convexity, the point $(A - B)/2$, which is not the origin and has integer coordinates, belongs to C.

THEOREM B.2. *Let $(u_{ij})_{1 \leq i,j \leq n}$ be a matrix with real coefficients and with determinant ± 1. Let $A_1, \ldots, A_n$ be positive real numbers satisfying $A_1 \ldots A_n = 1$. Then there exists a non-zero integer point $(x_1, \ldots, x_n)$ such* *that*

$$|u_{i1}x_1 + \ldots + u_{in}x_n| < A_i, \quad 1 \le i \le n-1,$$

and

$$|u_{n1}x_1 + \ldots + u_{nn}x_n| \le A_n.$$

PROOF. Since the absolute value of the determinant of the matrix $(u_{ij})_{1 \le i, j \le n}$ is equal to the product of the A_i's, we may assume that $A_1 = \cdots = A_n = 1$. For $i = 1, \ldots, n$ and $\underline{x} = (x_1, \ldots, x_n)$, we set

$$L_i(\underline{x}) := u_{i1}x_1 + \ldots + u_{in}x_n.$$

The linear system $|L_i(\underline{x})| \le 1$, $1 \le i \le n$, defines a symmetric, bounded, compact convex body of volume 2^n. By Theorem B.1, it contains a non-zero integer point. Classical topological arguments then allow us to replace $n-1$ of the large inequalities by strict inequalities.

We now state a deep generalization of Theorem B.2. Let $\mathcal{C}$ be a convex, compact body in $\mathbb{R}^n$, symmetric about the origin and of volume $\mathrm{vol}(\mathcal{C})$. Let $\lambda_1 := \lambda_1(\mathcal{C})$ be the infimum of the real numbers λ such that $\lambda\mathcal{C}$ contains an integer point other than the origin. Since $\mathcal{C}$ is compact, this infimum is a minimum. Further, we have $0 < \lambda_1 < +\infty$. Setting $\tilde{\lambda} := 2\,\mathrm{vol}(\mathcal{C})^{-1/n}$, the volume of the compact convex set $\tilde{\lambda}\mathcal{C}$ is equal to 2^n, and Theorem B.2 implies the upper bounds $\lambda_1 \le \tilde{\lambda}$ and

$$\lambda_1^n \,\mathrm{vol}(\mathcal{C}) \le 2^n. \tag{B.1}$$

Theorem B.3 allows us to strengthen (B.1). For any integer $j = 1, \ldots, n$, denote by $\lambda_j := \lambda_j(\mathcal{C})$ the infimum of the positive real numbers λ for which $\lambda\mathcal{C}$ contains j linearly independent integer points. Each of the λ_js is a minimum and we have

$$0 < \lambda_1 \le \lambda_2 \le \ldots \le \lambda_n < +\infty.$$

The real numbers $\lambda_1, \ldots, \lambda_n$ are called the *successive minima* of $\mathcal{C}$.

THEOREM B.3. *For a set $\mathcal{C}$ as above, we have*

$$\frac{2^n}{n!} \le \lambda_1 \ldots \lambda_n \mathrm{vol}(\mathcal{C}) \le 2^n.$$

Moreover, if the points $\underline{x}_1, \ldots, \underline{x}_n$ are linearly independent and realize the successive minima of $\mathcal{C}$, then we have

$$\det(\underline{x}_1, \ldots, \underline{x}_n) \le n!.$$

PROOF. This is Theorem V and Corollary, pages 218 and 219, of Cassels [157]. The proof of the upper bound for the product of the λ_i is difficult, while the lower bound is easier to establish.

Any hyperplane not containing the origin can be put in the form $y_1 x_1 + \ldots + y_n x_n = 1$, and so it may be represented as a point $\underline{y} = (y_1, \ldots, y_n)$ in $\mathbb{R}^n$. It turns out that the points $\underline{y}$ corresponding to hyperplanes having empty intersection with C form a convex set C^*, called the polar body of C.

THEOREM B.4. *Let $\lambda_1, \ldots, \lambda_n$ be the successive minima of a convex body C and let $\lambda_1^*, \ldots, \lambda_n^*$ be the successive minima of the polar body C^*. Then, for any integer $j = 1, \ldots, n$, we have*

$$1 \leq \lambda_j \lambda_{n+1-j}^* \leq n!.$$

PROOF. This is Theorem VI, page 219, of Cassels [157].

To conclude this Appendix, we state and prove Khintchine's Transference Theorem, which relates small values of linear forms to simultaneous Diophantine approximation. Let $\underline{x} = (x_1, \ldots, x_n)$ be a real n-tuple. We define $w(\underline{x})$ and $w'(\underline{x})$ to be the suprema of the real numbers w and w' for which the inequalities

$$|u_0 + u_1 x_1 + \ldots + u_n x_n| \leq \left(\max_{1 \leq i \leq n} |u_i| \right)^{-n-w}$$

and

$$\max_{1 \leq i \leq n} |v_0 x_i + v_i| \leq |v_0|^{-(1+w')/n}$$

have infinitely many integer solutions $(u_0, \ldots, u_n)$ and $(v_0, \ldots, v_n)$, with $v_0 \neq 0$. It follows from Theorem B.2 that $w(\underline{x})$ and $w'(\underline{x})$ are non-negative. Theorem B.5 shows how they are related.

THEOREM B.5. *With the above notation, we have*

$$\frac{w(\underline{x})}{n^2 + (n-1)w(\underline{x})} \leq w'(\underline{x}) \leq w(\underline{x}). \tag{B.2}$$

PROOF. We may assume that $n \geq 2$. We begin by treating some particular cases. If $w(\underline{x}) = 0$ (*resp.* $w'(\underline{x}) = 0$), then the left (*resp.* the right) inequality of (B.2) holds trivially. If $|u_0 + u_1 x_1 + \ldots + u_n x_n| = 0$ for some non-zero integer $(n+1)$-tuple $(u_0, \ldots, u_n)$, then $1, x_1, \ldots, x_n$ are linearly dependent over $\mathbb{Z}$ and there exists a positive constant c such that $\max_{1 \leq i \leq n} |v_0 x_i + v_i| \leq c|v_0|^{-1/(n-1)}$ has infinitely many integer solutions $(v_0, \ldots, v_n)$ with $v_0 \neq 0$. Consequently, we have $w(\underline{x}) = +\infty$ and $w'(\underline{x}) \geq 1/(n-1)$, hence, (B.2) holds. Furthermore, if $\max_{1 \leq i \leq n} |v_0 x_i + v_i| = 0$ for some non-zero integer $(n + 1)$-tuple $(v_0, \ldots, v_n)$, then $x_1, \ldots, x_n$ are rational numbers and $w(\underline{x}) = w'(\underline{x}) = +\infty$, hence, (B.2) holds.

We now assume that we are in none of these particular cases. Let $w > 0$ be a real number and let $u_0, \ldots, u_n$ be integers, not all zero, such that

$$0 < \rho := |u_0 + u_1 x_1 + \ldots + u_n x_n| \leq U^{-n-w},$$

$$\text{where } U = \max\{|u_0|, \ldots, |u_n|\}. \tag{B.3}$$

By Theorem B.2, there exist $v_0, \ldots, v_n$ integers, not all zero, such that

$$|v_0 x_i - v_i| \leq \rho^{1/n} \quad (i = 1, \ldots n),$$
$$|v_0 u_0 + v_1 u_1 + \ldots + v_n u_n| < 1, \tag{B.4}$$

since the system of linear forms appearing in the left-hand members of these inequalities has determinant $\pm \rho$. The u_is and the v_is are integers, thus we get

$$v_0 u_0 + v_1 u_1 + \ldots + v_n u_n = 0. \tag{B.5}$$

Combining (B.3) and (B.5), we obtain

$$|\rho v_0| = |(v_0 x_1 - v_1) u_1 + \ldots + (v_0 x_n - v_n) u_n| \leq n \, U \, \rho^{1/n}. \tag{B.6}$$

It follows from (B.3) that $U \leq \rho^{-1/(n+w)}$, thus (B.6) gives

$$|v_0| \leq n \rho^{1/n - 1 - 1/(n+w)} = n \rho^{-(n^2 + (n-1)w)/(n^2 + nw)},$$

which, combined with (B.4), yields the left inequality of (B.2).

As for the right inequality, let $w' > 0$ be a real number and let $v_0, \ldots, v_n$ be integers with $v_0 \neq 0$ and

$$0 < \rho := \max_{1 \leq i \leq n} |v_0 x_i + v_i| \leq |v_0|^{-(1+w')/n}.$$

Without any loss of generality, we assume that $\rho = |v_0 x_1 + v_1|$. By Theorem B.2, there exist $u_0, \ldots, u_n$ integers, not all zero, such that

$$|-u_0 + u_1 x_1 + \ldots + u_n x_n| \leq \rho^{(w'+n)/(w'+1)},$$
$$|v_0 u_0 + \ldots + v_n u_n| < 1, \tag{B.7}$$
$$|u_i| \leq \rho^{-1/(w'+1)} \quad (i = 2, \ldots, n),$$

since the system of linear forms appearing in the left-hand members of these inequalities has determinant $\pm \rho$. Arguing as previously, we get (B.5) and we deduce that $|u_1| \leq n \rho^{-1/(w'+1)}$, which, combined with (B.7), completes the proof of Theorem B.5.

Theorem B.5 is due to Khintchine [318, 320], and its proof has been simplified by Mahler [379] (see, for example, Gruber and Lekkerkerker [262], Section 45.3, Cassels [155], Chapter V, Theorem IV, or Schmidt [512], Chapter IV, Section 5). Inequalities (B.2) turn out to be best possible, as proved by Jarník [293, 295], who got some related results [294]. The generalization of Theorem B.5 to systems of linear forms is due to Dyson [224] and has been proved to be best possible by Jarník [299]. For other transference theorems, see Wang, Yu, and Zhu [594], Schmidt and Wang [516], and the survey of Xu [604].

References

[1]   Abercrombie, A. G. (1995). 'Badly approximable p-adic integers'. *Proc. Indian Acad. Sci. Math. Sci.*, **105**, 123–134. (Cited in Chapter 9.)

[2]   (1995). 'The Hausdorff dimension of some exceptional sets of p-adic integer matrices'. *J. Number Theory*, **53**, 311–341. (Cited in Chapter 9.)

[3]   Adhikari, S. D., Saradha, N., Shorey, T. N. & Tijdeman, R. (2001). 'Transcendental infinite sums'. *Indag. Math. (N. S.)*, **12**, 1–14. (Cited in Chapter 3.)

[4]   Ahlfors, L. V. (1966). *Complex Analysis: An Introduction of the Theory of Analytic Functions of One Complex Variable*. Second edition. McGraw-Hill Book Co., New York–Toronto–London. (Cited in Appendix A.)

[5]   Alessandri, P. & Berthé, V. (1998). 'Three distance theorems and combinatorics on words'. *Enseignement Math.*, **44**, 103–132. (Cited in Chapter 5.)

[6]   Allouche, J.-P., Davison, J. L., Queffélec, M. & Zamboni, L. Q. (2001). 'Transcendence of Sturmian or morphic continued fractions'. *J. Number Theory*, **91**, 39–66. (Cited in Chapter 2.)

[7]   Allouche, J.-P. & Shallit, J. (2003). *Automatic Sequences: Theory, Applications, Generalizations*. Cambridge University Press, Cambridge. (Cited in Chapter 2.)

[8]   Alniaçik, K. (1979). 'On the subclasses U_m in Mahler's classification of the transcendental numbers'. *Istanbul Üniv. Fen Fak. Mecm. Ser.*, **44**, 39–82. (Cited in Chapters 7 & 9.)

[9]   (1982). 'On U_m-numbers'. *Proc. Amer. Math. Soc.*, **85**, 499–505. (Cited in Chapter 7.)

[10] (1983). 'On Mahler's U-numbers'. *Amer. J. Math.*, **105**, 1347–1356. (Cited in Chapter 7.)

[11] (1986). 'On T-numbers'. *Glasnik Math.*, **21**, 271–282. (Cited in Chapter 7.)

[12] (1990). 'Representation of real numbers as sums of U_2-numbers'. *Acta Arith.*, **55**, 301–310. (Cited in Chapter 7.)

[13] (1991). 'On p-adic U_m-numbers'. *Istanbul Üniv. Fen Fak. Mat. Der.*, **50**, 1–7. (Cited in Chapter 9.)

[14] (1992). 'On p-adic T-numbers'. *Doğa Math.*, **16**, 119–128. (Cited in Chapter 9.)

[15] (1992). 'On semi-strong U-numbers'. *Acta Arith.*, **60**, 349–358. (Cited in Chapter 7.)

[16]   (1994). 'A note on some U-numbers'. *Indian J. Pure Appl. Math.*, **25**, 689–691. (Cited in Chapter 1.)

[17]   (1998). 'The points on curves whose coordinates are U-numbers'. *Rend. Mat. Roma*, Serie VII, **18**, 649–653. (Cited in Chapter 10.)

[18]   Alniaçik, K., Avci, Y. & Bugeaud, Y. (2003). 'On U_m-numbers with small transcendence measures'. *Acta. Math. Hungar.*, **99**, 271–277. (Cited in Chapter 7.)

[19]   Alniaçik, K. & Saias, É. (1994). 'Une remarque sur les G_δ-denses'. *Arch. Math.*, **62**, 425–426. (Cited in Chapter 1.)

[20]   Amice, Y. (1975). *Les nombres p-adiques*. Collection SUP: Le Mathématicien, No. 14, Presses Universitaires de France, Paris. (Cited in Chapter 9.)

[21]   Amoroso, F. (1988). 'On the distribution of complex numbers according to their transcendence types'. *Ann. Mat. Pura Appl.*, **151**, 359–368. (Cited in Chapter 8.)

[22]   (1996). 'Algebraic numbers close to 1 and variants of Mahler's measure'. *J. Number Theory*, **60**, 80–96. (Cited in Appendix A.)

[23]   (1998). 'Algebraic numbers close to 1: results and methods'. In *Number Theory* (Tiruchirapalli, 1996), pp. 305–316; *Amer. Math. Soc.*, Providence, RI. (Cited in Appendix A.)

[24]   (2001). 'Some metric results in transcendental numbers theory.' In *Introduction to algebraic independence theory*. Lecture Notes in Math. 1752, Springer, Berlin, 227–237. (Cited in Chapter 8.)

[25]   Amou, M. (1991). 'Approximation to certain transcendental decimal fractions by algebraic numbers'. *J. Number Theory*, **37**, 231–241. (Cited in Chapter 7.)

[26]   (1991). 'Algebraic independence of the values of certain functions at a transcendental number'. *Acta Arith.*, **59**, 71–82. (Cited in Chapter 8.)

[27]   (1992). 'An improvement of a transcendence measure of Galochkin and Mahler's S-numbers'. *J. Austral. Math. Soc. Ser.*, **52**, 130–140. (Cited in Chapter 3.)

[28]   (1996). 'On Sprindžuk's classification of transcendental numbers'. *J. reine angew. Math.*, **470**, 27–50. (Cited in Chapter 8 & Appendix A.)

[29]   (1996). 'Transcendence measures for almost all numbers'. In *Analytic Number Theory* (Kyoto, 1995). Sūrikaisekikenkyūsho Kōkyūroku, **961**, 112–116 (in Japanese). (Cited in Chapter 8.)

[30]   (1996). 'A metrical result on transcendence measures in certain fields'. *J. Number Theory*, **59**, 389–397. (Cited in Chapter 9.)

[31]   Amou, M. & Bugeaud, Y. (Preprint). 'Sur la séparation des racines des polynômes et une question de Sprindžuk'. (Cited in Chapter 8 & Appendix A.)

[32]   Arbour, B. & Roy, D. (2004). 'A Gel fond type criterion in degree two'. *Acta Arith.*, **111**, 97–103. (Cited in Chapters 3 & 8.)

[33]   Armitage, J. V. (1969). 'An analogue of a problem of Littlewood'. *Mathematika*, **16**, 101–105; Corrigendum and addendum (1970) in *Mathematika*, **17**, 173–178. (Cited in Chapter 10.)

[34]   Astels, S. (2000). 'Cantor sets and numbers with restricted partial quotients'. *Trans. Amer. Math. Soc.*, **352**, 133–170. (Cited in Chapter 1.)

[35]   (2001). 'Sums of numbers with small partial quotients II'. *J. Number Theory*, **91**, 187–205. (Cited in Chapter 1.)

[36]   (2002). 'Sums of numbers with small partial quotients II'. *Proc. Amer. Math. Soc.*, **130**, 637–642. (Cited in Chapter 1.)

[37] Baker, A. (1962). 'Continued fractions of transcendental numbers'. *Mathematika*, **9**, 1–8. (Cited in Chapters 2 & 3.)

[38] (1964). 'On Mahler's classification of transcendental numbers'. *Acta Math.*, **111**, 97–120. (Cited in Chapters 2 & 3.)

[39] (1964). 'On an analogue of Littlewood's Diophantine approximation problem'. *Michigan Math. J.*, **11**, 247–250. (Cited in Chapter 10.)

[40] (1964). 'Approximations to the logarithms of certain rational numbers'. *Acta Arith.*, **10**, 315–323. (Cited in Chapter 3.)

[41] (1966). 'On a theorem of Sprindžuk'. *Proc. Roy. Soc. London*, **292**, 92–104. (Cited in Chapters 4 & 9.)

[42] (1966/67). 'On Mahler's classification of transcendental numbers II: Simultaneous Diophantine approximation'. *Acta Arith.*, **12**, 281–288. (Cited in Chapter 3.)

[43] (1966/68). 'Linear forms in the logarithms of algebraic numbers I–IV'. *Mathematika*, **13** (1966), 204–216; **14** (1967), 102–107 & 220–224; **15** (1968), 204–216. (Cited in Chapter 2.)

[44] (1974). *Transcendental Number Theory*, Cambridge University Press. (Cited in Introduction, and Chapters 2, 3 & 4.)

[45] Baker, A. & Schmidt, W. M. (1970). 'Diophantine approximation and Hausdorff dimension'. *Proc. London Math. Soc.*, **21**, 1–11. (Cited in Chapters 3, 5 & 6.)

[46] Baker, A. & Stewart, C. L. (1988). 'On effective approximations to cubic irrationals'. In *New Advances in Transcendence Theory*. A. Baker (ed.), pp. 1–24, Cambridge University Press, Cambridge. (Cited in Chapter 2.)

[47] Baker, R. C. (1976). 'On approximation with algebraic numbers of bounded degree'. *Mathematika*, **23**, 18–31. (Cited in Chapters 7 & 9.)

[48] (1976). 'Sprindžuk's theorem and Hausdorff dimension'. *Mathematika*, **23**, 184–197. (Cited in Chapters 5 & 9.)

[49] (1976). 'Metric Diophantine approximation on manifolds'. *J. London Math. Soc.*, **14**, 43–48. (Cited in Chapter 3.)

[50] (1979). 'On numbers with many rational approximations'. *Math. Proc. Cambridge Philos. Soc.*, **86**, 25–27. (Cited in Chapter 1.)

[51] (1992). 'Singular *n*-tuples and Hausdorff dimension II'. *Math. Proc. Cambridge Phil. Soc.*, **111**, 577–584. (Cited in Chapter 5.)

[52] Baum, L. E. & Sweet, M. M. (1976). 'Continued fractions of algebraic power series in characteristic 2'. *Ann. of Math.*, **103**, 593–610. (Cited in Chapter 10.)

[53] Baxa, C. (1994). 'Fast growing sequences of partial denominators'. *Acta Math. Inform. Univ. Ostraviensis*, **2**, 81–84. (Cited in Chapter 3.)

[54] (1997). 'On the distribution of the sequence $(n\alpha)$ with transcendental α'. *Acta Arith.*, **81**, 357–363. (Cited in Chapters 3 & 7.)

[55] (To appear). 'Extremal values of continuants and transcendence of certain continued fractions. *Adv. in Appl. Math.* (Cited in Chapter 2.)

[56] Becker, P.-G. (1991). 'Effective measures for algebraic independence of the values of Mahler type functions'. *Acta Arith.*, **58**, 239–250. (Cited in Chapters 3 & 8.)

[57] (1994). 'Transcendence measures for the values of functions satisfying generalized Mahler type functional equations in arbitrary characteristic'. *Publ. Math. Debrecen*, **45**, 269–282. (Cited in Chapter 9.)

[58] Becker-Landeck, P.-G. (1986). 'Transcendence measures by Mahler's transcendence method'. *Bull. Austral. Math. Soc.*, **33**, 59–65. (Cited in Chapters 3 & 8.)

[59] Beckett, S. (1953). *L'Innommable*, Les éditions de Minuit.

[60] Beresnevich, V. (1996). 'Effective measure estimates for sets of real numbers with a given error of approximation by quadratic irrationalities'. *Vestsī Akad. Navuk Belarusī Ser. Fīz. Mat. Navuk*, **4**, 10–15 (in Russian). (Cited in Chapters 4, 6 & 7.)

[61] ——— (1999). 'On approximation of real numbers by real algebraic numbers'. *Acta Arith.*, **90**, 97–112. (Cited in Chapters 4, 5 & 6.)

[62] ——— (2000). 'Application of the concept of regular system of points in metric number theory'. *Vestsī Nats. Akad. Navuk Belarusī Ser. Fīz. Mat. Navuk*, **1**, 35–39 (in Russian). (Cited in Chapters 6 & 9.)

[63] ——— (2002). 'A Groshev type theorem for convergence on manifolds'. *Acta Math. Hungar.*, **94**, 99–130. (Cited in Chapters 4 & 5.)

[64] Beresnevich, V. & Bernik, V. (2000). 'A. Baker's conjecture and Hausdorff dimension'. *Publ. Math. Debrecen*, **56**, 263–269. (Cited in Chapter 5.)

[65] ——— (2002). 'Diophantine approximation on classical curves and Hausdorff dimension.' In *Analytic and Probabilistic Methods in Number Theory* (Palanga, 2001), 20–27, TEV, Vilnius. (Cited in Chapters 5 & 10.)

[66] Beresnevich, V., Bernik, V. I. & Dodson, M. M. (1997). 'Inhomogeneous nonlinear Diophantine approximation'. In *Papers in Honour of V. G. Sprindžuk's 60th. Birthday*, Inst. Math., Belarus Acad. Sci., 13–20. (Cited in Chapter 4.)

[67] ——— (2002). 'Regular systems, ubiquity and Diophantine approximation'. In *A Panorama of Number Theory or The View from Baker's Garden*. G. Wüstholz (ed.), Cambridge University Press, pp. 260–279. (Cited in Chapters 5 & 6.)

[68] Beresnevich, V., Bernik, V. I., Kleinbock, D. Y. & Margulis, G. A. (2002). 'Metric Diophantine approximation: the Khintchine–Groshev theorem for nondegenerate manifolds'. *Moscow Math. J.*, **2**, 203–225. (Cited in Chapters 4 & 5.)

[69] Beresnevich, V. V., Bernik, V. I. & Kovalevskaya, E. I. (Preprint). 'On approximation of *p*-adic numbers by *p*-adic algebraic numbers'. (Cited in Chapter 9.)

[70] Beresnevich, V. V., Dickinson, H. & Velani, S. L. (2001). 'Sets of exact 'logarithmic order' in the theory of Diophantine approximation'. *Math. Ann.*, **321**, 253–273. (Cited in Chapters 6 & 10.)

[71] ——— (Preprint). 'Measure theoretic laws for lim sup sets'. (Cited in Chapters 6 & 9.)

[72] Beresnevich, V. V. & Kovalevskaya, E. I. (2003). 'On Diophantine approximation of dependent quantities in the *p*-adic case'. *Mat. Zametki.* **73**, 22–37 (in Russian). English transl. in *Math. Notes*, **73** (2003), 21–35. (Cited in Chapter 9.)

[73] Bernhard, T. (1973). *Das Kalkwerk*, Suhrkamp Verlag.

[74] Bernik, V. I. (1977). 'Induced extremal surfaces'. *Mat. Sb. (N. S.)*, **103**, 480–489, 630 (in Russian). (Cited in Chapter 4.)

[75] ——— (1979). 'On the Baker–Schmidt hypothesis'. *Dokl. Akad. Nauk BSSR*, **23**, 392–395 (in Russian). (Cited in Chapter 5.)

[76] ——— (1980). 'A metric theorem on the simultaneous approximation of zero by the values of integral polynomials'. *Izv. Akad. Nauk SSSR Ser. Mat.* **44**, 24–45

(in Russian). English transl. in *Math. USSR Izv.*, **16** (1981), 21–40. (Cited in Chapters 4 & 9.)

[77]   (1983). 'Application of the Hausdorff dimension in the theory of Diophantine approximations'. *Acta Arith.*, **42**, 219–253 (in Russian). English transl. in *Amer. Math. Soc. Transl.*, **140** (1988), 15–44. (Cited in Chapter 5.)

[78]   (1984). 'A proof of Baker's conjecture in the metric theory of transcendental numbers'. *Dokl. Akad. Nauk. SSSR*, **277**, 1036–1039 (in Russian). English transl. in *Soviet Math. Dokl.*, **30** (1984), 186–189. (Cited in Chapter 4.)

[79]   (1986). 'A property of integer polynomials that realize Minkowski's theorem on linear forms'. *Dokl. Akad. Nauk BSSR*, **30**, 403–405, 477 (in Russian). (Cited in Chapter 6.)

[80]   (1988). 'Applications of measure theory and Hausdorff dimension to the theory of Diophantine approximation'. In *New Advances in Transcendence Theory* (Durham, 1986), 25–36, Cambridge University Press, Cambridge. (Cited in Chapter 9.)

[81]   (1989). 'On the best approximation of zero by values of integral polynomials'. *Acta Arith.*, **53**, 17–28 (in Russian). (Cited in Chapter 4.)

[82]   Bernik, V. I. & Borbat, V. N. (1997). 'Polynomials with overfalls in values of coefficients and a conjecture of A. Baker'. *Vestsī Nats. Akad. Navuk Belarusī Ser. Fīz. Mat. Navuk*, **3**, 5–8 (in Russian). (Cited in Chapter 4.)

[83]   (1997). 'Simultaneous approximation of zero by values of integer-valued polynomials'. *Tr. Mat. Inst. Steklova*, **218**, 58–73 (in Russian). English transl. in *Proc. Steklov Inst. Math.*, **218** (1997), 53–68. (Cited in Chapter 4.)

[84]   Bernik, V. I., Dickinson, H. & Dodson, M. M. (1998). 'Approximation of real numbers by values of integer polynomials'. *Dokl. Nats. Akad. Nauk Belarusi*, **42**, No.4, 51–54, 123 (in Russian). (Cited in Chapter 4.)

[85]   Bernik, V. I., Dickinson, H. & Yuan, J. (1999). 'Inhomogeneous Diophantine approximation on polynomials in $\mathbb{Q}_p$'. *Acta Arith.*, **90**, 37–48. (Cited in Chapter 9.)

[86]   Bernik, V. I. & Dodson, M. M. (1999). *Metric Diophantine approximation on manifolds*. Cambridge Tracts in Mathematics 137, Cambridge University Press. (Cited in Introduction, and Chapters 1, 4, 5, 6 & 9.)

[87]   Bernik, V. I. & Dombrovskiĭ, I. V. (1992). 'U_3-numbers on curves in $\mathbb{R}^2$'. *Vestsī Akad. Navuk Belarusī Ser. Fīz. Mat. Navuk*, **3/4**, 3–7, 123. (Cited in Chapter 10.)

[88]   (1994). 'Effective estimates for the measure of sets defined by Diophantine conditions'. *Trudy Mat. Inst. Steklov*, **207**, 35–41 (in Russian). English transl. in *Proc. Steklov Inst. Math.*, **207** (1995), 35–40. (Cited in Chapter 4.)

[89]   Bernik, V., Kleinbock, D. & Margulis, G. A. (2001). 'Khintchine-type theorems on manifolds: the convergence case for standard and multiplicative versions'. *Internat. Math. Res. Notices*, **9**, 453–486. (Cited in Chapter 4.)

[90]   Bernik, V. I. & Melnichuk, Y. I. (1988). *Diophantine approximation and Hausdorff dimension*. Akad. Nauk. BSSR, 1988 (in Russian). (Cited in Introduction and Chapter 5.)

[91]   Bernik V. I. & Morotskaya, I. L. (1986). 'Diophantine approximations in $\mathbb{Q}_p$ and Hausdorff dimension'. *Vestsī Nats. Akad. Navuk Belarusī Ser. Fīz. Mat. Navuk*, **3**, 3–9, 123 (in Russian). (Cited in Chapter 9.)

[92] Bernik, V. I. & Morozova, I. M. (1998). 'Diophantine approximations by lacunary polynomials and the Hausdorff dimension'. *Vestsī Nats. Akad. Navuk Belarusī Ser. Fīz. Mat. Navuk*, **4**, 47–51, 142 (in Russian). (Cited in Chapter 9.)

[93] ___ (1998). 'Diophantine approximations and zeros of lacunary polynomials'. *Dokl. Nats. Akad. Nauk Belarusī*, **42**, No.6, 5–8, 122 (in Russian). (Cited in Chapter 9.)

[94] Bernik, V. I. & Sakovich, N. V. (1994). 'Regular systems of complex algebraic numbers'. *Dokl. Akad. Nauk Belarusī*, **38**, 10–13, 122 (in Russian). (Cited in Chapter 9.)

[95] Bernik, V. I. & Tishchenko, K. (1993). 'Integral polynomials with an overfall of the coefficient values and Wirsing's problem'. *Dokl. Akad. Nauk Belarusī*, **37**, No.5, 9–11 (in Russian). (Cited in Chapter 3.)

[96] Bernik, V. I. & Vasiliev, D. V. (1999). 'A Khintchine-type theorem for integer-valued polynomials of a complex variable'. *Proceedings of the Institute of Mathematics*, **3**, 10–20. *Tr. Inst. Mat. (Minsk)*, **3**, *Natl. Akad. Nauk Belarusī, Inst. Mat. Minsk* (in Russian). (Cited in Chapter 9.)

[97] Bernik, V. I., Vasiliev, D. V. & Dodson, M. M. (2002). 'Metric theorems on approximation of real numbers by special real numbers'. *Dokl. Nats. Akad. Nauk Belarusī*, **46**, No.3, 60–63 (in Russian). (Cited in Chapter 4.)

[98] Bernstein, F. (1912). 'Über eine Anwendung der Mengenlehre auf ein aus der Theorie der säkularen Störungen herrührendes Problem'. *Math. Ann.*, **71**, 417–439. (Cited in Chapter 1.)

[99] ___ (1912). 'Über geometrische Wahrscheinlichkeit und über das Axiom der beschränkten Arithmetisierbarkeit der Beobachtungen'. *Math. Ann.*, **72**, 585–587. (Cited in Chapter 1.)

[100] Besicovitch, A. S. (1934). 'Sets of fractional dimension (IV); on rational approximation to real numbers'. *J. London Math. Soc.*, **9**, 126–131. (Cited in Chapter 5.)

[101] Bilu, Yu. & Bugeaud, Y. (2000). 'Démonstration du théorème de Baker–Feldman via les formes linéaires en deux logarithmes'. *J. Théor. Nombres Bordeaux*, **12**, 13–23. (Cited in Chapter 2.)

[102] Bluhm, C. (1998). 'On a theorem of Kaufman: Cantor-type construction of linear fractal Salem sets'. *Ark. Mat.*, **36**, 307–316. (Cited in Chapters 1 & 5.)

[103] ___ (2000). 'Liouville numbers, Rajchman measures, and small Cantor sets'. *Proc. Amer. Math. Soc.*, **128**, 2637–2640. (Cited in Chapter 1.)

[104] Bombieri, E. (1993). 'Effective Diophantine approximation on G_m'. *Ann. Scuola Norm. Sup. Pisa Cl. Sci.*, **20**, 61–89. (Cited in Chapter 2.)

[105] ___ (1997). 'The equivariant Thue–Siegel method'. In *Arithmetic Geometry* (Cortona, 1994), 70–86, *Sympos. Math.*, *XXXVII*, Cambridge University Press, Cambridge. (Cited in Chapter 2.)

[106] Bombieri, E. & Cohen, P. B. (1997). 'Effective Diophantine approximation on G_m, II'. *Ann. Scuola Norm. Sup. Pisa Cl. Sci.*, **24**, 205–225. (Cited in Chapter 2.)

[107] Bombieri, E. & Mueller, J. (1986). 'Remarks on the approximation to an algebraic number by algebraic numbers'. *Mich. Math. J.*, **33**, 83–93. (Cited in Chapter 2.)

[108] Bombieri, E., van der Poorten, A. J. & Vaaler, J. D. (1996). 'Effective measures of irrationality for cubic extensions of number fields'. *Ann. Scuola Norm. Sup. Pisa Cl. Sci.*, **23**, 211–248. (Cited in Chapter 2.)

[109] Borbat, V. N. (1995). 'Joint zero approximation by the values of integral polynomials and their derivatives'. *Vestsĭ Akad. Navuk BSSR Ser. Fĭz. Mat. Navuk*, **1**, 9–16, 123 (in Russian). (Cited in Chapter 6.)

[110] Borel, É. (1909). 'Les probabilités dénombrables et leurs applications arithmétiques'. *Rend. Circ. Math. Palermo*, **27**, 247–271. (Cited in Chapter 1.)

[111] (1912). 'Sur un problème de probabilités relatif aux fractions continues'. *Math. Ann.*, **72**, 578–584. (Cited in Chapter 1.)

[112] Borosh, I. & Fraenkel, A. S. (1972). 'A generalization of Jarník's theorem on Diophantine approximations'. *Indag. Math.*, **34**, 193–201. (Cited in Chapter 5.)

[113] (1975). 'A generalization of Jarník's theorem on Diophantine approximations to Ridout type numbers'. *Trans. Amer. Math. Soc.*, **211**, 23–38. (Cited in Chapter 5.)

[114] Borwein, P. & Erdélyi, T. (1995). *Polynomials and polynomial inequalities*. Graduate Texts in Mathematics 161, Springer-Verlag, New York. (Cited in Chapter 3.)

[115] Bovey, J. D. & Dodson, M. M. (1978). 'The fractional dimension of sets whose simultaneous rational approximations have errors with a small product'. *Bull. London Math. Soc.*, **10**, 213–218. (Cited in Chapter 5.)

[116] (1986). 'The Hausdorff dimension of systems of linear forms'. *Acta Arith.*, **45**, 337–358. (Cited in Chapters 5 & 9.)

[117] Braune, E. (1977). 'Über arithmetische Eigenschaften von Lückenreihen mit algebraischen Koeffizienten und algebraischem Argument'. *Monatsh. Math.*, **84**, 1–11. (Cited in Chapter 7.)

[118] Brownawell, D. (1974). 'Sequences of Diophantine approximations'. *J. Number Theory*, **6**, 11–21. (Cited in Chapter 8.)

[119] (1985). 'Effectivity in independence measures for values of E-functions'. *J. Austral. Math. Soc.*, **39**, 227–240. (Cited in Chapter 8.)

[120] Bugeaud, Y. (1998). 'Bornes effectives pour les solutions des équations en S-unités et des équations de Thue–Mahler'. *J. Number Theory*, **71**, 227–244. (Cited in Chapter 2.)

[121] (1998). 'Algebraic numbers close to 1 in non-Archimedean metrics'. *Ramanujan Math. J.*, **2**, 449–457. (Cited in Appendix A.)

[122] (2000). 'On the approximation by algebraic numbers with bounded degree'. *Algebraic Number Theory and Diophantine Analysis* (Graz, 1998), 47–53, de Gruyter, Berlin. (Cited in Chapter 3.)

[123] (2000). 'Approximation par des nombres algébriques'. *J. Number Theory*, **84**, 15–33. (Cited in Chapters 3 & 8.)

[124] (2002). 'Approximation par des nombres algébriques de degré borné et dimension de Hausdorff'. *J. Number Theory*, **96**, 174–200. (Cited in Chapter 6.)

[125] (2002). 'Approximation by algebraic integers and Hausdorff dimension'. *J. London Math. Soc.*, **65**, 547–559. (Cited in Chapters 3 & 10.)

[126] (2002). 'Nombres de Liouville et nombres normaux'. *C. R. Acad. Sci. Paris*, **335**, 117–120. (Cited in Chapter 1.)

[127]   (2003). 'A note on inhomogeneous Diophantine approximation'. *Glasgow Math. J.*, **45**, 105–110. (Cited in Chapter 5.)

[128]   (2003). 'Mahler's classification of numbers compared with Koksma's'. *Acta Arith.*, **110**, 89–105. (Cited in Chapters 3, 7 & 9.)

[129]   (2003). 'Sets of exact approximation order by rational numbers'. *Math. Ann.*, **327**, 171–190. (Cited in Chapters 6 & 10.)

[130]   (To appear). 'An inhomogeneous Jarník theorem'. *J. Analyse Math.* (Cited in Chapters 5 & 6.)

[131]   (2003). 'Approximation simultanée par des nombres algébriques'. *J. Théor. Nombres Bordeaux*, **15**, 665–672. (Cited in Chapter 3.)

[132]   (To appear). 'Intersective sets and Diophantine approximation'. *Mich. Math. J.* (Cited in Chapters 5 & 6.)

[133]   (Preprint). 'Mahler's classification of numbers compared with Koksma's, II'. (Cited in Chapters 7 & 9.)

[134]   Bugeaud, Y. & Chevallier, N. (Preprint). 'On simultaneous inhomogeneous Diophantine approximation'. (Cited in Chapter 5.)

[135]   Bugeaud, Y., Dodson, M. M. & Kristensen, S. (Preprint). 'Zero-infinity laws in Diophantine approximation'. (Cited in Chapters 6 & 9.)

[136]   Bugeaud, Y. & Győry, K. (1996). 'Bounds for the solutions of Thue–Mahler equations and norm form equation'. *Acta Arith.*, **74**, 273–292. (Cited in Chapter 2.)

[137]   Bugeaud, Y. & Laurent, M. (Preprint). 'Exponents of Diophantine Approximation and Sturmian Continued Fractions'. (Cited in Chapter 3.)

[138]   (Predprint). 'Exponents of Inhomogeneous Diophantine Approximation'. (Cited in Chapter 3.)

[139]   Bugeaud, Y & Mignotte, M. (To appear). 'On the distance between roots of integer polynomials'. *Proc. Edinburgh Math. Soc.* (Cited in Appendix A.)

[140]   Bugeaud, Y. & Teulié, O. (2000). 'Approximation d'un nombre réel par des nombres algébriques de degré donné'. *Acta Arith.*, **93**, 77–86. (Cited in Chapters 3 & 10.)

[141]   Bundschuh, P. (1978). 'Transzendenzmasse in Körpern formaler Laurentreihen'. *J. reine angew. Math.* **299/300**, 411–432. (Cited in Chapter 9.)

[142]   (1980) 'Über eine Klasse reeller transzendenter Zahlen mit explicit angebbarer g-adischer und Kettenbruch-Entwicklung'. *J. reine angew. Math.*, **318**, 110–119. (Cited in Chapter 3.)

[143]   Burger, E. B. (1992). 'Homogeneous Diophantine approximation in S-integers'. *Pacific J. Math.*, **152**, 211–253. (Cited in Chapter 9.)

[144]   (1996). 'On Liouville decompositions in local fields'. *Proc. Amer. Math. Soc.*, **124**, 3305–3310. (Cited in Chapters 1 & 9.)

[145]   (1996). 'Uniformly approximable numbers and the uniform approximation spectrum'. *J. Number Theory*, **61**, 194–208. (Cited in Chapter 1.)

[146]   (1999). 'On real quadratic fields and simultaneous Diophantine approximation'. *Monatsh. Math.*, **128**, 201–209. (Cited in Chapter 9.)

[147]   (2000). 'On simultaneous Diophantine approximation in the vector space $\mathbb{Q}+\mathbb{Q}\alpha$'. *J. Number Theory*, **82**, 12–24. (Cited in Chapter 9.)

[148]   (2001). 'Diophantine inequalities and irrationality measures for certain transcendental numbers'. *Indian J. Pure Appl. Math.*, **32**, 1591–1599. (Cited in Chapter 7.)

[149] Burger, E. B. & Dubois, E. (1994). 'Sur les quotients partiels de U-nombres dans un corps de séries formelles'. *C. R. Acad. Sci. Paris*, **319**, 421–426. (Cited in Chapter 9.)

[150] Burger, E. B. & Struppeck, T. (1993). 'On frequency distributions of partial quotients of U-numbers'. *Mathematika*, **40**, 215–225. (Cited in Chapter 7.)

[151] Cantor, D. G. (1965). 'On the elementary theory of Diophantine approximation over the ring of adeles. I'. *Illinois J. Math.*, **9**, 677–700. (Cited in Chapter 9.)

[152] Cantor, G. (1874). 'Über eine Eigenschaft des Inbegriffs aller reellen algebraischen Zahlen'. *J. reine angew. Math.*, **77**, 258–262. (Cited in Chapter 1.)

[153] Cassels, J. W. S. (1951). 'Some metrical theorems in Diophantine approximation, V: On a conjecture of Mahler'. *Proc. Cambridge Philos. Soc.*, **47**, 18–21. (Cited in Chapter 4.)

[154] ——— (1955). 'Simultaneous Diophantine approximation, II'. *Proc. London Math. Soc.*, **5**, 435–448. (Cited in Chapter 3.)

[155] ——— (1957). *An introduction to Diophantine approximation*. Cambridge Tracts in Math. and Math. Phys., vol. 99, Cambridge University Press. (Cited in Chapters 1 & 6, and Appendix B.)

[156] ——— (1986). *Local fields*. London Mathematical Society Student Texts 3, Cambridge University Press, Cambridge. (Cited in Chapter 3 and Appendix A.)

[157] ——— (1997). *An Introduction to the Geometry of Numbers*. Springer Verlag. (Cited in Appendix B.)

[158] Cassels, J. W. S. & Swinnerton-Dyer, H. P. F. (1955). 'On the product of three homogeneous linear forms and indefinite ternary quadratic forms'. *Philos. Trans. Roy. Soc. London, Ser. A.*, **248**, 73–96. (Cited in Chapter 10.)

[159] Caveny, D. (1993). 'U-numbers and T-numbers: Some elementary transcendence and algebraic independence results'. *Number Theory with an Emphasis on the Markoff Spectrum* (Provo, UT, 1991), 43–52, Dekker, New York. (Cited in Chapter 3.)

[160] Caveny, D. & Tubbs, R. (1993). 'The arithmetic of well-approximated numbers'. *Number Theory with an Emphasis on the Markoff Spectrum* (Provo, UT, 1991), 53–59, Dekker, New York. (Cited in Chapter 3.)

[161] Černy, K. (1952). 'Sur les approximations diophantiennes'. *Čehoslovack. Mat. Ž.*, **2**, 191–220. (Cited in Chapter 6.)

[162] Chowla, S. & DeLeon, M. J. (1968). 'On a conjecture of Littlewood'. *Norske Vid. Selsk. Forh. (Trondheim)*, **41**, 45–47. (Cited in Chapter 10.)

[163] Chudnovsky, G. V. (1980). 'Algebraic independence of the values of elliptic functions at algebraic points. Elliptic analogue of the Lindemann–Weierstrass theorem'. *Invent. Math.*, **61**, 267–290. (Cited in Chapters 3 & 8.)

[164] ——— (1984). *Contributions to the theory of transcendental numbers*. (Translated from the Russian by J. A. Kandall.) Math. Surveys Monographs, 19, Amer. Math. Soc., Providence, R. I. (Cited in Chapters 3 & 8, and Appendix A.)

[165] Chung, K. L. & Erdős, P. (1952). 'On the application of the Borel–Cantelli lemma'. *Trans. Amer. Math. Soc.*, **72**, 179–186. (Cited in Chapter 6.)

[166] Cijsouw, P. L. (1972). *Transcendence Measures*. Doctoral dissertation, University of Amsterdam. (Cited in Chapters 3 & 8.)

[167] Cohn, H. (1946). 'Note on almost-algebraic numbers'. *Bull. Amer. Math. Soc.*, **52**, 1042–1045. (Cited in Appendix A.)

[168] Collins, G. E. (2001). 'Polynomial minimum root separation'. *J. Symbol. Comp.*, **32**, 467–473. (Cited in Appendix A.)

[169] Collins, G. E. & Horowitz, E. (1974). 'The minimum root separation of a polynomial'. *Math. Comp.*, **28**, 589–597. (Cited in Appendix A.)

[170] Cusick, T. W. (1967). 'Littlewood's Diophantine approximation problem for series'. *Proc. Amer. Math. Soc.*, **18**, 920–924. (Cited in Chapter 10.)

[171] (1971). 'Lower bound for a Diophantine approximation function'. *Monatsh. Math.*, **75**, 398–401. (Cited in Chapter 10.)

[172] (1990). 'Hausdorff dimensions of sets of continued fractions'. *Quart. J. Math. Oxford*, **41**, 277–286. (Cited in Chapter 5.)

[173] Cusick, T. W. & Flahive, M. E. (1989). *The Markoff and Lagrange spectra.* Mathematical Surveys and Monographs, 30. American Mathematical Society, Providence, RI. (Cited in Chapter 1.)

[174] Dajani, K. & Kraaikamp, C. (2002). *Ergodic theory of numbers.* Carus Mathematical Monographs, 29. Mathematical Association of America, Washington, DC. (Cited in Chapter 1.)

[175] Davenport, H. (1954). 'Simultaneous Diophantine approximation'. *Mathematika*, **1**, 51–72. (Cited in Chapters 1, 3 & 10.)

[176] (1961). 'A note on binary cubic forms'. *Mathematika*, **8**, 58–62. (Cited in Chapter 4.)

[177] Davenport, H., Erdős, P. & LeVeque, W. J. (1963). 'On Weyl's criterion for uniform distribution'. *Michigan Math. J.*, **10**, 311–314. (Cited in Chapter 1.)

[178] Davenport, H. & Lewis, D. J. (1963). 'An analogue of a problem of Littlewood'. *Michigan Math. J.*, **10**, 157–160. (Cited in Chapter 10.)

[179] Davenport, H. & Roth, K. F. (1955). 'Rational approximations to algebraic numbers'. *Mathematika*, **2**, 160–167. (Cited in Chapter 2.)

[180] Davenport, H. & Schmidt, W. M. (1967). 'Approximation to real numbers by quadratic irrationals'. *Acta Arith.*, **13**, 169–176. (Cited in Chapters 3 & 10.)

[181] (1967/68). 'A theorem on linear forms'. *Acta Arith.*, **14**, 209–223. (Cited in Chapter 3.)

[182] (1969). 'Approximation to real numbers by algebraic integers'. *Acta Arith.*, **15**, 393–416. (Cited in Chapters 2, 3, 8 & 9.)

[183] (1970). 'Dirichlet's theorem on Diophantine approximation'. *Symposia Mathematica*, Vol. IV (INDAM, Rome, 1968/69), pp. 113–132. Academic Press, London. (Cited in Chapters 1 & 3.)

[184] (1970). 'Dirichlet's theorem on Diophantine approximation, II'. *Acta Arith.*, **16**, 413–423. (Cited in Chapters 1 & 5.)

[185] Davison, J. L. (1989). 'A class of transcendental numbers with bounded partial quotients'. In *Number Theory and Applications*, R. A. Mollin, (ed.) pp. 365–371. Kluwer Academic Publishers. (Cited in Chapter 2.)

[186] (2002). 'Continued fractions with bounded partial quotients'. *Proc. Edinburgh Math. Soc.*, **45**, 653–671. (Cited in Chapter 2.)

[187] Diaz, G. (1993). 'Une nouvelle minoration de $|\log \alpha - \beta|$, $|\alpha - \exp \beta|$, α et β algébriques'. *Acta Arith.*, **64**, 43–57. (Cited in Chapters 3 & 8.)

[188]   (1997). 'Une nouvelle propriété d'approximation diophantienne'. *C. R. Acad. Sci. Paris,* **324**, 969–972. (Cited in Chapter 8 and Appendix A.)

[189]   Diaz, G. & Mignotte, M. (1991). 'Passage d'une mesure d'approximation à une mesure de transcendence'. *C. R. Math. Rep. Acad. Sci. Canada,* **13**, 131–134. (Cited in Chapter 8 and Appendix A.)

[190]   Dickinson, H. (1993). 'The Hausdorff dimension of systems of simultaneously small linear forms'. *Mathematika,* **40**, 367–374. (Cited in Chapter 5.)

[191]   (1994). 'The Hausdorff dimension of sets arising in metric Diophantine approximation'. *Acta Arith.,* **68**, 133–140. (Cited in Chapter 5.)

[192]   (1997). 'A remark on the Jarník–Besicovitch theorem'. *Glasgow Math. J.,* **39**, 233–236. (Cited in Chapters 5 & 6.)

[193]   Dickinson, H. & Dodson, M. M. (2001). 'Diophantine approximation and Hausdorff dimension on the circle'. *Math. Proc. Cambridge Phil. Soc.,* **130**, 515–522. (Cited in Chapters 4 & 5.)

[194]   Dickinson, H., Dodson, M. M. & Yuan, J. (1999). 'Hausdorff dimension and *p*-adic Diophantine approximation'. *Indag. Math. (N. S.),* **10**, 337–347. (Cited in Chapter 9.)

[195]   Dickinson, H. & Velani, S. L. (1997). 'Hausdorff measure and linear forms'. *J. reine angew. Math.,* **490**, 1–36. (Cited in Chapter 6.)

[196]   Dirichlet, L. G. P. (1842). 'Verallgemeinerung eines Satzes aus der Lehre von den Kettenbrüchen nebst einige Anwendungen auf die Theorie der Zahlen'. *S.-B. Preuss. Akad. Wiss.,* 93–95. (Cited in Chapter 1.)

[197]   Diviš, B. (1972). 'An analog to the Lagrange numbers'. *J. Number Theory,* **4**, 274–285. (Cited in Chapter 1.)

[198]   Dodson, M. M. (1984). 'A note on the Hausdorff–Besicovitch dimension of systems of linear forms'. *Acta Arith.,* **44**, 87–98. (Cited in Chapter 5.)

[199]   (1991). 'Star bodies and Diophantine approximation'. *J. London Math. Soc.,* **44**, 1–8. (Cited in Chapter 5.)

[200]   (1992). 'Hausdorff dimension, lower order and Khintchine's theorem in metric Diophantine approximation'. *J. reine angew. Math.,* **432**, 69–76. (Cited in Chapter 6.)

[201]   (1993). 'Geometric and probabilistic ideas in the metrical theory of Diophantine approximation'. *Tsp. Mat. Nauk,* **48**, 77–106 (in Russian). English transl. in *Russian Math. Surveys,* **48** (1993), 73–102. (Cited in Chapter 5.)

[202]   (1997). 'A note on metric inhomogeneous Diophantine approximation'. *J. Austral. Math. Soc.,* **62**, 175–185. (Cited in Chapter 5.)

[203]   (2002). 'Exceptional sets in dynamical systems and Diophantine approximation'. *Rigidity in dynamics and geometry* (Cambridge, 2000), 77–98, Springer, Berlin. (Cited in Chapter 5.)

[204]   Dodson, M. M. & Hasan, S. (1992). 'Systems of linear forms and covers for star bodies'. *Acta Arith.,* **61**, 119–127. (Cited in Chapter 5.)

[205]   Dodson, M. M. & Kristensen, S. (To appear). 'Hausdorff dimension and Diophantine approximation'. *Fractal Geometry and Applications: A Jubilee of Benoit Mandelbrot,* Proceedings of Symposia in Pure Mathematics, American Mathematical Society. (Cited in Chapters 5, 6 & 9.)

[206]   Dodson, M. M., Rynne, B. P. & Vickers, J. A. G. (1990). 'Diophantine approximation and a lower bound for Hausdorff dimension'. *Mathematika,* **37**, 59–73. (Cited in Chapters 5 & 6.)

[207] Dombrovskiĭ, R. I. (1989). 'Simultaneous approximations of real numbers by algebraic numbers of bounded degree'. *Dokl. Akad. Nauk BSSR*, **33**, 205–208, 283 (in Russian). (Cited in Chapter 5.)

[208] Dress, A., Elkies, N. & Luca, F. (Preprint). 'A characterization of Mahler's generalized Liouville numbers by simultaneous rational approximation'. (Cited in Chapter 3.)

[209] Dubickas, A. (1992). 'An estimation of the difference between two zeros of a polynomial'. *New Trends in Probability and Statistics*, Vol. 2 (Palanga, 1991), 17–21, VSP, Utrecht. (Cited in Appendix A.)

[210] (1998). 'On algebraic numbers close to 1'. *Bull. Austral. Math. Soc.*, **58**, 423–434. (Cited in Appendix A.)

[211] Dubois, E. (1977). 'Application de la méthode de W. M. Schmidt à l'approximation de nombres algébriques dans un corps de fonctions de caractéristique zéro'. *C. R. Acad. Sci. Paris*, **284**, 1527–1530. (Cited in Chapter 9.)

[212] (1996). 'On Mahler's classification in Laurent series fields'. *Rocky Mountain J.*, **26**, 1003–1016. (Cited in Chapter 9.)

[213] Duffin, R. J. & Schaeffer, A. C. (1941). 'Khintchine's problem in metric Diophantine approximation'. *Duke J.*, **8**, 243–255. (Cited in Chapter 1.)

[214] Dumas, G. (1906). 'Sur quelques cas d'irréductibilité des polynomes à coefficients rationnels'. *J. Math. pures appl.*, (6) **2**, 191–258. (Cited in Chapter 7.)

[215] Durand, A. (1973/74). 'Un critère de transcendence'. *Séminaire Delange–Pisot–Poitou: Groupe d'étude de théorie des nombres*, 15ème année, numéro G11. (Cited in Chapter 8.)

[216] (1974). 'Quatre problèmes de Mahler sur la fonction ordre d'un nombre transcendant'. *Bull. Soc. Math. France*, **102**, 365–377. (Cited in Chapter 8.)

[217] (1975). 'Fonctions Θ-ordre et classification de $\mathbb{C}_p$'. *C. R. Acad. Sci. Paris*, **280**, 1085–1088. (Cited in Chapter 8.)

[218] (1977). 'Indépendance algébrique de nombres complexes et critères de transcendance'. *Compositio Math.*, **35**, 259–267. (Cited in Chapter 8.)

[219] (1978). 'Approximations algébriques d'un nombre transcendant'. *C. R. Acad. Sci. Paris*, **287**, 595–597. (Cited in Chapter 8.)

[220] (1990). 'Quelques aspects de la théorie analytique des polynômes I, II'. *Cinquante ans de polynômes* (Paris, 1988), pp. 1–42, 43–85, *Lecture Notes in Math.*, **1415**, Springer, Berlin. (Cited in Appendix A.)

[221] (1990). 'Approximations algébriques d'un nombre transcendant'. *Cinquante ans de polynômes* (Paris, 1988), pp. 94–96, *Lecture Notes in Math.*, **1415**, Springer, Berlin. (Cited in Chapter 8.)

[222] Duverney, D. (2001). 'Transcendence of a fast converging series of rational numbers'. *Math. Proc. Cambridge Phil. Soc.*, **130**, 193–207. (Cited in Chapter 3.)

[223] Dyson, F. J. (1947). 'The approximation to algebraic numbers by rationals'. *Acta Math.*, **79**, 225–240. (Cited in Chapter 2.)

[224] (1947). 'On simultaneous Diophantine approximations'. *Proc. London Math. Soc.*, **49**, 409–420. (Cited in Appendix B.)

[225] Eggleston, H. G. (1951). 'Sets of fractional dimension which occur in some problems in number theory'. *Proc. London Math. Soc.*, **54**, 42–93. (Cited in Chapter 5.)

[226] Erdős, P. (1959). 'Some results on Diophantine approximation'. *Acta Arith.*, **5**, 359–369. (Cited in Chapter 1.)

[227]   (1962). 'Representations of real numbers as sums and products of Liouville numbers'. *Michigan Math. J.*, **9**, 59–60. (Cited in Chapters 1 & 9.)

[228]   (1970). 'On the distribution of convergents of almost all real numbers'. *J. Number Theory*, **2**, 425–441. (Cited in Chapter 1.)

[229] Erdős, P. & Mahler, K. (1939). 'Some arithmetical properties of the convergents of a continued fraction'. *J. London Math. Soc.*, **14**, 12–18. (Cited in Chapters 5 & 10.)

[230] Euler, L. (1737). 'De fractionibus continuis'. *Commentarii Acad. Sci. Imperiali Petropolitanae*, **9**. (Cited in Chapter 1.)

[231] Everest, G. & Ward, T. (1999). *Heights of polynomials and entropy in algebraic dynamics*. Universitext, Springer-Verlag London, Ltd, London. (Cited in Appenidx A.)

[232] Evertse, J-H. (2000). 'Symmetric improvements of Liouville's inequality: A survey'. In *Algebraic Number Theory and Diophantine Analysis* (Graz, 1998), 129–141, de Gruyter, Berlin. (Cited in Chapter 10.)

[233]   'Approximation of complex algebraic numbers by algebraic numbers of bounded degree'. www.math.leidenuniv.nl/evertse/publicaties.shtml (Cited in Chapter 9.)

[234] Falconer, K. (1985). *The geometry of fractal sets*. Cambridge Tracts in Mathematics, 85, Cambridge University Press. (Cited in Chapter 5.)

[235]   (1985). 'Classes of sets with large intersections'. *Mathematika*, **32**, 191–205. (Cited in Chapter 6.)

[236]   (1990). *Fractal Geometry: Mathematical Foundations and Applications*. John Wiley & Sons. (Cited in Chapters 5 & 6.)

[237]   (1994). 'Sets with large intersection properties'. *J. London Math. Soc.*, **49**, 267–280. (Cited in Chapters 5 & 6.)

[238]   (2000). 'Representation of families of sets by measures, multifractal analysis and Diophantine approximation'. *Math. Proc. Cambridge Philos. Soc.*, **128**, 111–121. (Cited in Chapter 5.)

[239] Fatou, P. (1904). 'Sur l'approximation des incommensurables et des séries trigonométriques'. *C. R. Acad. Sci. Paris*, **139**, 1019–1021. (Cited in Chapter 1.)

[240] Feldman, N. I. (1962). 'On transcendental numbers having an approximation of given type'. *Uspehi mat. Nauk*, **17**, No.5 (107), 145–151 (in Russian). (Cited in Chapter 7.)

[241]   (1971). 'An effective refinement of the exponent in Liouville's theorem'. *Izv. Akad. Nauk*, **35**, 973–990 (in Russian). English transl. in *Math. USSR Izv.*, **5** (1971), 985–1002. (Cited in Chapter 2.)

[242]   (1981). *Approximation of algebraic numbers*. Moskov. Gos. Univ., Moscow (in Russian). (Cited in Chapter 2.)

[243]   (1982). *Hilbert's seventh problem*. Moskov. Gos. Univ., Moscow (in Russian). (Cited in Appenidx A.)

[244] Feldman, N. I. & Nesterenko, Yu. V. (1998). *Number Theory IV: Transcendental Numbers*. Encyclopaedia of Mathematical Sciences 44. Springer-Verlag, Berlin. (Cited in Chapters 2, 3 & 8.)

[245] Feng, D.-J., Wu, J., Liang, J.-C. & Tseng, S. (1997). 'Appendix to the paper by T. Łuczak – a simple proof of the lower bound'. *Mathematika*, **44**, 54–55. (Cited in Chapter 5.)

[246] Ferenczi, S. & Mauduit, C. (1997). 'Transcendence of numbers with a low complexity expansion'. *J. Number Theory*, **67**, 146–161. (Cited in Chapter 2.)

[247] Ford, L. R. (1925). 'On the closeness of approach of complex rational fractions to a complex irrational number'. *Trans. Amer. Math. Soc.*, **27**, 146–154. (Cited in Chapter 9.)

[248] Forder, H. G. (1963). 'A simple proof of a result on Diophantine approximation'. *Math. Gaz.*, **47**, 237–238. (Cited in Chapter 1.)

[249] Fraenkel, A. S. (1964). 'Transcendental numbers and a conjecture of Erdős and Mahler'. *J. London Math. Soc.*, **39**, 405–416. (Cited in Chapter 5.)

[250] Fraenkel, A. S. & Borosh, I. (1965). 'Fractional dimension of a set of transcendental numbers'. *Proc. London Math. Soc.*, **15**, 458–470; corrigendum: *Proc. London Math. Soc.*, **16** (1966), 192. (Cited in Chapters 5 & 10.)

[251] Frostman, O. (1935). 'Potentiel d'équilibre et capacité des ensembles avec quelques applications à la théorie des fonctions'. *Meddel. Lunds Univ. Math. Sem.*, **3**, 1–118. (Cited in Chapter 5.)

[252] Fuchs, M. (2002). 'On metric Diophantine approximation in the field of formal Laurent series'. *Finite Fields Appl.*, **8**, 343–368. (Cited in Chapter 9.)

[253] Gallagher, P. (1962). 'Metric simultaneous Diophantine approximation'. *J. London Math. Soc.*, **37**, 387–390. (Cited in Chapter 4.)

[254] Galočkin, A. I. (1980). 'A transcendence measure for the values of functions satisfying certain functional equations'. *Mat. Zametki.*, **27**, 175–183, 317 (in Russian). English transl. in *Math. Notes*, **27**, 83–88. (Cited in Chapters 3 & 8.)

[255] Ganesa Moorthy, C. (1992). 'A problem of Good on Hausdorff dimension'. *Mathematika*, **39**, 244–246. (Cited in Chapter 5.)

[256] Gelfond, A. O. (1948). 'Approximation of algebraic irrationalities and their logarithms'. *Vest. Mosk. Univ.*, **9**, 3–25 (in Russian). (Cited in Chapter 2.)

[257] (1952). *Transcendental and algebraic numbers*. Gosudartsv. Izdat. Tehn.-Teor. Lit., Moscow (in Russian). Translated from the first Russian edition by Leo F. Boron, Dover publ., New York, 1960. (Cited in Chapter 8 and Appenidx A.)

[258] Good, I. J. (1941). 'The fractional dimensional theory of continued fractions'. *Proc. Cambridge Phil. Soc.*, **37**, 199–228. (Cited in Chapter 5.)

[259] Grace, J. H. (1918). 'The classification of rational approximations'. *Proc. London Math. Soc.*, **17**, 247–258. (Cited in Chapter 1.)

[260] Groshev, A. V. (1938). 'A theorem on systems of linear forms'. *Dokl. Akad. Nauk SSSR*, **19**, 151–152 (in Russian). (Cited in Chapter 1.)

[261] Gruber, P. M. (1983). 'In most cases approximation is irregular'. *Rend. Sem. Mat. Univers. Politecn. Torino*, **41**, 19–33. (Cited in Chapter 1.)

[262] Gruber, P. M. & Lekkerkerker, C. G. (1987). *Geometry of numbers*. Series Bibliotheca Mathematica, 8, North-Holland, Amsterdam. (Cited in Chapter 1 and Appendix B.)

[263] Guntermann, N. (1996). 'Approximation durch algebraische Zahlen beschränkten Grades im Körper der formalen Laurentreihen'. *Monatsh. Math.*, **122**, 345–354. (Cited in Chapter 9.)

[264] Gürses, H. (2000). 'Über verallgemeinerte Lückenreihen'. *Istanbul Üniv. Fen Fak. Mat. Derg.*, **59**, 59–88. (Cited in Chapters 3 & 7.)

[265] (2000). 'Über eine Klasse verallgemeinerter Lückenreihen mit algebraischen Koeffizienten'. *Istanbul Üniv. Fen Fak. Mat. Derg.*, **59**, 89–102. (Cited in Chapter 7.)

[266] Güting, R. (1963). 'On Mahler's function θ_1'. *Mich. Math. J.*, **10**, 161–179. (Cited in Chapter 5.)

[267] (1965). 'Über den Zusammenhang zwischen rationalen Approximationen und Kettenbruchentwicklungen'. *Math. Z.*, **90**, 382–387. (Cited in Chapter 3.)

[268] (1967). 'Polynomials with multiple zeros'. *Mathematika*, **14**, 181–196. (Cited in Appendix A.)

[269] (1968). 'Zur Berechnung der Mahlerschen Funktionen w_n'. *J. reine angew. Math.*, **232**, 122–135. (Cited in Chapters 3 & 7.)

[270] Hall, M. (1947). 'On the sum and product of continued fractions'. *Ann. of Math.*, **48**, 966–993. (Cited in Chapter 1.)

[271] Hardy, G. H. & Wright, E. M. (1979). *An Introduction to the Theory of Numbers*, 5th. Edition. Clarendon Press. (Cited in Chapter 1.)

[272] Harman, G. (1988). 'Metric Diophantine approximation with two restricted variables, III. Two prime numbers'. *J. Number Theory*, **29**, 364–375. (Cited in Chapter 1.)

[273] (1998). *Metric Number Theory*. LMS Monographs New Series, vol. 18. Clarendon Press. (Cited in Introduction, and Chapters 1, 4 & 5.)

[274] (2000). 'Variants of the second Borel–Cantelli lemma and their applications in metric number theory'. *Number Theory*, 121–140, Trends Math., Birkhäuser, Basel. (Cited in Chapters 1 & 6.)

[275] (2003). 'Simultaneous Diophantine approximation and asymptotic formulae on manifolds'. *Acta Arith.*, **108**, 379–389. (Cited in Chapter 4.)

[276] Hausdorff, F. (1919). 'Dimension und äusseres Maß'. *Math. Ann.*, **79**, 157–179. (Cited in Chapter 5.)

[277] Hensley, D. (1992). 'Continued fractions Cantor sets, Hausdorff dimension and functional analysis'. *J. Number Theory*, **40**, 336–358. (Cited in Chapter 5.)

[278] (1996). 'A polynomial time algorithm for the Hausdorff dimension of continued fractions Cantor sets'. *J. Number Theory*, **58**, 9–45; erratum *J. Number Theory*, **59** (1996), 419. (Cited in Chapter 5.)

[279] Hernández, S. (Preprint). 'Comparing classifications of p-adic transcendental numbers'. (Cited in Chapter 9.)

[280] Hightower, C. J. (1975). 'Approximation to complex numbers by certain biquadratic numbers'. *J. Number Theory*, **7**, 293–309. (Cited in Chapter 9.)

[281] Hinokuma, T. & Shiga, H. (1992). 'A remark on theorem of Jarník'. *Ryukyu Math. J.*, **5**, 1–6. (Cited in Chapter 5.)

[282] (1996). 'Hausdorff dimension of sets arising in Diophantine approximation'. *Kodai Math. J.*, **19**, 365–377. (Cited in Chapters 5 & 6.)

[283] Hirst, K. E. (1970). 'A problem in the fractional dimension theory of continued fractions'. *Quart. J. Math. Oxford*, **21**, 29–35. (Cited in Chapter 5.)

[284] Hurwitz, A. (1891). 'Über die angenäherte Darstellung der Irrationalzahlen durch rationale Brüche'. *Math. Ann.*, **39**, 279–284. (Cited in Chapter 1.)

[285]   Inoue, K. & Nakada, H. (2003). 'On metric Diophantine approximation in positive characteristic'. *Acta Arith.*, **110**, 205–218. (Cited in Chapter 9.)

[286]   Iosifescu, M. & Kraaikamp, C. (2002). *Metrical Theory of Continued Fractions*. Mathematics and its Applications 547. Kluwer Academic Publishers, Dordrecht. (Cited in Chapter 1.)

[287]   Ivanov, V. A. (1978). 'A theorem of Dirichlet in the theory of Diophantine approximations'. *Mat. Zametki*, **24**, 459–474, 589 (in Russian). English transl. in *Math. Notes*, **24** (1978), 747–755. (Cited in Chapter 1.)

[288]   Jarník, V. (1928/29). 'Zur metrischen Theorie der diophantischen Approximationen'. *Prace Mat. Fíz.*, **36**, 91–106. (Cited in Chapters 5 & 9.)

[289]   (1929). 'Diophantischen Approximationen und Hausdorffsches Mass'. *Mat. Sbornik*, **36**, 371–382. (Cited in Chapters 5 & 9.)

[290]   (1930). 'Několik poznámek o Hausdorffově míře'. *Rozpravy Tř. České Akad.*, **40**, c. 9, 8 p. (Cited in Chapter 6.)

[291]   (1930). 'Quelques remarques sur la mesure de M. Hausdorff'. *Bull. Int. Acad. Sci. Bohême*, 1–6. (Cited in Chapter 6.)

[292]   (1931). 'Über die simultanen Diophantische Approximationen'. *Math. Z.*, **33**, 505–543. (Cited in Chapters 1, 5 & 6.)

[293]   (1935). 'Über einen Satz von A. Khintchine'. *Práce Mat. Fíz.*, **43**, 1–16. (Cited in Chapter 10 and Appendix B.)

[294]   (1936). 'O simultánnich diofantických approximaciích'. *Rozpravy Tř. České Akad.*, **45**, c. 19, 16 p. (Cited in Appendix B.)

[295]   (1936). 'Über einen Satz von A. Khintchine, 2. Mitteilung'. *Acta Arith.*, **2**, 1–22. (Cited in Chapter 10 and Appendix B.)

[296]   (1939). 'Über einen *p*-adischen Übertragungssatz'. *Monatsh. Math. Phys.*, **48**, 277–287. (Cited in Chapter 9.)

[297]   (1945). 'Sur les approximations diophantiennes des nombres *p*-adiques'. *Revista Ci. Lima*, **47**, 489–505. (Cited in Chapter 9.)

[298]   (1954). 'Contribution to the metric theory of continued fractions'. *Czechoslovak Math. J.*, **4** (79), 318–329 (in Russian). (Cited in Chapter 5.)

[299]   (1959). 'Eine Bemerkung zum Übertragungssatz'. *Bŭlgar. Akad. Nauk Izv. Mat. Inst.*, **3**, 169–175. (Cited in Appendix B.)

[300]   (1969). 'Un théorème d'existence pour les approximations diophantiennes'. *Enseignement Math.*, **15**, 171–175. (Cited in Chapter 6.)

[301]   Jenkinson, O. & Pollicott, M. (2001). 'Computing the dimension of dynamically defined sets: E_2 and bounded continued fractions'. *Ergod. Th. Dynam. Sys.*, **21**, 1429–1445. (Cited in Chapter 5.)

[302]   Jones, H. (2001). *Contributions to Metric Number Theory*. Ph.D. thesis, Cardiff. (Cited in Chapter 1.)

[303]   (2001). 'Khintchine's theorem in k dimensions with prime numerator and denominator'. *Acta Arith.*, **99**, 205–225. (Cited in Chapter 1.)

[304]   Kahane, J.-P. & Katznelson, Y. (1973). 'Sur les ensembles d'unicité $U(\varepsilon)$ de Zygmund'. *C. R. Acad. Sci. Paris*, **227**, 893–895. (Cited in Chapter 7.)

[305]   Kahane, J.-P. & Salem, R. (1963). *Ensembles Parfaits et Séries Trigonométriques*. Herrmann. (Cited in Chapter 7.)

[306]   Kaindl, G. (1976). 'Simultane Diophantische Approximationen'. *Österreich. Akad. Wiss. Math. Naturwiss. Kl. S.-B. II*, **185**, 387–404. (Cited in Chapter 1.)

[307] Kalosha, N. I. (2001). 'On the approximation properties of algebraic numbers of the third degree'. *Vestsī Nats. Akad. Navuk Belarusī Ser. Fīz. Mat. Navuk*, **3**, 12–17, 140 (in Russian). (Cited in Chapter 4.)

[308] ——— (2003). 'Simultaneous approximations of zero by values of integral polynomials in $\mathbb{R} \times \mathbb{Q}_p$'. *Dokl. Nats. Akad. Nauk Belarusī*, **47**, No.2, 19–22 (in Russian). (Cited in Chapter 9.)

[309] Kargaev, P. & Zhigljavsky, A. (1996). 'Approximation of real numbers by rationals: some metric theorems'. *J. Number Theory*, **61**, 209–225. (Cited in Chapter 1.)

[310] Kasch, F. (1958). 'Über eine metrische Eigenschaft der S-Zahlen'. *Math. Z.*, **70**, 263–270. (Cited in Chapters 4 & 9.)

[311] Kasch, F. & Volkmann, B. (1958). 'Zur Mahlerschen Vermutung über S-Zahlen'. *Math. Ann.*, **136**, 442–453. (Cited in Chapters 3, 4, 5 & 7.)

[312] ——— (1959/60). 'Metrische Sätze über transzendente Zahlen in P-adischen Körpern'. *Math. Z.*, **72**, 367–378. (Cited in Chapter 9.)

[313] ——— (1962). 'Metrische Sätze über transzendente Zahlen in P-adischen Körpern, II'. *Math. Z.*, **78**, 171–174. (Cited in Chapter 9.)

[314] Kaufman, R. (1980). 'Continued fractions and Fourier transforms'. *Mathematika*, **27**, 262–267. (Cited in Chapter 7.)

[315] ——— (1981). 'On the theorem of Jarník and Besicovitch'. *Acta Arith.*, **39**, 265–267. (Cited in Chapters 1, 5 & 7.)

[316] Kechris, A. S. (1995). *Classical Descriptive Set Theory*. Springer-Verlag, Berlin and New York. (Cited in Chapter 7.)

[317] Khintchine, A. Ya. (1924). 'Einige Sätze über Kettenbrüche, mit Anwendungen auf die Theorie der diophantischen Approximationen'. *Math. Ann.*, **92**, 115–125. (Cited in Chapters 1 & 6.)

[318] ——— (1925). 'Zwei Bemerkungen zu einer Arbeit des Herrn Perron'. *Math. Z.*, **22**, 274–284. (Cited in Chapters 3 & 10, and Appendix B.)

[319] ——— (1926). 'Zur metrischen Theorie der diophantischen Approximationen'. *Math. Z.*, **24**, 706–714. (Cited in Chapter 1.)

[320] ——— (1926). 'Über eine Klasse linearer diophantischer Approximationen'. *Rendiconti Circ. Mat. Palermo*, **50**, 170–195. (Cited in Chapter 1 and Appendix B.)

[321] ——— (1935). 'Metrische Kettenbruchprobleme'. *Compositio Math.*, **1**, 361–382. (Cited in Chapter 3.)

[322] ——— (1936). 'Zur metrischen Kettenbrüchetheorie'. *Compositio Math.*, **3**, 275–285. (Cited in Chapter 1.)

[323] ——— (1964). *Continued Fractions*. The University of Chicago Press. (Cited in Chapters 1 & 3.)

[324] Haseo Ki. (1995). 'The Borel classes of Mahler's A, S, T and U numbers'. *Proc. Amer. Math. Soc.*, **123**, 3197–3204. (Cited in Chapter 7.)

[325] Kleinbock, D. (1999). 'Badly approximable systems of affine forms'. *J. Number Theory*, **79**, 83–102. (Cited in Chapter 5.)

[326] ——— (2001). 'Some applications of homogeneous dynamics to number theory'. *Smooth Ergodic Theory and its Applications* (Seattle, WA, 1999). Proceedings of Symposia in Pure Mathematics, **69**, Amer. Math. Soc., Providence. (Cited in Chapter 4.)

[327]   (2003). 'Extremal subspaces and their submanifolds'. *Geom. Funct. Anal.*, **13**, 437–466. (Cited in Chapter 4.)

[328]   (To appear). 'Baker–Sprindžuk conjectures for complex analytic manifolds.' *Algebraic groups and Arithmetic.* T.I.F.R., India. (Cited in Chapter 9.)

[329]   Kleinbock, D., Lindenstrauss, E. & Weiss, B. (To appear). 'On fractal measures and Diophantine approximation'. *Selecta Math.* (Cited in Chapters 4, 5 & 10.)

[330]   Kleinbock, D. & Margulis, G. A. (1998). 'Flows on homogeneous spaces and Diophantine approximation on manifolds'. *Ann. of Math.*, **148**, 339–360. (Cited in Chapters 4 & 5.)

[331]   Kleinbock, D. & Tomanov, G. (Preprint). 'Flows on S-arithmetic homogeneous spaces and applications to metric Diophantine approximation'. (Cited in Chapter 9.)

[332]   Koksma, J. F. (1936). *Diophantische Approximationen.* Ergebnisse d. Math. u. ihrer Grenzgebiete, 4, Springer. (Cited in Introduction and Chapter 9.)

[333]   (1939). 'Über die Mahlersche Klasseneinteilung der transzendenten Zahlen und die Approximation komplexer Zahlen durch algebraische Zahlen'. *Monatsh. Math. Phys.*, **48**, 176–189. (Cited in Chapters 3, 4, 6, 8, 9 & 10.)

[334]   Koksma, J. F. & Popken, J. (1932). 'Zur Transzendenz von e'. *J. reine angew. Math.*, **168**, 211–230. (Cited in Appendix A.)

[335]   Kopetzky, H. G. (1989). 'Some results on Diophantine approximation related to Dirichlet's theorem'. *Number Theory* (Ulm, 1987), 137–149, Lecture Notes in Math., **1380**, Springer, New York. (Cited in Chapter 1.)

[336]   Kopetzky, H. G. & Schnitzer, F. J. (1986). 'Quadratische Approximationen am Einheitskreis'. *Arch. Math.*, **46**, 144–147. (Cited in Chapter 3.)

[337]   Korkine, A. & Zolotareff, G. (1873). 'Sur les formes quadratiques'. *Math. Ann.*, **6**, 366–389. (Cited in Chapter 1.)

[338]   Kovalevskaya, E. I. (1999). 'Simultaneous approximations of real numbers by algebraic numbers of a given type'. *Vestsī Nats. Akad. Navuk Belarusī Ser. Fīz. Mat. Navuk*, **3**, 5–9, 138 (in Russian). (Cited in Chapter 5.)

[339]   (1999). 'A metric theorem on the exact order of approximation of zero by values of integer polynomials in $\mathbb{Q}_p$'. *Dokl. Nats. Akad. Nauk Belarusī*, **43**, No.5, 34–36, 123 (in Russian). (Cited in Chapter 9.)

[340]   (2000). 'The convergence Khintchine theorem for polynomials and planar p-adic curves'. Number theory (Liptovský Ján, 1999), *Tatra Mt. Math. Publ.* **20**, 163–172. (Cited in Chapter 9.)

[341]   Kristensen, S. (2003). 'On well-approximable matrices over a field of formal series'. *Math. Proc. Cambridge Philos. Soc.*, **135**, 255–268. (Cited in Chapter 9.)

[342]   (Preprint). 'Badly approximable systems of linear forms over a field of formal series'. (Cited in Chapter 9.)

[343]   (2003). 'Diophantine approximation and the solubility of the Schrödinger equation'. *Phys. Lett. A*, **314**, 15–18. (Cited in Chapter 5.)

[344]   Kubilius, I. (1949). 'On the application of I. M. Vinogradov's method to the solution of a problem of the metric theory of numbers'. *Doklady Akad. Nauk SSSR (N.S.)*, **67**, 783–786 (in Russian). (Cited in Chapter 4.)

[345]   Kühnlein, S. (1996). 'On a measure theoretic aspect of Diophantine approximation'. *J. reine angew. Math.*, **477**, 117–127. (Cited in Chapter 1.)

[346] Lagarias, J. C. (1981). 'A complement to Ridout's p-adic generalization of the Thue–Siegel–Roth theorem'. *Analytic Number Theory* (Philadelphia, 1980), 264–275, *Lecture Notes in Math.*, **899**, Springer, Berlin–New York. (Cited in Chapters 2 & 10.)

[347] Lagrange, J. L. (1770). 'Additions au mémoire sur la résolution des équations numériques'. *Mém. Berl.*, **24**. (Cited in Chapter 1.)

[348] Lang, S. (1962). 'A transcendence measure for E-functions'. *Mathematika*, **9**, 157–161. (Cited in Chapters 3 & 8.)

[349] (2002). *Algebra*, revised third edition. *Graduate Texts in Mathematics*, **211**. Springer-Verlag, New York. (Cited in Chapter 3 and Appendix A.)

[350] Langmayr, F. (1980). 'On Dirichlet's approximation theorem'. *Monatsh. Math.*, **90**, 229–232. (Cited in Chapter 1.)

[351] Lasjaunias, A. (2000). 'A survey of Diophantine approximation in fields of power series'. *Monatsh. Math.*, **130**, 211–229. (Cited in Chapters 9 & 10.)

[352] Laurent, M. (1999). 'Diophantine approximation and algebraic independence'. *Colloque Franco–Japonais: Théorie des Nombres Transcendants* (Tokyo, 1998), 75–89. *Sem. Math. Sci.*, **27**, Keio Univ. Yokohama. (Cited in Chapter 8.)

[353] (1999). 'Some remarks on the approximation of complex numbers by algebraic numbers'. *Proceedings of the 2nd Panhellenic Conference in Algebra and Number Theory* (Thessaloniki, 1998). *Bull. Greek Math. Soc.*, **42**, 49–57. (Cited in Chapter 8.)

[354] (2003). 'Simultaneous rational approximation to the successive powers of a real number'. *Indag. Math. (N.S.)*, **11**, 45–53. (Cited in Chapter 3.)

[355] Laurent, M. & Poulakis, D. (2004). 'On the global distance between two algebraic points on a curve'. *J. Number Theory*, **104**, 210–254. (Cited in Chapter 7.)

[356] Laurent, M. & Roy, D. (1999). 'Criteria of algebraic independence with multiplicities and interpolation determinants'. *Trans. Amer. Math. Soc.*, **351**, 1845–1870. (Cited in Chapter 8 and Appendix A.)

[357] (1999). 'Sur l'approximation algébrique en degré de transcendance un'. *Ann. Inst. Fourier (Grenoble)*, **49**, 27–55. (Cited in Chapter 8.)

[358] (2001). 'Criteria of algebraic independence with multiplicities and approximation by hypersurfaces'. *J. reine angew. Math.*, **536**, 65–114. (Cited in Chapter 8.)

[359] Legendre, A.-M. (1830). *Théorie des Nombres*, troisième édition, Tome 1. Paris. (Cited in Chapter 1.)

[360] LeVeque, W. J. (1952). 'Continued fractions and approximations in $k(i)$, I and II'. *Indag. Math.*, **14**, 526–545. (Cited in Chapter 9.)

[361] (1953). 'On Mahler's U-numbers'. *J. London Math. Soc.*, **28**, 220–229. (Cited in Chapters 3, 7 & 9.)

[362] (1953). 'Note on S-numbers'. *Proc. Amer. Math. Soc.*, **4**, 189–190. (Cited in Chapter 4.)

[363] (1956). *Topics in Number Theory*, Vols. 1 & 2. Addison-Wesley Publishing Co., Inc., Reading, MA. (Cited in Chapter 2.)

[364] (1960). 'On the frequency of small fractional parts in certain real sequences, II'. *Trans. Amer. Math. Soc.*, **94**, 130–149. (Cited in Chapter 1.)

[365] Levesley, J. (1998). 'A general inhomogeneous Jarník–Besicovitch theorem'. *J. Number Theory*, **71**, 65–80. (Cited in Chapter 5.)

[366] Lévy, P. (1936). 'Sur le devéloppement en fraction continue d'un nombre choisi au hasard'. *Compositio Math.*, **3**, 286–303. (Cited in Chapter 1.)

[367] Liardet, P. & Stambul, P. (2000). 'Séries de Engel et fractions continuées'. *J. Théor. Nombres Bordeaux*, **12**, 37–68. (Cited in Chaprer 2.)

[368] Liouville, J. (1844). 'Remarques relatives à des classes très-étendues de quantités dont la valeur n'est ni algébrique, ni même réductible à des irrationnelles algébriques'. *C. R. Acad. Sci. Paris*, **18**, 883–885. (Cited in Chapter 1.)

[369] (1844). 'Nouvelle démonstration d'un théorème sur les irrationnelles algébriques'. *C. R. Acad. Sci. Paris*, **18**, 910–911. (Cited in Chapter 1.)

[370] (1851). 'Sur des classes très-étendues de quantités dont la valeur n'est ni algébrique, ni même réductible à des irrationnelles algébriques'. *J. Math. pures appl.*, (1) **16**, 133–142. (Cited in Chapter 1.)

[371] Lock, D. J. (1947). *Metrisch-Diophantische Onderzoekingen in $K(P)$ en $K^n(P)$'*. Dissertation, Vrije Univ., Amsterdam. (Cited in Chapter 9.)

[372] Łuczak, T. (1997). 'On the fractional dimension of sets of continued fractions'. *Mathematika*, **44**, 50–53. (Cited in Chapter 5.)

[373] Lutz, É. (1955). *Sur les Approximations Diophantiennes Linéaires et p-adiques*. Hermann. (Cited in Chapter 9.)

[374] Lützen, J. (1990). *Joseph Liouville 1809–1882: Master of Pure and Applied Mathematics*. Studies in the History of Mathematics and Physical Sciences, **15**, Springer-Verlag, New York. (Cited in Chapter 1.)

[375] Mahler, K. (1930). 'Über Beziehungen zwischen der Zahl e und den Liouvilleschen Zahlen'. *Math. Z.*, **31**, 729–732. (Cited in Chapter 3.)

[376] (1932). 'Zur Approximation der Exponentialfunktionen und des Logarithmus, I, II'. *J. reine angew. Math.*, **166**, 118–150. (Cited in Chapters 3, 7, 8 & 9.)

[377] (1932). 'Über das Mass der Menge aller S-Zahlen'. *Math. Ann.*, **106**, 131–139. (Cited in Chapters 3, 4 & 9.)

[378] (1934/35). 'Über eine Klasseneinteilung der p-adischen Zahlen'. *Mathematica Leiden*, **3**, 177–185. (Cited in Chapter 9.)

[379] (1936). 'Neuer Beweis einer Satz von A. Khintchine'. *Mat. Sbornik*, **43**, 961–962. (Cited in Appendix B.)

[380] (1937). 'Arithmetische Eigenschaften einer Klasse von Dezimalbrüchen'. *Proc. Akad. v. Wetensch., Amsterdam*, **40**, 421–428. (Cited in Chapters 2 & 7.)

[381] (1940). 'On a geometrical representation of p-adic numbers'. *Ann. of Math.*, **41**, 8–56. (Cited in Chapter 9.)

[382] (1941). 'An analogue to Minkowski's geometry of numbers in a field of series'. *Ann. of Math.*, **42**, 488–522. (Cited in Chapter 9.)

[383] (1949). 'On the continued fraction of quadratic and cubic irrationals'. *Ann. Mat. Pura Appl.*, (4) **30**, 147–172. (Cited in Chapter 2.)

[384] (1953). 'On the approximation of π'. *Indag. Math.*, **15**, 30–42. (Cited in Chapter 3.)

[385] (1953). 'On the approximation of logarithms of algebraic numbers'. *Philos. Trans. Roy. Soc. London. Ser. A*, **245**, 371–398. (Cited in Chapter 3.)

[386] (1957). 'On the fractional parts of the powers of a rational number, II'. *Mathematika*, **4**, 122–124. (Cited in Chapter 2.)

[387]   (1960). 'An application to Jensen's formula to polynomials'. *Mathematika*, **7**, 98–100. (Cited in Appendix A.)

[388]   (1961). *Lectures on Diophantine approximation, Part 1: g-adic numbers and Roth's theorem*. University of Notre Dame, Ann Arbor. (Cited in Chapter 2.)

[389]   (1962). 'On some inequalities for polynomials in several variables'. *J. London Math. Soc.*, **37**, 341–344. (Cited in Appendix A.)

[390]   (1963). 'On the approximation of algebraic numbers by algebraic integers'. *J. Austral. Math. Soc.*, **3**, 408–434. (Cited in Chapter 2.)

[391]   (1964). 'An inequality for the discriminant of a polynomial'. *Michigan Math. J.*, **11**, 257–262. (Cited in Appendix A.)

[392]   (1965). 'Arithmetic properties of lacunary power series with integral coefficients'. *J. Austral. Math. Soc.*, **5**, 56–64. (Cited in Chapter 7.)

[393]   (1971). 'On the order function of a transcendental number'. *Acta Arith.*, **18**, 63–76. (Cited in Chapter 8.)

[394]   (1971). 'Lectures on transcendental numbers'. 1969 Number Theory Institute (Proc. Sympos. Pure Math., Vol. XX, State Univ. New York, Stony Brook, NY, 1969), pp. 248–274. *Amer. Math. Soc.*, Providence, RI. (Cited in Chapters 8 & 10.)

[395]   (1973). 'The classification of transcendental numbers'. *Analytic Number Theory* (Proc. Sympos. Pure Math., Vol. XXIV, St. Louis Univ., St. Louis, 1972), pp. 175–179. Amer. Math. Soc., Providence, RI. (Cited in Chapter 8.)

[396]   (1975). 'A necessary and sufficient condition for transcendency'. *Math. Comp.*, **29**, 145–153. (Cited in Chapter 8.)

[397]   (1976). 'On a class of transcendental decimal fractions'. *Comm. Pure Appl. Math.*, **29**, 717–725. (Cited in Chapter 2.)

[398]   (1981). p-*adic Numbers and their Functions*, second edition. Cambridge Tracts in Mathematics, **76**, Cambridge University Press. (Cited in Chapter 9.)

[399]   (1982). 'Fifty years as a mathematician'. *J. Number Theory*, **14**, 121–155. (Cited in Chapter 10.)

[400]   (1984). 'Some suggestions for further research'. *Bull. Austral. Math. Soc.*, **29**, 101–108. (Cited in Chapter 10.)

[401]   (1986). 'The successive minima in the geometry of numbers and the distinction between algebraic and transcendental numbers'. *J. Number Theory*, **22**, 147–160. (Cited in Chapters 3 & 9.)

[402]   Mahler, K. & Szekeres, G. (1966/67). 'On the approximation of real numbers by roots of integers'. *Acta Arith.*, **12**, 315–320. (Cited in Chapter 3.)

[403]   Maillet, E. (1906). *Introduction à la Théorie des Nombres Transcendants et des Propriétés Arithmétiques des Fonctions*. Gauthier-Villars, Paris. (Cited in Chapters 2, 3, 7 & 10.)

[404]   (1906). 'Sur la classification des irrationnelles'. *C. R. Acad. Sci. Paris*, **143**, 26–28. (Cited in Chapter 3.)

[405]   Margulis, G. (2002). 'Diophantine approximation, lattices and flows on homogeneous spaces'. In: *A Panorama of Number Theory or The View from Baker's Garden*, G. Wüstholz (ed.), Cambridge University Press, pp. 280–310. (Cited in Chapter 4.)

[406] Markoff, A. A. (1879). 'Sur les formes quadratiques binaires indéfinies'. *Math. Ann.*, **15**, 381–407. (Cited in Chapter 1.)

[407] Markovich, N. I. (1986). 'Approximations of zero by values of quadratic polynomials'. *Vestsī Akad. Navuk BSSR Ser. Fīz. Mat. Navuk*, **4**, 18–22, 123 (in Russian). (Cited in Chapter 9.)

[408] Martin, G. (2001). 'Absolutely abnormal numbers'. *Amer. Math. Monthly*, **108**, 746–754. (Cited in Chapter 1.)

[409] Mashanov, V. I. (1987). 'On a problem of Baker in the metric theory of Diophantine approximations'. *Vestsī Akad. Navuk BSSR Ser. Fīz. Mat. Navuk*, **1**, 34–38, 125 (in Russian). (Cited in Chapter 4.)

[410] de Mathan, B. (1970). 'Approximations diophantiennes dans un corps local'. *Bull. Soc. Math. France Suppl. Mém.*, **21**. (Cited in Chapter 9.)

[411] ——— (1991). 'Simultaneous Diophantine approximation for algebraic functions in positive characteristic'. *Monatsh. Math.*, **111**, 187–193. (Cited in Chapter 10.)

[412] ——— (2001). 'A remark about Peck's method in positive characteristic'. *Monatsh. Math.*, **133**, 341–345. (Cited in Chapter 9.)

[413] ——— (2003). 'Conjecture de Littlewood et récurrences linéaires'. *J. Théor. Nombres Bordeaux*, **15**, 249–266. (Cited in Chapter 10.)

[414] de Mathan, B. & Teulié, O. (To appear). 'Problèmes diophantiens simultanés'. *Monatsh. Math.*. (Cited in Chapter 10.)

[415] Mattila, P. (1995). *Geometry of Sets and Measures in Euclidean Spaces*. Cambridge University Press. (Cited in Chapters 3, 5, 6 & 7.)

[416] Melián, M. V. & Pestana, D. (1993). 'Geodesic excursions into cusps in finite volume hyperbolic manifolds'. *Mich. Math. J.*, **40**, 77–93. (Cited in Chapter 9.)

[417] Melničuk, Ju. V. (1980). 'Hausdorff dimension in Diophantine approximation of p-adic numbers'. *Ukrain. Mat. Zh.*, **32**, 118–124, 144 (in Russian). (Cited in Chapter 9.)

[418] Menken, H. (2000). 'An investigation on p-adic U-numbers'. *Istanbul Üniv. Fen Fak. Mat. Derg.*, **59**, 111–143. (Cited in Chapter 9.)

[419] Mignotte, M. (1976). 'Sur la complexité de certains algorithmes où intervient la séparation des racines d'un polynôme'. *RAIRO Informatique Théorique*, **10**, 51–55. (Cited in Appendix A.)

[420] ——— (1979). 'Approximation des nombres algébriques par des nombres algébriques de grand degré'. *Ann. Fac. Sci. Toulouse Math.*, (5) **1**, 165–170. (Cited in Appendix A.)

[421] ——— (1982). 'Some useful bounds'. In: B. Buchberger, G. E. Collins & R. Loos (eds.), *Computer Algebra*. Springer-Verlag, pp. 259–263. (Cited in Chapter 7.)

[422] ——— (1995). 'On the distance between the roots of a polynomial'. *Applic. Alg. Eng. Comput. Comm*, **6**, 327–332. (Cited in Appendix A.)

[423] Mignotte, M. & Payafar, M. (1979). 'Distance entre les racines d'un polynôme'. *RAIRO Anal. Numér.*, **13**, 181–192. (Cited in Appendix A.)

[424] Miller, W. (1982). 'Transcendence measures by a method of Mahler'. *J. Austral. Math. Soc.*, **32**, 68–78. (Cited in Chapter 8.)

[425] Mkaouar, M. (2001). 'Continued fractions of transcendental numbers'. *Bull. Greek Math. Soc.*, **45**, 79–85. (Cited in Chapter 3.)

[426] Molchanov, S. M. (1983). 'On the p-adic transcendence measure for the values of functions which satisfy some functional equations'. *Vestnik Moskov. Univ. Ser. I Mat. Mekh.*, **2**, 31–37 (in Russian). English transl. in *Moscow Univ. Math. Bull.*, **38** (1983), No.2, 36–43. (Cited in Chapter 9.)

[427] Montgomery, H. M. (1994). *Ten Lectures on the Interface between Analytic Number Theory and Harmonic Analysis.* American Mathematical Society, Providence, RI. (Cited in Chapters 1 & 10.)

[428] Moran, W., Pearce, C. E. M. & Pollington, A. D. (1992). 'T-numbers form an M_0 set'. *Mathematika*, **39**, 18–24. (Cited in Chapter 7.)

[429] Mordell, L. J. (1934). 'On some arithmetical results in the geometry of numbers'. *Compositio Math.*, **1**, 248–253. (Cited in Appendix B.)

[430] Morduchai-Boltovskoj, D. (1934). 'On transcendental numbers with successive approximations defined by algebraic equations'. *Rec. Math. Moscou*, **41**, 221–232 (in Russian). (Cited in Chapter 3.)

[431] Morotskaya, I. L. (1987). 'Hausdorff dimension and Diophantine approximations in $\mathbb{Q}_p$'. *Dokl. Akad. Nauk BSSR*, **31**, No.7, 597–600, 668 (in Russian). (Cited in Chapter 9.)

[432] Morrison, J. F. (1978). 'Approximation of p-adic numbers by algebraic numbers of bounded degree'. *J. Number Theory*, **10**, 334–350. (Cited in Chapter 9.)

[433] Müller, P. (2002). 'Finiteness results for Hilbert's irreducibility theorem'. *Ann. Inst. Fourier (Grenoble)*, **52**, 983–1015. (Cited in Chapter 7.)

[434] Nakada, H. (1988). 'On metrical theory of Diophantine approximation over imaginary quadratic field'. *Acta Arith.*, **51**, 393–403. (Cited in Chapter 9.)

[435] Narkiewicz, W. (1990). *Elementary and Analytic Theory of Algebraic Numbers.* Springer-Verlag, Berlin. (Cited in Chapter 2.)

[436] Nesterenko, Yu. V. (1974). 'An order function for almost all numbers'. *Mat. Zametki*, **15**, 405–414 (in Russian). English transl. in *Math. Notes*, **15** (1974), 234–240. (Cited in Chapter 8.)

[437] Nesterenko, Yu. V. & Waldschmidt, M. (1996). 'On the approximation of the values of exponential function and logarithm by algebraic numbers'. *Mat. Zapiski 2*, Diophantine approximations. *Proceedings of papers dedicated to the memory of Prof N.I. Feldman.* Yu. V. Nesterenko (ed.), Centre for applied research under Mech Math. Faculty of MSU, Moscow, 23–42 (in Russian). English transl. available at www.math.jussieu.fr/miw/articles/ps/Nesterenko.ps (Cited in Chapter 8.)

[438] Nishioka, Ku. (1991). 'Algebraic independence measures of the values of Mahler functions'. *J. reine angew. Math.*, **420**, 203–214. (Cited in Chapters 3, 7 & 8.)

[439] ———— (1996). *Mahler functions and transcendence.* Lecture Notes in Math., 1631, Springer-Verlag, Berlin. (Cited in Chapters. 3, 7 & 8.)

[440] Nishioka, Ku. & Töpfer, T. (1991). 'Transcendence measures and non-linear functional equations of Mahler type'. *Arch. Math.*, **57**, 370–378. (Cited in Chapters 3 & 8.)

[441] Obreškov, N. (1950). 'On Diophantine approximations of linear forms for positive values of the variables'. *Dokl. Akad. Nauk SSSR*, **73**, 21–24 (in Russian). (Cited in Chapter 1.)

[442] Oryan, M. H. (1980). 'Über gewisse Potenzreihen, die für algebraische Argumente Werte aus den Mahlerschen Unterklassen U_m nehmen'. *Istanbul Üniv Fen Fak. Mec. Seri*, **45**, 1–42. (Cited in Chapter 7.)

[443] (1980). 'Über die Unterklassen U_m der Mahlerschen Klasseneinteilung der transzendenten formalen Laurentreihen'. *Istanbul Üniv Fen Fak. Mec. Seri*, **45**, 43–63. (Cited in Chapter 9.)

[444] (1983–86). 'Über gewisse Potenzreihen, deren Funktionswerte für Argumente aus der Menge der Liouvilleschen Zahlen U-Zahlen vom Grad $\leq m$ sind'. *Istanbul Üniv Fen Fak. Mec. Seri*, **47**, 15–34. (Cited in Chapter 7.)

[445] (1983–86). 'Über gewisse Potenzreihen, deren Funktionswerte für Argumente aus der Menge der p-adischen Liouvilleschen Zahlen p-adische U-Zahlen vom Grad $\leq m$ sind'. *Istanbul Üniv Fen Fak. Mec. Seri*, **47**, 53–67. (Cited in Chapter 9.)

[446] (1989). 'On power series and Mahler's U-numbers'. *Math. Scand.*, **65**, 143–151. (Cited in Chapters 3 & 7.)

[447] (1990). 'On the power series and Liouville numbers'. *Doğa Math.*, **14**, 79–90. (Cited in Chapter 7.)

[448] Özdemir, A. S. (1996). 'On the partial generalization of the measure of transcendence of some formal Laurent series'. *Math. Comput. Appl.*, **1**, 100–110. (Cited in Chapter 9.)

[449] (2000). 'On the measure of transcendence of certain formal Laurent series'. *Bull. Pure Appl. Sci. Sect. E Math. Stat.*, **19**, 541–550. (Cited in Chapter 9.)

[450] Peck, L. G. (1961). 'Simultaneous rational approximations to algebraic numbers'. *Bull. Amer. Math. Soc.*, **67**, 197–201. (Cited in Chapters 9 & 10.)

[451] Pereverzeva, N. A. (1987). 'A lower bound for the Hausdorff dimension of some sets of the Euclidean plane'. *Vestsī Akad. Navuk BSSR Ser. Fīz Mat. Navuk*, **6**, 102–104, 128 (in Russian). (Cited in Chapter 5.)

[452] (1989). 'A set-theoretic characteristic of sets in $\mathbb{R}^2$ that admit a given order of approximation by algebraic numbers'. *Vestsī Akad. Navuk BSSR Ser. Fīz Mat. Navuk*, **4**, 34–38, 124 (in Russian). (Cited in Chapter 5.)

[453] Perna, A. (1914). 'Sui numeri transcendenti in generale e sulla loro costruzione in base al criterio di Liouville'. *Giorn. Mat. Battaglini*, **52**, 305–365. (Cited in Chapter 3.)

[454] Perron, O. (1929). *Die Lehre von den Ketterbrüchen*. Teubner, Leipzig. (Cited in Chapters 1, 2 & 7.)

[455] Petruska, G. (1992). 'On strong Liouville numbers'. *Indag. Math. (N. S.)*, **3**, 211–218. (Cited in Chapter 7.)

[456] Philippon, P. (1994). 'Classification de Mahler et distances locales'. *Bull. Austral. Math. Soc.*, **49**, 219–238. (Cited in Chapter 8.)

[457] (2000). 'Approximation algébrique des points dans les espaces projectifs, I'. *J. Number Theory*, **81**, 234–253. (Cited in Chapter 8.)

[458] Pollington, A. D. (1979). 'The Hausdorff dimension of a set of non-normal well approximable numbers'. *Number Theory*, Carbondale (1979), *Lectures Notes in Math.*, **751**, Springer-Verlag, Berlin, 256–264. (Cited in Chapter 5.)

[459] (1993). 'Sum sets and U-numbers.' *Number theory with an emphasis on the Markoff spectrum.* (Provo, UT, 1991), 207–214, Dekker, New York. (Cited in Chapter 7.)

[460] Pollington, A. D. & Velani, S. L. (2000). 'On a problem in simultaneous Diophantine approximation: Littlewood's conjecture'. *Acta Math.*, **185**, 287–306. (Cited in Chapters 7 & 10.)

[461] (2002). 'On simultaneously badly approximable numbers'. *J. London Math. Soc.*, **66**, 29–40. (Cited in Chapter 10.)

[462] van der Poorten, A. J. (1986). 'An introduction to continued fractions.' *Diophantine analysis.* (Kensington, 1985), 99–138, London Math. Soc. Lecture Note Ser., **109**, Cambridge University Press (Cited in Chapter 1.)

[463] Popken, J. (1929). 'Zur Transzendenz von *e*'. *Math. Z.*, **29**, 525–541. (Cited in Chapter 3.)

[464] Queffélec, M. (1998). 'Transcendance des fractions continues de Thue–Morse'. *J. Number Theory*, **73**, 201–211. (Cited in Chapter 2.)

[465] Queffélec, M. & Ramaré, O. (2003). 'Analyse de Fourier des fractions continues à quotients restreints'. *Enseign. Math.*, **49**, 335–356. (Cited in Chapter 1.)

[466] Raisbeck, G. (1950). 'Simultaneous Diophantine approximation'. *Canadian J. Math.*, **2**, 283–288. (Cited in Chapter 1.)

[467] Ramharter, G. (1983). 'Some metrical properties of continued fractions'. *Mathematika*, **30**, 117–132. (Cited in Chapter 5.)

[468] (1994). 'On the fractional dimension theory of a class of expansions'. *Quart. J. Math. Oxford*, **45**, 91–102. (Cited in Chapter 5.)

[469] Ratliff, M. (1978). 'The Thue–Siegel–Roth–Schmidt theorem for algebraic functions'. *J. Number Theory*, **10**, 99–126. (Cited in Chapter 9.)

[470] Ridout, D. (1957). 'Rational approximations to algebraic numbers'. *Mathematika*, **4**, 125–131. (Cited in Chapter 2.).

[471] Rieger, G.-J. (1975). 'Über die Lösbarkeit von Gleichungssystemen durch Liouville Zahlen'. *Arch. Math.*, **26**, 40–43. (Cited in Chapter 1.)

[472] Riesz, F. & Sz.-Nagy, B. (1990). *Functional Analysis.* Translated from the second French edition by Leo F. Boron. Reprint of the 1955 original. Dover Publications Inc., New York. (Cited in Chapters 3 & 6.)

[473] Robert, A. M. (2000). *A Course in* p-adic *Analysis.* Graduate Texts in Mathematics, **198**, Springer-Verlag, New York. (Cited in Chapter 9.)

[474] Rockett, A. M. & Szüsz, P. (1992). *Continued Fractions.* World Scientific, Singapore. (Cited in Chapter 1.)

[475] Rogers, C. A. (1951). 'The signatures of the errors of simultaneous Diophantine approximations'. *Proc. London Math. Soc.*, **52**, 186–190. (Cited in Chapter 1.)

[476] (1970). *Hausdorff Measures.* Cambridge University Press. (Cited in Chapter 5.)

[477] Roth, K. F. (1955). 'Rational approximations to algebraic numbers'. *Mathematika*, **2**, 1–20; corrigendum, 168. (Cited in Chapter 2.)

[478] Roy, D. (2001). 'Approximation algébrique simultanée de nombres de Liouville'. *Canad. Math. Bull.*, **44**, 115–120. (Cited in Chapter 8.)

[479] (2003). 'Approximation simultanée d'un nombre et son carré'. *C. R. Acad. Sci. Paris*, **336**, 1–6. (Cited in Chapter 3.)

[480] (2004). 'Approximation to real numbers by cubic algebraic numbers, I'. *Proc. London Math. Soc.*, **88**, 42–62. (Cited in Chapters 3 & 10.)

[481] (2003). 'Approximation to real numbers by cubic algebraic numbers, II'. *Ann. of Math.*, **158**, 1081–1087. (Cited in Chapters 3 & 10.)

[482] (To appear). 'Diophantine approximation in small degree'. In: Proceedings of the seventh conference of the Canadian number theory association. (Cited in Chapter 3.)

[483] Roy, D. & Waldschmidt, M. (1997). 'Approximation diophantienne et indépendance algébrique de logarithmes'. *Ann. Sci. École Norm. Sup.*, **30**, 753–796. (Cited in Chapter 8.)

[484] (To appear). 'Diophantine approximation by conjugate algebraic integers'. *Compositio Math.* (Cited in Chapters 3, 8 & 9.)

[485] Rump, S. M. (1979). 'Polynomial real root separation'. *Math. Comp.*, **33**, 327–336. (Cited in Appendix A.)

[486] Rynne, B. P. (1992). 'The Hausdorff dimension of certain sets arising from Diophantine approximation by restricted sequences of integer vectors'. *Acta Arith.*, **61**, 69–81. (Cited in Chapter 5.)

[487] (1992). 'Regular and ubiquitous systems and $\mathcal{M}_\infty^s$-dense sequences'. *Mathematika*, **39**, 234–243. (Cited in Chapters 5 & 6.)

[488] (1998). 'The Hausdorff dimension of sets arising from Diophantine approximation with a general error function'. *J. Number Theory*, **71**, 166–171. (Cited in Chapters 5 & 6.)

[489] (1998). 'Hausdorff dimension and generalised simultaneous Diophantine approximation'. *Bull. London Math. Soc.*, **30**, 365–376. (Cited in Chapter 5.)

[490] (2003). 'Simultaneous Diophantine approximation on manifolds and Hausdorff dimension'. *J. Number Theory*, **98**, 1–9. (Cited in Chapter 5.)

[491] Rynne, B. P. & Dickinson, H. (2000). 'Hausdorff dimension and a generalized form of simultaneous Diophantine approximation'. *Acta Arith.*, **93**, 21–36. (Cited in Chapter 5.)

[492] Sakovich, N. V. (1995). 'The Hausdorff dimension and the distribution of values of integral polynomials over $\mathbb{C}$'. *Vestsī Akad. Navuk Belarusī Ser. Fīz Mat. Navuk*, **2**, 39–46, 124 (in Russian). (Cited in Chapter 9.)

[493] Saradha, N. & Tijdeman, R. (2003). 'On the transcendence of infinite sums of values of rational functions'. *J. London Math. Soc.*, **67**, 580–592. (Cited in Chapter 3.)

[494] Schikhof, W. H. (1984). *Ultrametric Calculus: An Introduction to p-adic Analysis.* Cambridge Studies in Advanced Mathematics **4**, Cambridge University Press. (Cited in Chapter 9.)

[495] Schlickewei, H.-P. (1976). 'On products of special linear forms with algebraic coefficients'. *Acta Arith.*, **31**, 389–398. (Cited in Chapter 9.)

[496] (1981). '*p*-adic *T*-numbers do exist'. *Acta Arith.*, **39**, 181–191. (Cited in Chapter 9.)

[497] Schmeling, J. & Troubetzkoy, S. (2003). 'Inhomogeneous Diophantine approximation and angular recurrence for polygonal billiards'. *Mat. Sbornik*, **194**, 295–309. (Cited in Chapter 5.)

[498] Schmidt, A. L. (1974). 'On the approximation of quaternions'. *Math. Scand.*, **34**, 184–186. (Cited in Chapter 9.)

[499] (1975). 'Diophantine approximation of complex numbers'. *Acta Math.*, **134**, 1–85. (Cited in Chapter 9.)

[500] Schmidt, W. M. (1960). 'A metrical theorem in Diophantine approximation'. *Canad. J. Math.*, **12**, 619–631. (Cited in Chapter 1.)

[501]   (1961). 'Bounds for certain sums; a remark on a conjecture of Mahler'. *Trans. Amer. Math. Soc.*, **101**, 200–210. (Cited in Chapter 4.)

[502]   (1964). 'Metrical theorems on fractional parts of sequences'. *Trans. Amer. Math. Soc.*, **110**, 493–518. (Cited in Chapter 1.)

[503]   (1966). 'On badly approximable numbers and certain games'. *Trans. Amer. Math. Soc.*, **123**, 178–199. (Cited in Chapters 5 & 9.)

[504]   (1967). 'On simultaneous approximations of two algebraic numbers by rationals'. *Acta Math.*, **119**, 27–50. (Cited in Chapter 2.)

[505]   (1969). 'Badly approximable systems of linear forms'. *J. Number Theory*, **1**, 139–154. (Cited in Chapters 5 & 9.)

[506]   (1970). 'Simultaneous approximations to algebraic numbers by rationals'. *Acta Math.*, **125**, 189–201. (Cited in Chapter 2.)

[507]   (1970). '*T*-numbers do exist'. Symposia Math. IV, Inst. Naz. di Alta Math., Rome 1968, Academic Press, pp. 3–26. (Cited in Chapters 1, 3, 7 & 9.)

[508]   (1971). 'Mahler's *T*-numbers'. 1969 Number Theory Institute, (Proc. of Symposia in Pure Math., Vol. XX, State Univ. New York, Stony Brook, NY, 1969), pp. 275–286, *Amer. Math. Soc.*, Providence, RI. (Cited in Chapters 3, 7 & 9.)

[509]   (1971). 'Diophantine approximation and certain sequences of lattices'. *Acta Arith.*, **18**, 165–178. (Cited in Chapter 1.)

[510]   (1971). *Approximation to algebraic numbers*. Monographie de l'Enseignement Mathématique, 19, Genève. (Cited in Introduction and Chapter 2.)

[511]   (1976). 'Two questions in Diophantine approximation'. *Monatsh. Math.*, **82**, 237–245. (Cited in Chapters 1 & 10.)

[512]   (1980). *Diophantine approximation*. Lecture Notes in Math., 785, Springer, Berlin. (Cited in Introduction, in Chapters 1, 2, 3, 5, & 10, and in Appendices A & B.)

[513]   (1983). 'Open problems in Diophantine approximation'. In: *Approximations Diophantiennes et Nombres Transcendants*. Luminy, 1982. Progress in Mathematics, Birkhäuser. (Cited in Chapter 10.)

[514]   (1983). 'Diophantine approximation properties of certain infinite sets'. *Trans. Amer. Math. Soc.*, **278**, 635–645. (Cited in Chapter 1.)

[515]   (2000). 'On continued fractions and Diophantine approximation in power series fields'. *Acta Arith.*, **95**, 139–166. (Cited in Chapters 9 & 10.)

[516]   Schmidt, W. M. & Wang, Y. (1979). 'A note on a transference theorem of linear forms'. *Sci. Sinica*, **22**, 276–280. (Cited in Chapter 4 and Appendix B.)

[517]   Schneider, T. (1957). *Einführung in die Transzendenten Zahlen*. Springer-Verlag, Berlin–Göttingen–Heidelberg. (Cited in Introduction, and Chapters 3 & 9.)

[518]   Schwarz, W. (1977). 'Liouville Zahlen und der Satz von Baire'. *Math. Phys. Semesterber.*, **24**, 84–87. (Cited in Chapter 1.)

[519]   Schweiger, F. (1995). *Ergodic Theory of Fibred Systems and Metric Number Theory*. Oxford Science Publications, The Clarendon Press, Oxford University Press, New York. (Cited in Chapter 1.)

[520]   Serret, J.-A. (1877). *Cours d'Algèbre supérieure*, quatrième edition. Gauthiers-Villars, Paris. (Cited in Chapter 1.)

[521] Shallit, J. (1992). 'Real numbers with bounded partial quotients: a survey'. *Enseign. Math.*, **38**, 151–187. (Cited in Chapter 10.)

[522] Shidlovskii, A. B. (1989). *Transcendental Numbers*. Translated from the Russian by Neal Koblitz, de Gruyter Studies in Mathematics 12, Walter de Gruyter & Co., Berlin. (Cited in Chapters 3 & 8.)

[523] Shorey, T. N. (1983). 'Divisors of convergents of a continued fraction'. *J. Number Theory*, **17**, 127–133. (Cited in Chapter 2.)

[524] Siegel, C. L. (1921). 'Approximation algebraischer Zahlen'. *Math. Z.*, **10**, 173–213. (Cited in Chapter 2.)

[525] Slater, N. B. (1967). 'Gaps and steps for the sequence $n\theta$ mod 1'. *Proc. Cambridge Philos. Soc.*, **63**, 1115–1123. (Cited in Chapter 5.)

[526] Slesoraĭtene, R. (1973). 'An analogue of a theorem of Mahler and Sprindžuk for polynomials of degree two in two p-adic variables'. *Litovsk. Mat. Sb.*, **13**, 177–188 (in Russian). English transl. in *Lithuanian Math. J.*, **13** (1973), 291–299. (Cited in Chapter 9.)

[527] ——— (1987). 'An analogue of the Mahler–Sprindžuk theorem for polynomials of fourth degree in two variables'. *Litovsk. Mat. Sb.*, **27**, 165–171 (in Russian). English transl. in *Lithuanian Math. J.*, **27** (1987), 94–99. (Cited in Chapter 4.)

[528] Sorokin, V. N. (1991). 'Hermite–Padé approximations of sequential powers of a logarithm and their arithmetic applications'. *Izv. Vyssh. Uchebn. Zaved. Mat.*, No. 11, 66–74 (in Russian). English transl. in *Soviet Math. (Iz. VUZ)*, **35** (1991), No. 11, 67–74. (Cited in Chapter 3.)

[529] Sós, V. T. (1957). 'On the theory of Diophantine approximations, I'. *Acta Math. Hung.*, **8**, 461–472. (Cited in Chapter 5.)

[530] Sprindžuk, V. G. (1962). 'On some general problems of approximating numbers by algebraic numbers'. *Litovsk. Mat. Sb.*, **2**, 129–145 (in Russian). (Cited in Chapter 3.)

[531] ——— (1962). 'On theorems of Khintchine and Kubilius'. *Litovsk. Mat. Sb.*, **2**, 147–152 (in Russian). (Cited in Chapter 4.)

[532] ——— (1962). 'On a classification of transcendental numbers'. *Litovsk. Mat. Sb.*, **2**, 215–219 (in Russian). (Cited in Chapter 8.)

[533] ——— (1962). 'On a conjecture of K. Mahler concerning the measure of the set of S-numbers'. *Litovsk. Mat. Sb.*, **2**, 221–226 (in Russian). (Cited in Chapter 4.)

[534] ——— (1963). 'On algebraic approximations in the field of power series'. *Vestnik Leningrad Univ. Ser. Mat. Meh. Astronom.*, **18**, 130–134 (in Russian). (Cited in Chapter 9.)

[535] ——— (1963). 'On the measure of the set of S-numbers in the p-adic field'. *Dokl. Akad. Nauk SSSR.*, **151**, 1292 (in Russian). English transl. in *Soviet Math. Dokl.*, **4** (1963), 1201–1202. (Cited in Chapter 9.)

[536] ——— (1964). 'On Mahler's conjecture'. *Dokl. Akad. Nauk SSSR.*, **154**, 783–786 (in Russian). English transl. in *Soviet Math. Dokl.*, **5** (1964), 183–187. (Cited in Chapter 4.)

[537] ——— (1964). 'More on Mahler's conjecture'. *Dokl. Akad. Nauk SSSR.*, **155**, 54–56. (in Russian). English transl. in *Soviet Math. Dokl.*, **5** (1964), 361–363. (Cited in Chapter 4.)

[538]   (1965). 'A proof of Mahler's conjecture on the measure of the set of
          S-numbers'. *Izv. Akad. Nauk SSSR. Ser. Mat.*, **29**, 379–436 (in Russian). English
          transl. in *Amer. Math. Soc. Transl.*, **51** (1966), 215–272. (Cited in Chapters 3, 4
          & 9.)
[539]   (1967). *Mahler's Problem in Metric Number Theory*. Izdat. "Nauka i
          Tehnika", Minsk, (in Russian). English translation by B. Volkmann, *Transla-
          tions of Mathematical Monographs*, **25**, American Mathematical Society, Prov-
          idence, RI, 1969. (Cited in Introduction, and Chapters 3, 4, 6 & 9.)
[540]   (1977). *Metric Theory of Diophantine Approximation*. Izdat. "Nauka",
          Moscow, (in Russian). English translation by R. A. Silverman, *Scripta Series
          in Mathematics*, John Wiley & Sons, 1979. (Cited in Introduction, and Chapters
          1 & 4.)
[541]   (1980). 'Achievements and problems of the theory of Diophantine approx-
          imations'. *Uspekhi Mat. Nauk*, **35**, 3–68, 248 (in Russian). English transl. in
          Russ. Math. Survey, **35** (1980), 1–80. (Cited in Chapter 9.)
[542]   Starkov, A. N. (2000). 'Dynamical systems on homogeneous spaces'. *Trans-
          lations of Mathematical Monographs*, **190**, American Mathematical Society,
          Providence, RI. (Cited in Chapter 4.)
[543]   Steinig, J. (1992). 'A proof of Lagrange's theorem on periodic continued frac-
          tions'. *Arch. Math. (Basel)*, **59**, 21–23. (Cited in Chapter 1.)
[544]   Stewart, C. L. (1984). 'A note on the product of consecutive integers'. *Topics in
          classical number theory*, Vol. I, II (Budapest, 1981), 1523–1537, Colloq. Math.
          Soc., János Bolyai 34, North-Holland, Amsterdam. (Cited in Chapter 2.)
[545]   Stokolos, A. M. (2001). 'Some applications of Gallagher's theorem in harmonic
          analysis'. *Bull. London Math. Soc.*, **33**, 210–212. (Cited in Chapter 4.)
[546]   Stolarsky, K. B. (1974). *Algebraic numbers and Diophantine approximation*.
          Pure and Applied Mathematics, No. 26. Marcel Dekker Inc., New York. (Cited
          in Chapter 3.)
[547]   Sullivan, D. (1982). 'Disjoint spheres, approximation by imaginary numbers,
          and the logarithm law for geodesics'. *Acta Math.*, **149**, 215–237. (Cited in Chap-
          ters 1 & 9.)
[548]   Teulié, O. (2001). *Approximation de nombres réels et p-adiques par des nom-
          bres algébriques*. Thèse, Université de Bordeaux. (Cited in Chapter 9.)
[549]   (2001). 'Approximation d'un nombre réel par des unités algébriques'.
          Monatsh. Math., **133**, 169–176. (Cited in Chapters 3 & 10.)
[550]   (2002). 'Approximation d'un nombre p-adique par des nombres algébriques'.
          Acta Arith., **102**, 137–155. (Cited in Chapter 9.)
[551]   (Preprint). 'Approximation d'un nombre p-adique par des nombres quadra-
          tiques'. (Cited in Chapter 9.)
[552]   (2002). 'Approximations simultanées de nombres algébriques de $\mathbb{Q}_p$ par des
          rationnels'. *Monatsh. Math.*, **137**, 313–324. (Cited in Chapter 9.)
[553]   Thue, A. (1909). 'Über Annäherungswerte algebraischer Zahlen'. *J. reine
          angew. Math.*, **135**, 284–305. (Cited in Chapter 2.)
[554]   Thurnheer, P. (1981). 'Un raffinement du théorème de Dirichlet sur
          l'approximation diophantienne'. *C. R. Acad. Sci. Paris Ser. I Math.*, **293**, 623–
          624. (Cited in Chapters 1 & 10.)
[555]   (1982). 'Eine Verschärfung des Satzes von Dirichlet über diophantische Ap-
          proximation'. *Comment. Math. Helv.*, **57**, 60–78. (Cited in Chapters 1 & 10.)

[556]   (1984). 'Zur diophantischen Approximation von zwei reellen Zahlen'. *Acta Arith.*, **44**, 201–206. (Cited in Chapters 1 & 10.)

[557]   (1985). 'Approximation diophantienne par certains couples d'entiers'. *C. R. Math. Rep. Acad. Sci. Canada*, **7**, 51–53. (Cited in Chapters 1 & 10.)

[558]   (1990). 'On Dirichlet's theorem concerning Diophantine approximation'. *Acta Arith.*, **54**, 241–250. (Cited in Chapters 1 & 10.)

[559]   Tichy, R. F. (1979). 'Zum Approximationssatz von Dirichlet'. *Monatsh. Math.*, **88**, 331–333. (Cited in Chapter 1.)

[560]   Tishchenko, K. I. (1996). 'Systems of linearly independent polynomials with height gaps and the Wirsing problem'. *Vestsī Nats. Akad. Navuk Belarusī Ser. Fīz Mat. Navuk*, **4**, 16–22, 140 (in Russian). (Cited in Chapter 3.)

[561]   (2000). 'On approximation to real numbers by algebraic numbers'. *Acta Arith.*, **94**, 1–24. (Cited in Chapter 3.)

[562]   (2000). 'On the approximation of complex numbers by algebraic numbers of bounded degree'. *Dokl. Nats. Akad. Nauk Belarusī*, **44**, No.3, 25–28 (in Russian). (Cited in Chapter 9.)

[563]   (2000). 'On the application of linearly independent polynomials in the Wirsing problem'. *Dokl. Nats. Akad. Nauk Belarusi*, **44**, No.5, 34–36 (in Russian). (Cited in Chapter 3.)

[564]   (2000). 'On some special cases of the Wirsing conjecture'. *Vestsī Nats. Akad. Navuk Belarusī Ser. Fīz. Mat. Navuk*, **3**, 47–52, 140 (in Russian). (Cited in Chapter 3.)

[565]   (2000). 'On new methods in the problem of the approximation of real numbers by algebraic numbers of degree at most three'. *Vestsī Nats. Akad. Navuk Belarusī Ser. Fīz. Mat. Navuk*, **4**, 26–31, 141 (in Russian). (Cited in Chapter 3.)

[566]   (2001). 'On the problem of the approximation of p-adic numbers by algebraic numbers'. *Vestsī Nats. Akad. Navuk Belarusī Ser. Fīz. Mat. Navuk*, **4**, 129–130, 144 (in Russian). (Cited in Chapter 9.)

[567]   (2001). 'On the approximation of real numbers by algebraic integers of the third degree'. *Dokl. Nats. Akad. Nauk Belarusī*, **45**, No.1, 17–19 (in Russian). (Cited in Chapter 3.)

[568]   (2002). 'On simultaneous approximation of two real numbers by real numbers'. *Dokl. Nats. Akad. Nauk Belarusī*, **46**, No.2, 29–33 (in Russian). (Cited in Chapter 3.)

[569]   Töpfer, T. (1995). 'Algebraic independence of the values of generalized Mahler functions'. *Acta Arith.*, **70**, 161–181. (Cited in Chapters 3 & 8.)

[570]   (1995). 'An axiomatization of Nesterenko's method and applications on Mahler functions, II'. *Compositio Math.*, **95**, 323–342. (Cited in Chapters 3 & 8.)

[571]   Tricot, C. (1980). 'Rarefaction indices'. *Mathematika*, **27**, 46–57. (Cited in Chapter 5.)

[572]   (1982). 'Two definitions of fractional dimension'. *Math. Proc. Cambridge Philos. Soc.*, **91**, 54–74. (Cited in Chapter 5.)

[573]   Turkstra, H. (1936). *Metrische bijdragen tot de theorie der Diophantische approximaties in het lichaam der p-adische getallen*. Dissertation, Vrije Univ., Amsterdam. (Cited in Chapter 9.)

[574]   Väänänen, K. (1989). 'Application of Hermite–Mahler polynomials to the approximation of the values of p-adic exponential function'. *Monatsh. Math.*, **108**, 219–227. (Cited in Chapter 9.)

[575]   Vahlen, K. Th. (1895). 'Über Näherungswerte und Kettenbrüche'. *J. reine angew. Math.*, **115**, 221–233. (Cited in Chapter 1.)

[576]   Vallée, B. (1998). 'Dynamique des fractions continues à contraintes périodiques'. *J. Number Theory*, **72**, 183–235. (Cited in Chapter 5.)

[577]   Vilchinskiĭ, V. T. (1990). 'On the metric theory of non-linear Diophantine approximation'. *Dokl. Akad. Nauk BSSR*, **34**, No.8, 677–680 (in Russian). (Cited in Chapter 5.)

[578]   Vilchinskii, V. T. & Dombrovskiĭ, I. R. (1990). 'The Hausdorff dimension and estimates for trigonometric sums over prime numbers'. *Vestsī Akad. Navuk BSSR Ser. Fīz. Mat. Navuk*, **1**, 3–6, 123 (in Russian). (Cited in Chapter 5.)

[579]   Vinogradov, A. I. & Chudnovsky, G. V. (1984). 'The proof of extremality of certain manifolds'. Appendix in: G.V. Chudnovsky *Contributions to the Theory of Transcendental Numbers*. Translated from the Russian by J. A. Kandall, *Math. Surveys Monographs*, **19**, Amer. Math. Soc., Providence, RI. (Cited in Chapter 4.)

[580]   Viola, C. (1979). 'Diophantine approximation in short intervals'. *Ann. Scuola Norm. Sup. Pisa Cl. Sci.*, (4) **6**, 703–717. (Cited in Chapter 1.)

[581]   Volkmann, B. (1960). Zur Mahlerschen Vermutung im Komplexen'. *Math. Ann.*, **140**, 351–359. (Cited in Chapter 9.)

[582]   ——— (1961). 'The real cubic case of Mahler's conjecture'. *Mathematika*, **8**, 55–57. (Cited in Chapter 4.)

[583]   ——— (1962). 'Zur metrischen Theorie der S-Zahlen'. *J. reine angew. Math.*, **209**, 201–210. (Cited in Chapter 4.)

[584]   ——— (1963/64) 'Zur metrischen Theorie der S-Zahlen, II'. *J. reine angew. Math.*, **213**, 58–65. (Cited in Chapter 4.)

[585]   van der Waerden, B. L. (1953). *Modern Algebra*, Vol. I, Revised English edition. Ungar. (Cited in Chapter 7.)

[586]   ——— (1991). *Algebra*, Vol. I. Springer-Verlag, New York (Cited in Chapter 3 and Appendix A.)

[587]   Waldschmidt, M. (1971). 'Indépendance algébrique des valeurs de la fonction exponentielle'. *Bull. Soc. Math. France*, **99**, 285–304. (Cited in Chapter 8.)

[588]   ——— (1974). *Nombres transcendants*. Lecture Notes in Mathematics, 402, Springer-Verlag, Berlin–New York. (Cited in Chapter 8.)

[589]   ——— (1978). 'Transcendence measures for exponentials and logarithms'. *J. Austral. Math. Soc. (Series A)*, **25**, 445–465. (Cited in Chapter 8.)

[590]   ——— (2000). 'Conjectures for large transcendence degree'. *Algebraic Number Theory and Diophantine Analysis* (Graz, 1998), 497–520, de Gruyter, Berlin. (Cited in Chapter 8.)

[591]   ——— (2000). *Diophantine Approximation on Linear Algebraic Groups. Transcendence properties of the exponential function in several variables*. Grundlehren der Mathematischen Wissenschaften, 326, Springer-Verlag, Berlin. (Cited in Chapters 2 & 8, and Appendix A.)

[592]   ——— (2004). 'Open Diophantine Problems'. *Moscow Math. J.*, **4**. (Cited in Chapter 10.)

[593] Wang, Y. & Yu, K. (1981). 'A note on some metrical theorems in Diophantine approximation'. *Chinese Ann. Math.*, **2**, 1–12. (Cited in Chapter 4.)

[594] Wang, Y., Yu, K. & Zhu, Y. C. (1979). 'A transfer theorem for linear forms'. *Acta Math. Sinica*, **22**, 237–240 (in Chinese). (Cited in Chapter 4 and Appendix B.)

[595] Wass, N. Ch. (1990). 'Algebraic independence of the values at algebraic points of a class of functions considered by Mahler'. *Dissertationes Math. (Rozprawy Mat.)*, **303**, 61 pp. (Cited in Chapters 3 & 8.)

[596] Weiss, B. (2001). 'Almost no points on a Cantor set are very well approximable'. *R. Soc. Lond. Proc. Ser. A Math. Phys. Eng. Sci.*, **457**, 949–952. (Cited in Chapter 5.)

[597] ——— (2002). 'Dynamics on parameter spaces: submanifold and fractal subset questions'. *Rigidity in Dynamics and Geometry* (Cambridge, 2000), 425–440, Springer, Berlin. (Cited in Chapter 5.)

[598] Wirsing, E. (1961). 'Approximation mit algebraischen Zahlen beschränkten Grades'. *J. reine angew. Math.*, **206**, 67–77. (Cited in Chapters 2, 3, 8, 9 & 10, and Appendix A.)

[599] ——— (1971). 'On approximations of algebraic numbers by algebraic numbers of bounded degree'. *1969 Number Theory Institute* (Proc. of Symposia in Pure Math., Vol. XX, State Univ. New York, Stony Brook, NY, 1969), pp. 213–247, Amer. Math. Soc., Providence, RI. (Cited in Chapter 2.)

[600] Wolfskill, J. (1983). 'A growth bound on the partial quotients of cubic numbers'. *J. reine angew. Math.*, **346**, 129–140. (Cited in Chapter 2.)

[601] Worley, R. T. (1981). 'Estimating $|\alpha - p/q|$'. *J. Austral. Math. Soc. (Series A)*, **31**, 202–206. (Cited in Chapter 1.)

[602] Xin, X. L. (1989). 'Mahler's classification of p-adic numbers'. *Pure Appl. Math.*, **5**, 73–80 (in Chinese). (Cited in Chapter 9.)

[603] Xu, G. S. (1986). 'On the arithmetic properties of the values of p-adic E-functions'. *Acta Math. Sinica*, **29**, 444–453 (in Chinese). (Cited in Chapter 9.)

[604] ——— (1988). 'Diophantine approximation and transcendental number theory'. *Number Theory and its Applications in China*, 127–142, Contemp. Math., **77**, Amer. Math. Soc., Providence, RI. (Cited in Chapter 4 and Appendix B.)

[605] Yanchenko, A. Ya. (1982). 'The measure of mutual transcendence of almost all pairs of p-adic numbers'. *Vestnik Moskov. Univ. Ser. I Mat. Mekh.*, **3**, 62–65, 111 (in Russian). English transl. in *Moscow Univ. Math. Bull.*, **37**, No.3, 71–74. (Cited in Chapter 9.)

[606] Yilmaz, G. (1996/97). 'Arithmetical properties of the values of some power series with algebraic coefficients taken for U_m-numbers arguments'. *Istanbul Üniv. Fen Fak. Mat. Derg.*, **55/56**, 111–144. (Cited in Chapters 3, 7 & 9.)

[607] Yu, K. (1981). 'Hausdorff dimension and simultaneous rational approximation'. *J. London Math. Soc.*, **24**, 79–84. (Cited in Chapter 5.)

[608] ——— (1981). 'A note on a problem of Baker in metrical number theory'. *Math. Proc. Cambridge Philos. Soc.*, **90**, 215–227. (Cited in Chapters 4, 5 & 9.)

[609] ——— (1987). 'A generalization of Mahler's classification to several variables'. *J. reine angew. Math.*, **377**, 113–126. (Cited in Chapter 3.)

[610] Zeren, B. M. (1980). 'Über einige komplexe und p-adische Lückenreihen mit Werten aus den Mahlerschen Unterklassen U_m'. *Istanbul Üniv. Fen Fak. Mec. Seri.*, **45**, 89–130. (Cited in Chapters 7 & 9.)

[611]   (1991). 'Über eine Klasse von verallgemeinerten Lückenreihen, deren Werte für algebraische Argumente transzendent, aber keine U-Zahlen sind, I, II, III'. *Istanbul Üniv. Fen Fak. Mat. Der.*, **50**, 79–99; 101–114; 147–158. (Cited in Chapters 3 & 7.)

[612]   Zheludevich, F. F. (1983). 'Simultaneous approximations of numbers by roots of integral polynomials'. *Vestsī Akad. Navuk BSSR Ser. Fīz. Mat. Navuk*, **4**, 14–18 (in Russian). (Cited in Chapter 9.)

[613]   (1984). 'The simultaneous approximation of numbers by the roots of integer polynomials'. *Vestsī Akad. Navuk BSSR Ser. Fīz. Mat. Navuk*, **2**, 14–20 (in Russian). (Cited in Chapter 9.)

[614]   (1986). 'Simultane diophantische Approximationen abhängiger Grössen in mehreren Metriken'. *Acta Arith.*, **45**, 87–98. (Cited in Chapter 9.)

[615]   (1987). 'Approximations of S-adèles by algebraic numbers of bounded degree'. *Vestsī Akad. Navuk BSSR Ser. Fīz. Mat. Navuk*, **6**, 19–22, 124 (in Russian). (Cited in Chapter 9.)

[616]   Zhu, Y. C. (1981). 'On a problem of Mahler concerning the transcendence of a decimal fraction'. *Acta Math. Sinica*, **24**, 246–253 (in Chinese). (Cited in Chapter 2.)

[617]   Zong, C. M. (1994). 'Analogues of K. Mahler's theory'. *Acta Math. Sinica (N. S.)*, **10**, Special Issue, 141–154. (Cited in Chapter 9.)

Index